W9-CGL-356

THIS FASCINATING OIL BUSINESS

THIS FASCINATING
OIL BUSINESS

by

Max W. Ball

Douglas Ball

Daniel S. Turner

**Maps and Illustrations by
Laurence N. Watts**

THE BOBBS-MERRILL COMPANY, INC.
Indianapolis / New York

Copyright 1940 by The Bobbs-Merrill Company
Copyright © 1965 by The Bobbs-Merrill Company, Inc.
All rights reserved

First paperback printing 1979

PRINTED IN THE UNITED STATES OF AMERICA

Library of Congress Cataloging in Publication Data

Ball, Max Waite, 1885–1954.

This fascinating oil business, by Max W. Ball, Douglas Ball
[and] Daniel S. Turner. Illustrations and maps by Laurence N.
Watts. Indianapolis, Bobbs-Merrill [©1965]

xiv, 464 p. illus., maps. 22 cm.

Completely revised and up-dated edition of the work by M. W.
Ball, first published in 1940.

Bibliography: p. 434.

1. Petroleum industry and trade. I. Ball, Douglas. II. Turner,
Daniel S. III. Title.

TN870.B3   1965   553.282   64–15660
ISBN 0–672–50829–X
ISBN 0–672–52584–4 paperback

Dedications

In the United States alone, the oil industry and its suppliers now employ more than fourteen million people. Billions of dollars are invested each year, and the production statistics of the industry have risen to astronomical figures. But the human story is often lost, and newspaper and technical accounts seem distant from the concerns of people in their day-to-day lives. This edition is dedicated to everyone who desires a comprehensive treatment of a subject that touches them every day—the search for oil.

DOUGLAS BALL
February 1979

My grandmother used to tell me about a proud mother who said that her son had made a fiddle right out of his head and had wood enough left for another. All of the wood in this book came out of my own head, but much information, many suggestions, and much helpful criticism have come from the heads of others. To them, in deep gratitude, I dedicate this book.

MAX W. BALL
November 1939

Since Max Ball's readers toured through geologic aeons and thousands of miles the oil business has fueled a war and enough automobile miles for a round trip to the moon for every family in the United States.

But the oil business is more remarkable as an instigator of friendships than as a generator of statistics, and we dedicate this new edition to the many friends who have helped it to press.

DOUGLAS BALL
DANIEL S. TURNER
March 1964

Contents

royalty. Is an eighth enough? Minor provisions of the lease. Provisions by law or custom. Cash for the landowner. Mutual advantages of the lease. Producers form 88. Rival lessors. When will drilling start?

production problems. Refinery research. Improvement in quality. New products and new uses. Jet fuels and lubricants. Alcohols and antifreezes. Lacquers, paints, varnishes, and solvents. Nylons, rayons, and plastics. Dyestuffs, textile oils, and leather oils. Synthetic rubbers. Medicines, poisons, and toiletries. Explosives and poison gases. Anesthetics. Even a vitamin. Food fats. Fruit and vegetable uses. Fertilizers. Petroleum mulch. Odds and ends. Competitive sources. Your day. Prepare for a tour.

Appendix D

Some Definitions and Usages

Before we start let's agree on the meanings of a few words and phrases that we may hear frequently:

Oil to us will mean petroleum or "rock oil." Linseed oil, castor oil, banana oil, and other oils have their uses, but not in this discussion. Most of the time *oil* will mean the whole petroleum group, including natural gas, paraffin, and asphalt.

Gallon means the wine or U.S. liquid gallon. It is one-sixth smaller than the Imperial gallon used in Canada and Great Britain.

Barrel means the standard oil barrel, which equals 42 U. S. gallons (35 Imperial gallons).

Temperatures are in degrees Fahrenheit.

Pressures are in pounds per square inch. In case you have forgotten, the pressure of the atmosphere at sea level is 14.7.

A number of intransitive verbs become transitive in the oil business. Thus to *flow* a well or *produce* a well means to let or cause the well to flow or produce its oil.

Operator means the individual or company that conducts the operation, whether the drilling of a well or the production of the oil from a tract or field. The operator is usually the lessee rather than the owner of the land on which the operation takes place. If drilling is done by contract, the operator is the man or company for whom, rather than by whom, the drilling is done.

The adjective *proven*, meaning known to be capable of producing oil or gas, is common usage among oil men. The dictionaries may say that "proven" is archaic and has been replaced by "proved," but the men who wrote the dictionaries didn't live in the oil country. In any case it doesn't matter much whether a lease is called "proven" or "proved"; the important question is whether it can honestly be called either.

Wildcat and *wildcatter* are not opprobrious terms in the oil business. Any well drilled in unproven territory is a wildcat. The man or company that drills it is a wildcatter and, if otherwise worthy, is held in high honor by the industry.

The *play* is where the action is when oil exploration comes to an area. The play is also the action itself, and there is nothing playful about the intensely competitive *leaseplays* and *market plays* described in Chapters 5 and 14 respectively.

A *dry hole* may be exceedingly wet. Any well that fails to produce oil or gas in commercial quantities is called a dry hole, even though it flows a stream of water, as many dry holes do.

Other new terms and usages will come to our attention as we go along. Not all of them will be defined at the first encounter, but all of them will be explained sooner or later. If a word or phrase puzzles you, and you don't want to wait until you come to a later explanation, the index may help you to locate the explanation forthwith.

Now, if you are ready, we'll amplify our definition of oil.

THIS FASCINATING OIL BUSINESS

CHAPTER 1

What Is Oil?

Crude Oil and Some of Its Products

Oil, as everyone knows, is a greasy liquid found in the rocks at or beneath the earth's surface. It is a mineral of organic origin, the result of the slow distillation of animal or vegetable remains during millions of years. A chemist can simulate the conditions of its formation and could make a small amount of oil out of a tree trunk, or out of any reader, no matter how emaciated. Oil men call it "crude oil," or simply "crude," to distinguish it from the refined oils manufactured from it.

Crude oil is the raw material of the oil industry. From it are made gasoline and kerosene, naphtha and other cleaning fluids, motor oils and cup grease and sewing machine oil and axle grease and a host of other lubricants, candles and paraffin and laundry wax and many other waxes and wax products, Vaseline and other ointments and jellies, nasal sprays and intestinal lubricants and other medicinal products, ethers and other anesthetics, "wood" and "grain" and other alcohols, paint and varnish constituents, fly killers and mosquito exterminators and other insecticides, disinfectants, dyes, chemical and commercial solvents, oil for making artificial gas, oils for making candy and cosmetics and for packing fruit and preserving eggs, gases and oils for ripening fruits and for accelerating plant growth, flotation oils and reducing agents for metallurgical use, roofing materials and timber preservatives, rubber and plastics, paving and road oils, fuel oil for steamships and boilers and diesel engines and household furnaces and kitchen ranges and all sorts of heating and power plants, coke, and a multitude of other useful things.

USERS OF OIL PRODUCTS

These people are using some of the 1,000 or more products made from oil. Most of us use several of these products every day. They are essential to modern life as we know it.

Hydrocarbons and More Hydrocarbons[1]

Every substance on or in the earth belongs to one of two great classes. If it is or has been part of living organisms, whether animal or vegetable, it is an *organic* substance. All other substances are classed as *inorganic*. Oil, though a mineral, is an organic substance, derived from parts of living organisms. It and coal and certain other organic mineral substances are composed mainly of hydrogen and carbon and are called "hydrocarbons." Hydrogen in its free state is a lightweight gas. Carbon, when not combined with other elements, is a solid that takes such diverse forms as charcoal, graphite, and diamonds. Atoms of hydrogen and atoms of carbon combine in different proportions to form the molecules of thousands of different hydrocarbons, each with its own chemical and physical characteristics and its own chemical formula, and each differing from all the others.

Oil is not a simple hydrocarbon with a definite chemical formula. On the contrary, it is an extremely complex mixture of hydrocarbons that differ widely in their formulas and characteristics. Natural gas is nearly always associated with it, and gas and oil together form a continuous succession of hydrocarbons. Some of the hydrocarbons in the succession contain, in addition to their carbon and hydrogen atoms, an atom or more of oxygen or sulphur or nitrogen.

At one end of the succession is a very lightweight, odorless, heavy flammable gas known as methane or marsh gas; toward the other end are solids, such as the paraffin that makes candles and the asphalt that makes pavements. In the succession of gases, liquids, semi-solids, and solids, from methane through gasoline and kerosene and fuel oil and paraffin and asphalt and on down to the solid compounds, each compound is a little heavier than the last one.

The lightest gases are known as *fixed gases* or *dry gases* because no part of them is liquefied in ordinary commercial practice. Next are slightly heavier gases known as *wet gases*; out of these, by chilling, compression, or absorption, a small percentage of very light, very volatile gasoline is made. This is known as "natural gasoline."

Next heavier than the wet gases are the volatile liquids that we know as gasoline, and next come kerosene, diesel oil, lubricating oils, wax

---

[1] Much of the material under this and the next heading is explained in greater detail in Chapter 11. If you find it heavy going, skip it here and try it there.

oils, and paraffins or asphalts. *Naphtha* includes the heavier part of the gasoline group and the lighter part of the kerosene group. Fuel oil may include any part of the series heavier than kerosene.

Proper use of heat and pressure will transform almost any of these products into others, and will make the hundreds of derived products to which we have briefly referred. The oil business is no longer content with dividing crude oil into its natural constituents; it rearranges and rebuilds the natural constituents into nearly any combination of hydrocarbons that the world will buy.

As Diverse as a Subway Crowd

No one should think from what has been said that all crude oils are alike, or contain the same hydrocarbons or contain them in the same proportions. Oils from different fields are almost as individual in their characteristics as the same number of human beings. An oil may be high in gasoline or low. It may have a large kerosene cut or none. It may be rich or wanting in lubricants. It may have an asphalt base or a paraffin base or a mixture of both. Oil is oil in the sense that "pigs is pigs," but there is more variation in oils than in pigs!

Many differences in color, weight, viscosity, and other physical characteristics result from the variations in composition. Most crude oils look black at first glance, but only the heavier asphaltic oils are really black. The *paraffin-base oils* and some of the lighter *mixed-base oils* are greenish if you look directly at them and dark reddish if you look through them. A few oils are even straw-colored. As a rule, the asphaltic oils are heavier and more viscous, the paraffin oils lighter and more fluid.

These and many other variations in characteristics result in notable variations in value. In general, the lighter oils, most of them of paraffin or mixed base, command higher prices than the heavier oils, because more gasoline can be made more easily from the lighter oils by simple refining methods. But gasoline from the heavier oils usually has a higher anti-knock value or *octane number* than that from the lighter oils. This fact tends to lessen the differences in price.

An Important Substance

Such is oil—complex, variable, and useful. Try to imagine present-day America without automobiles or asphalt streets or dry cleaning or plastics, or think what would happen to the machinery of the world

MORE USERS OF OIL PRODUCTS

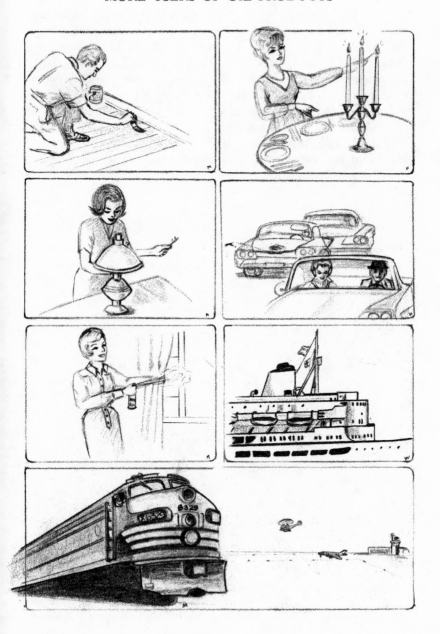

without lubricants, and you will realize that the oil industry is important. The American public thinks so; it pays over ten billion dollars to service stations each year despite, or perhaps because of, the fact that petroleum products have increased less in price than almost anything else the public buys. The Oil Business and its products pay more taxes than any other industry, probably more than any two or three combined. Besides all this, the industry is intrinsically fascinating; in almost every phase there is a sense of lilt and adventure that few industries have in such degree.

If these things move you, or if you'd like to know why your next tankful of gasoline is so easy to obtain, come along and we'll see something of how it's done.

CHAPTER 2

Meet the Oil Men

Many Men of Many Functions

THE business of the oil industry is to supply the motorist with gasoline, to lubricate the machinery of this machine age, and to provide the world with more than a thousand other necessities. To do this the industry must find oil, acquire title to it, bring it to the surface, carry it to refineries, manufacture it into the products desired, transport the products to their points of use, and sell them.

These are widely diverse functions, as diverse as buying land, raising wheat, transporting the wheat to the mill, making flour, baking bread, and selling the bread over the counter. The *geologist* who locates the well, the *driller* on the derrick floor, the *pumper* on the pipeline, the *stillman* in the refinery, the *service station attendant*—these men differ widely in training and function and know little of one another's work, yet all are called *oil men*, as though the farmer, the railroader, the miller, the baker, and the corner grocer were all called wheat men.

For this there is a reason. Much of the oil business is carried on by integrated units which engage in all branches of the industry—from finding the oil to distributing the finished products. There are thousands of smaller units, each confining itself to some one branch—production, transportation, refining, or marketing—and most of them are efficient, useful, and in normal times profitable, but despite their numbers they do the smaller part of the industry's business. The better known companies are complete units, competing with one another over the whole range of the industry, each producing, processing, and distributing its own products.

The various employees of an integrated company may be engaged in different types of work in different branches of the business, but they consider their jobs as parts of a common effort. The men working for individuals and for non-integrated companies have come to have a similar feeling. Thus every man connected with the industry, in whatever branch or calling, considers himself—and is—an oil man.

Would you like to meet these various oil men and become acquainted with their respective functions? It will give us a good start to our study of the oil business.

The Geologic Staff

If you own a farm in prospective oil territory, you may some day notice a group of surveyor-looking men working quietly in the neighborhood. They are a geologic crew, the vanguard in the search for oil.

The geologist does not find the oil; he merely recommends where to drill for it. He may be part of the geologic staff of a company, or he may be in business for himself as a consulting geologist, or he may be an oil producer on his own account. He has a degree in engineering or science, perhaps two or three of them, but there is nothing pedantic about him. He is an explorer, with his feet on the ground and his mind ranging the rocks for thousands of feet below.

You may watch him and his instrument man surveying with plane table and stadia rod, mapping the formations that outcrop at the surface and measuring the thickness of exposed strata. You may see a *geophysicist* with or without half-a-dozen strange trucks making delicate measurements with some strange device such as a *gravity meter* or *magnetometer* or a *seismograph*. In the head office you may find a *subsurface geologist* determining the character and attitude of the rocks by studying the logs of wells, or a *photogeologist* poring over pairs of pictures to outline the contacts between types of rocks, or a *paleontologist* determining the age of certain beds from a study of their fossils, or a *geochemist* studying the cuttings from deeply buried sands, or a *microscopic lithologist* correlating strata by their mineral constituents. All of them are devoted to a single end, the finding and the production of the greatest amount of oil.

The geologist's duties do not end when oil is found. He advises when and where more wells should be drilled and whether to drill deeper sands in the hope of finding additional production. Many geologists become *valuation engineers* and specialize in estimating the value of properties to be bought or sold by estimating the amount of oil that may be produced from them.

The field of geology is wide, and in it are many specialists. The oil man irreverently calls them all *rock hounds* and their assistants *pebble pups*.

The Land Man

After the geologists have left your territory a friendly but businesslike fellow may call on you. He is the *land man* or *lease man*, and his mission is to lease the areas recommended by the geologists. His prime qualifications are a knowledge of human nature, a sound trading instinct, and a keen understanding of values. If he and his principals are reputable he will not try to trick or outtrade you, for a resentful lessor is no asset to his lessee. The land man never knows, moreover, how soon he may have occasion to take other leases in the neighborhood, and he therefore wants your good will as well as a lease on your land. The first-class land man has learned—if he needed to—that honesty is the best policy.

The Scout

While you are negotiating with the land man you may notice another keen-eyed chap in the neighborhood, the *scout* for a rival company. His business is to keep his employer informed of the activities of other oil operators. If someone starts a well, he must keep in touch with its progress and make frequent reports on it. If strange geologists start work in his territory he must try to learn what they are doing and why. If someone launches a leasing campaign he must report it before the campaign is well under way. He is his company's field intelligence service.

The scout is a versatile individual. He must know something about geology, something about drilling, and much about drillers. He must have a working knowledge of lease practice and a persuasive personality. He must be able to gain a driller's confidence, and then be able to check what the driller tells him. The number of *stands* of drill pipe in the derrick tells him the depth of the well; a glance at the *shale shaker* tells him the geologic formations being penetrated; and the look in the driller's eye usually tells him when he is being lied to.

In most of the active districts scouting practice has been greatly improved by the substitution of co-operation for secrecy. Once, scouts swapped information secretly, chiseled facts where they could, and guessed at the rest. Now they hold weekly meetings and exchange well logs and other factual data, but the agenda do not include geologic

SOME OIL MEN

The rockhound reeks with his lore profound
Of domes and salt cellars underground;
Mysterious depths pretends to see--
Then guesses at where some oil may be.

The leaseman lives with a sole design;
His heart is glued to a dotted line.
His goal is single, his purpose pure;
All he wants is your signature.

The rig builder hopes to end on high
As the derrick grows from earth to sky;
But a slight misstep on an upper girt
And he ends below when he hits the dirt.

ideas and leasing information, which the scouts must learn as best they can. Scouting continues to call for alertness.

The Drilling Crew

Some day, long after you and your neighbors have signed your leases and perhaps have begun to doubt the good intentions of the company, behold! work starts on a well. It probably will not be on your land—fate is seldom so kind—but it may be in your neighbor's pasture where you can see the derrick. Here comes a fleet of trucks, bogging down in the slough, roaring away up the hill, and finally unloading at the selected spot a stack of ungainly machinery and steel parts. With them or shortly behind them comes the *drilling crew*. With the apparently casual efficiency of experts they unload, erect, and assemble derrick and engines, pumps and hoists, and may be "cutting hole" the same day. At the head of the drilling crew are the *drillers*, proudest men in the oil business, seasoned veterans of the derrick floor.

There are two methods of drilling wells, and consequently two types of drillers. In the *cable-tool method* a heavy *bit* suspended from a cable is raised and dropped in the hole, pounding its way down through the rocks. In the *rotary method* a bit is rotated on the end of a string of *drill pipe* and grinds its way down.

A cable-tool driller must know by the feel of an inch-thick steel cable how his two-ton tools are performing at the bottom of the hole— a hole that may be only six inches in diameter and a mile or two miles deep. A rotary driller must tell what is happening down the hole by the sound and speed of whirring machinery and an array of gauges and recording instruments. In either case the driller must be cool, vigilant, and resourceful, and must be something of a steam fitter, plumber, blacksmith, carpenter, and stationary engineer. He has usually worked in a dozen oil fields in the United States and perhaps in two or three in foreign countries. He is a rover, with none of the rover's irresponsibility. There are three or four drillers in each crew, one for each *tour*. The word rhymes with hour and would be called a shift in most businesses.

The cable-tool driller's assistant and right-hand man is the *tool dresser* or *toolie*. When the heavy bit is worn, the toolie and the driller heat it, then "dress" it by hammering it with a sixteen-pound sledge, but this light exercise is only one of the tool dresser's many duties. For smooth and instant co-ordination, a good driller and tool dresser are a pleasure to watch. A score of intricate operations, varied to meet the conditions of the moment and requiring perfect team work, are performed with never a word and scarcely a signal, as though controlled by telepathy. The tool dresser, of course, expects to become a driller. He is usually sure that he is already better than any driller alive except the one with whom he is drilling.

The rotary driller has no tool dresser, but has four or five helpers, skilled men who perform the intricate and strenuous duties that are incident to rotary drilling. In most crews they are designated as the *derrick man, lead-tong man, back-up man,* and *motor man,* according to the principal functions they perform when the drill pipe is being "pulled" or "run." In perfect co-ordination, a rotary crew surpasses even a cable-tool crew, and the term *roughneck* by which rotary helpers are commonly called is a proud title devoid of stigma.

The *roustabout* does the common labor of the field and the odd jobs about the rig; he aspires to be a roughneck or tool dresser.

The Tool Pusher

Over the drilling crew is the *tool pusher.* He has charge of one or two or three or four wells, and sees that they have supplies and equipment and keep going diligently. The statement that "Harry Gray is pushing Prairie tools in Ada" does not depict Harry at the handles of a plow; it means that Harry is overseeing the drilling of Prairie Drilling Company's wells in the Ada district of Oklahoma.

The Completion, Logging, and Testing Specialists

At last comes the fateful day when the well reaches the *sand* in which oil is expected. Everyone holds his breath while the drilling crew "drills-in" the well and specialists "core," "log," or "drillstem test" it. The well may be dry—at least four out of five wildcat wells are, you know—and that will be that. It may be an oil well or a gas well, to everyone's joy. Or it may be a *teaser,* with a *show* of oil but not enough to be of commercial value.

In recent years science and the oil industry have developed more and more special techniques and tools for measuring the characteristics of a sand and its contained fluids. Contractors (called *service* companies) now will, upon the tool pusher's call, drillstem test his well, or log it, by one or more of several methods. The service company men arrive at the well as fast as their cars and trucks can get them there, and within minutes the *drillstem test engineer* may be supervising the drilling crew in lowering several thousand dollars' worth of his company's special pressure gauges and packers into the hole at the bottom of the drill pipe to sample the well's fluids and pressures. The loggers will lower their *electrodes* and *sondes* into the well on a cable, and deliver a record of the well's electrical or radio-active characteristics, temperature, and diameter. Other common logs are records of sound-wave velocities and "dips" or tilts of the rocks the well penetrates. The loggers and testers have come to be indispensable to well drilling and completing, although their contribution to a well may be made in a few hours.

The logs and the drillstem test may show that the sand is "tight," which means lacking in porosity. If so, a *well-shooter* may arrive in short order in an *Explosives* labeled truck. He makes his living by exploding nitroglycerine in wells to shatter the producing sands and improve the productivities. The truck may contain enough "soup" to level a hill.

Well-shooting was formerly one of the most dangerous occupations. A hard bump in the road or a slip at the well head, and the well-shooter could not be found or assembled for burial. Improved technique and materials have lessened the hazards, and shooters now can give thought to such things as arteriosclerosis and old-age pensions.

Many tight formations can be made to yield their oil or gas by *fracing* (it rhymes with cracking), in which case we see several trucks drive up loaded with high-pressure pumps and a ton or so of special materials. The *fracer* creates cracks in the oil-bearing formations with a particular mixture of sand and oil or other ingredients injected at high pressure into the well, thus providing access to the well through the erstwhile-tight formations.

Where nature has cemented the oil-bearing zones with limestone, the *acidizer* may be called on. He comes with truckfuls of tubing, carboys, and pumps and dissolves the lime out of the sand with hydrochloric acid. Fracing and acidizing require skill and training, and the fracer and the acidizer are usually chemists or engineers or both.

The Production Crew

If the well is dry you have probably seen all the oil men you will see for some time. But let's suppose that this is a producer, that it "makes a well," as the oil men say. In due time, the neighboring landscape is cluttered with derricks, tanks, pipes, warehouses, and what not. These are serviced by a small army of men known as the *production crew*, including the roustabouts, the *bull gang*, the *connection gang*, the *clean-out crew*, and the *pumpers*. The bull gang and the roustabouts dig ditches, wrestle pipe and lumber, and perform the other less-skilled jobs about the field. The connection gang connects the pipes and valves and installs much of the equipment. The clean-out crew has a light drilling rig and cleans out wells that accumulate mud or sand or paraffin wax. In a few months or years, when the *flush* production is over and the gas pressure will no longer "flow" the wells, the field will fall to the pumpers, who through the many years of its slowly declining life will see that the wells are pumped efficiently. The pumper is the lingering rear guard of the oil business; not infrequently he does a little farming on the side.

The Production Engineer

Watching the seeming chaos while the field is developing, you may wonder who does the thinking for all these men. The answer is three-fold: The production crew are experienced men and do most of the thinking for themselves; over them are foremen and field superintendents who do the larger planning; and back of these are production engineers whose business it is to see that the most oil is produced for the least money. The *production engineer* is a rapidly increasing member of the oil fraternity. He may be part geologist; his primary function is to prevent waste, whether of oil, of the gas that produces the oil, of materials, or of labor. He advises where casing should be set to keep water from flooding the oil sand, how much gas should be allowed to flow from the well for each barrel of oil produced, what pumping method and equipment should be used, and a multitude of similar things. He is the conservation engineer of the production department.

The Pipeliners

After the first few wells have been completed, a new gang of men appears in the field. This is the *pipeline crew*—tough, hard-boiled, and efficient. They will lay the pipeline, a dozen or a hundred miles long, to carry the oil from the field to the nearest existing pipeline or to the refinery. They could be arrested for their songs and jokes, they'll get out of hand if given half a chance, but they'll lay a good line from here to there and vanish like smoke when the job is over.

The lordly driller begins to smile
When his tools get down past the second mile,
And asserts, in words that would shock the ladies,
That he'll drill you a round-trip well to Hades.

The pumper nurses the pumping wells,
Sits up nights with their sinking spells,
Gives them shots when their output drops--
And the rest of the time he tends his crops.

The pipe line crew are a hardy lot,
Their heads are cool while their joints are hot.
They struggle and sweat till the job is right--
Then promptly bury it out of sight.

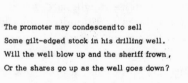

The promoter may condescend to sell
Some gilt-edged stock in his drilling well.
Will the well blow up and the sheriff frown,
Or the shares go up as the well goes down?

Their place will be taken by the *pipeline operating crew: plant men* who tend the big pumps that push the oil through the line, *tank gaugers* who measure oil in the tanks to determine the amount being run from each lease, and *line patrolmen* who walk, ride, or fly along the line at frequent intervals to detect leaks.

The Refinery Crew

If we want to see more oil men we will have to follow the pipeline to the refinery. There we will find *pipe gangs* who have the skill of the best steam fitters and who install equipment for pressures of which ordinary steam fitters never dream, *stillmen* who have charge of the great stills in which the oil is refined, *lube-house men* who work with the oils and greases, *wax-house men* who work among the filter presses that extract the paraffin, roustabouts who do the unskilled labor, and many others. In the laboratory are the chemists, constantly checking and testing the crude oil coming in and the refined products going out, and doing research on possible new products and on better ways to make those now being manufactured.

Over them all are other petroleum engineers who design, build, and often operate the refineries. So inventive are they that refinery design and technique have been advancing with disconcerting speed for many years. One of their specialties is labor-saving efficiency, and a modern refinery uses less labor per unit of value of output than almost any other manufacturing unit.

The Traffic Men

At the refinery we are likely to see a whole fleet of tank trucks or perhaps a string of tank cars being loaded with gasoline and other refined products. On our way home, the tank trucks remind us to be thankful for highway passing lanes, and we can see the tank cars on sidings here and there, waiting to be unloaded and sent back for more. The traffic department has been at work. The *traffic manager* and his assistants sit in the office like spiders at the center of a web, and shuttle tank trucks and tank cars to the four corners of the territory, a truckload of lube oil here, six cars of gasoline there, ten trailerloads of jet fuel yonder. They may send even more barrels of products to market through pipelines. On them depends the prompt delivery of the company's products, with the most efficient use of trucks and tank cars.

The Wholesaler

Notice the warehouse and storage tanks beside which that string of tank cars is standing? That's a bulk station, from which gasoline and lubricating oils are distributed to service stations. Its owner may be the company whose refinery we have just left, or a wholesaler or *jobber* not connected with any producing or refining company. He may distribute through his own service stations or he may sell to service stations owned by others. Some wholesalers and jobbers are big, some are small. Some serve only a few stations in one locality; some distribute to hundreds of stations in several states. Big or little, on his own or employed by a refiner, the bulk station operator is an indispensable link between the refinery and the service stations.

The Retailer

Halfway home we stop for "five of gas and a quart of oil." There we make contact with the final department in the great business of making a hundred million cars go. Everybody knows all about service stations, or should—at any rate there are too many of them. If all business operated on the same scale of service, the butcher would bathe your dog whenever you buy a pound of scraps. The service station attendant is the contact man between the industry and the public. His first name is service, his second celerity, and his surname is courtesy.

The Sales Staff

Back of the service station attendant and the owner or lessee of the station is the sales force of the manufacturer or jobber whose products are sold.

On the road, we passed one of them bringing a truckload of gasoline from the bulk station to the service station. He is called a *tank-truck driver*, and closely related to his job is that of the *route salesman*, the man who delivers heating oil to your home. Each of them is the man who, in spite of snow or heat, flat tires or motor trouble, must deliver the goods where and when they are wanted, and who by his efficiency and courtesy must keep the service station men and the homeowners on friendly terms with the refiner or jobber.

Back of the tank-truck driver are the refiner's or jobber's local agents, district supervisors, traveling representatives and inspectors,

The skillful stillman is seldom still;
Much depends on the stillman's skill;
If he opens a valve that he ought to close,
Up with the still the stillman goes.

The tank-truck man has
      a one-track ear,
If you honk to pass him
      he just can't hear;
But whisper a station is
      short of gas
And he's gone before
      you can try to pass.

The station attendant, with zealous care,
Will check the water and check the air,
Wash all the glassware round and about--
And your B.V.D.'s, if you don't watch out.

the sales engineers who advise customers regarding fuel and lubricating requirements, the advertising men who proclaim that "Grumble Gas Never Gripes," the district sales managers, the general sales managers, and so on up to the vice-presidents in charge of distribution.

The men in the marketing department contend that theirs is the most interesting and essential part of the business, that they are the front-line troops—the producers, transporters, and refiners acting merely as the service of supply. That's one thing that gives zest to the oil business, the fact that each department considers its work the most exciting and important of all.

The Promoter

Before we go home I want you to meet another important oil man, the *promoter*. He's drilling a well over in the next county, or perhaps he's assembling a block of leases on which he hopes to get a test drilled, or he may be building a refinery somewhere. He's rushed to death, but he'll be glad to stop long enough to sell you a few shares in his company or a spread of acreage around his well or an interest in his block of leases.

Let's not get this man wrong. The term "promoter" seems to arouse suspicion in some circles, but the promoter plays a valuable role in the oil industry. His name has been besmirched by fake stock peddlers who never promoted anything but their own pocketbooks; but the legitimate promoter who has a vision and is trying to translate it into an accomplishment is likely to be a pretty straight shooter. He may be a geologist or a driller or a lease hound or a refinery man who prefers the risks and rewards of his own business to the so-called "security"[1] of a company payroll; he may be a businessman with oil experience who is attracted by the speculative possibilities of oil development. He may drill the well or build the plant himself or he may do the preliminary work and then turn the deal over to one of the big companies. He is not always popular, nor even recognized by some companies, but the discovery of a reasonable proportion of new fields is due to his activities. If he is honest and intelligent, he is worth listening

---

[1] The "security of tenure" about which company executives used to talk to ambitious employees proves to be something of a grim joke during depressions, although a few companies make it a reality.

to, bearing in mind always that his enterprises are strictly speculative, and assuring yourself that you are listening to a real promoter and not a mere slicker.

A Word of Caution

Your new acquaintanceship with oil men may be of value here. The men we have met are all specialists in some department of the business and may know little or nothing about other departments. Keep that fact in mind. When asked to invest in such and such a wildcat because it has been located by "an old Standard Oil man," or by "the man who drilled the discovery well in Seminole," find out whether the man is a geologist. Or, when told that a production superintendent recommends building a refinery, check up on his knowledge of refining and marketing. You would not take farming lessons from a baker nor let a farmer design your flour mill, though both could appropriately be called wheat men. Use a little caution when you deal with oil men; philanthropists are as rare among oil promoters and oil investment bankers as they are in most other businesses.

Now for Some Geology

Now we have met most of the men in the oil business, except the executives, lawyers, bankers, accountants, and clerks without whom no business can run. If you have enjoyed meeting these men, perhaps you would like to watch them at their work. If so, let's begin at the beginning; let's pick a good geologist and go with him while he hunts for a promising place to drill.

CHAPTER 3

Picking the Place
To Drill—The Theory

What the Geologist Seeks

Oil is not a quick distillate resulting from sudden heat such as that of a volcano; it comes from the slow distillation of organic material in *source rocks* that were laid down as sediment on the bottom of ancient seas. It is not found in underground lakes filling vast caverns or in streams flowing in subterranean channels; it is found in the minute pores or voids of *reservoir rocks* such as sandstones and some limestones. It does not collect in the troughs and low places of such rocks; it accumulates in *traps* in their higher parts, held there by water that fills the pore spaces of the rest of the reservoir rock.

The three essentials for a commercial accumulation of oil are, therefore, source beds, reservoir beds, and traps, and these are the things for which an oil geologist looks. If there has been no source, there will be no oil—so he looks for beds that seem to have contained enough organic material to have furnished some oil. If there is no porous or reservoir bed the oil will still be disseminated through the source beds, so he looks for a rock stratum porous enough to hold fluids and pervious (the oil man's term is "permeable") enough to permit their relatively free movement and accumulation. If there is no trap the oil will be distributed through the reservoir bed in quantities too small to be useful, so he looks for a place where the folding or the termination or interruption of the reservoir bed may have caused a concentration of the oil.

The source beds will probably be shales or limestones which, when they were deposited as mud or ooze on the sea floor millions of years ago, contained an abundance of organic remains, either animal or vegetable or both. The reservoir bed will probably be a sandstone, but may be one of the porous varieties of limestone or its cousin dolomite. The structure or trap will probably be an arch or upfold in the strata, formed by the folding or settling of the earth's crust in the course of some crustal readjustment, or a place where the continuity of the reservoir bed is interrupted or terminated.

There is nothing mysterious about all this. Geologists have observed what is happening on the earth's surface today and have projected their observations back through the eons of geologic time. We can do the same, with no more equipment than our eyes, open minds, and a little imagination.

Degrading Forces

The hills and valleys you crossed last Sunday, the canyons and mountains that awed you on your vacation, were not carved out by a single stroke, to endure for all eternity. The face of nature is not a fixed and static thing. Rain, snow, frost, wind, and vegetation, the plastic surgeons of geologic time, are busy remolding it. Frost and roots shatter the rocks within their reach. Wind carries sand and acts as a sand blast, wearing away exposed surfaces. Percolating waters, supplied with humic acids from vegetable decay, complete the disintegration, and what were solid rocks become clay and sand. Rain and rivulets wash away this soil, exposing fresh rock surfaces to frost, roots, and percolating waters. Wind aids and abets them. The rivulets and the wind deliver the mud and sand to the creeks, the creeks deliver it to the rivers, and the rivers carry it to the sea where waves and tides spread it out over the floor of the ocean.

These forces of erosion and deposition, wearing away the land and dumping their materials into the sea, work slowly but unceasingly. There is no haste. They have been steadily at work for hundreds of millions of years of the earth's history; they work as steadily today; ahead lie other hundreds of millions of years. What has been is and will be; the surface of the earth has been, is being, and will be shaped by infinitesimal tools working through near-infinite time.

Do you doubt that frost and raindrops can wear a mountain down? Watch a hard rain making gullies on a bare hillside. Dip a pail of water from a muddy creek, put it on the stove until the water is gone, and see how much sand and soil was on its way somewhere. Multiply this pail by the water in all the creeks and rivers, and see whether your imagination can comprehend how much of the land is traveling to

The valley has been made by the stream that flows through it. Rain, frost, wind, and vegetation have shaped it and its enclosing hills. The solid rocks that once filled the valley and covered the hills are now sand and mud and ooze on the bottom of the sea to which the water in the stream flows.

the sea. Then perhaps you can realize that on what is now the Piedmont Plateau, from New York City to Atlanta, once stood a mountain range more majestic than the present Appalachians, to which the Piedmont is now only a low and insignificant footstool; or that more of the Rockies has been removed, perhaps, than still remains. The eternal hills are eternal only when compared to the brevity of man's existence.

Crustal Movements

If these forces of degradation held sway unchecked, the earth's surface would long ago have been reduced to flat, low, monotonous land areas, washed by the waves of shallow seas. Fortunately, counterbalancing forces are at work. The earth's surface is not static or stable; great segments of it are slowly rising, other great segments are slowly sinking. Rising movements, long continued, result in new land areas —lowlands like the Mississippi Valley, high plains like those east of the Rockies, and mountain ranges like the Rockies themselves. Despite many theories, we know little of the causes of these crustal movements. One rule seems to obtain: The sea floor most heavily loaded with sediments is most likely to be the locus of the next great uplift.

Whatever may be the causes, the earth's crust moves. The east coast of North America, from Cape Hatteras northward, has been slowly sinking; note how far the sea has encroached up old river valleys like those of the Potomac, the Delaware, the Hudson, and the St. Lawrence. The west coast of South America is probably slowly rising; note its lack of indentations and the steep gradients by which its rivers drop to the sea.

Time and Mountains

These are not sudden cataclysmic things, these movements that hoist up continents and buckle them into mountain ranges. They are slow, scarcely faster than the forces working to counteract them; sometimes they are even slower, and the land wears down faster than it is being raised. For the degrading forces do not wait until the uplifting movement is complete; as soon as the land area emerges from the sea they attack it and begin to wear it down. Thus the two sets of forces work simultaneously, one uplifting, one degrading—with sometimes one, sometimes the other, gaining in the endless competition.

Men used to look at mountain ranges and speak of "convulsions of nature." Some still think of them in terms of sudden outbursts like those of a volcano. Such a concept is misleading. Volcanic eruptions are a result, not a cause, of mountain uplifts; and though nature may have her convulsions, most of her shudders are far too slow for human observation.

When you drove through Wind River Canyon on your way from Casper to Yellowstone Park, or down the Hudson Gorge through the Catskills, or along the Columbia above Portland, you may have asked how the river found such a convenient chasm through the mountains. The answer is simple: The river did not find the chasm; it made it. The river was there before the mountains. As the mountains grew the river maintained its channel, cutting its groove deeper through them as they rose. Now the mountains tower above the river, and may—who knows?—be still growing. Thus, slowly—inconceivably slowly—are mountain ranges uplifted.

Rocks Formed Beneath the Sea

What happens to the great masses of sediment that are carried into the sea? When the streams that carry them enter the quieter waters of the sea, the currents of the streams are checked and begin to drop their load. The coarser sediments such as gravel and sand are dropped first; the muds settle farther out in quieter waters; the lime and other soluble materials, carried in solution instead of in suspension, are slowly precipitated still farther out or become the shells and skeletal structures of myriads of minute organisms, and sink to the bottom as these myriads die. The waves play an important part, especially near the shore. They distribute the sand along the coast and move the finer sediments out to deeper waters.

As the sediments pile up to thicknesses of thousands of feet, those at the bottom are weighted down by those above; much of the water is squeezed out and compaction into rocks begins. Then sometime, when a crustal movement starts and what was sea floor gradually becomes an upland, the sediments are subjected to great and shifting pressures through hundreds of thousands of years. Under these pressures, compaction is completed and the sediments are transformed into solid rocks. The sands become sandstones, the muds become shales, and the limey oozes becomes limestones. The substance of the mighty cliffs that look down on Canada's Lake Louise was once ooze on an ocean floor; bit by bit it is on its way to becoming ooze again.

Streams carry the disintegrated fragments of the rocks to the sea, and waves, currents, and tides spread them over the sea bottom. The sand forms beaches and bars along the shore, the mud settles farther out, and the lime dissolved in the water is taken into the shells and skeletons of sea organisms which lie and leave their remains as ooze on the bottom, usually still farther from shore. Shiftings of the shoreline, either by erosion or by movements of the earth's crust, result in sand, mud, and ooze being piled up in alternate layers or beds, which in time compact and solidify into sandstones, shales, and limestones.

Other Sedimentary Rocks

Not all of the soil formed by the disintegration of the rocks reaches the sea. Some may be deposited in inland lakes and ponds. Some may be dropped in the channels of the streams or may form broad flats and terraces in valley bottoms. Torrential streams in steep mountain channels may dump their loads when they emerge from the mountains, forming *alluvial fans* and *outwash plains*. Sand may be blown by the wind and piled up into dunes. Clay may be carried by the wind and spread over wide areas, perhaps forming deposits of great extent and many feet in thickness like the famous *loess beds* of China. Any or all of the beds so laid down, in lakes or valleys or on dry land, may eventually become solid rocks. Those laid down in lakes are called *lacustrine* deposits; the others are called *terrestrial* deposits. Together they are known as *non-marine beds* to distinguish them from the marine beds laid down beneath the sea.

Rocks formed from sediments, whether marine or non-marine, are naturally referred to as *sedimentary* rocks because they are usually laid down in layers or beds or strata. They are also called *stratified* rocks.

Igneous and Metamorphic Rocks

Some rocks were not formed from sediments but have been melted and have solidified from a molten state. These are the *igneous* rocks. Those that have reached or nearly reached the surface while still molten are called *lavas*; they form *volcanic cones*, or spread out in *flows* or *sheets*, or insert themselves as *sills* between beds of other rocks, or squeeze and melt their way up through other rocks to solidify as *dikes*. Rocks that have solidified far beneath the surface are called *plutonic*. Granite is the most widespread igneous rock; it is usually a plutonic rock rather than a lava. Basalt and rhyolite are common lavas.

When either igneous or sedimentary rocks are subjected to enough heat and pressure their character and appearance are changed, chiefly by rearranging their mineral constituents along planes or zones roughly perpendicular to the direction of the pressure. The deposition of mineral matter by water being squeezed out of the rocks also plays an important part. Either granitic rocks or masses of interbedded shales and sandstones become gneisses and schists, sandstones become quartz-

Sedimentary rocks are composed of sediments deposited under water or laid down on the land by streams or winds. Igneous rocks have solidified from a molten state. Metamorphic rocks were once sedimentary or igneous, but have been changed by heat or pressure. All three classes of rocks—sedimentary, metamorphic, and igneous—can be seen in this cliff.

ites, shales become slates, and limestone becomes marble. All of the rocks thus formed by the metamorphosis of other rocks are called *metamorphic* rocks.

Igneous and metamorphic rocks form the cores of most mountain ranges. In some ranges the core is still deeply buried beneath sedimentary rocks, but as a rule erosion has removed the overlying sediments and cut deeply into the igneous and metamorphic core. The scenery of the American Rockies, the Sierra Nevadas, and the Alps, for example, is mainly sculptured in granite, gneiss, schist, and marble.

Rocks Favorable for Oil

Igneous and metamorphic rocks are important to the oil geologist only because they form the *basement complex* beneath the sedimentary rocks in which oil may be found, and because the debris from their erosion has furnished a large part of the sediments from which sedimentary rocks are formed. Otherwise, igneous and metamorphic rocks are anathema in oil geology; except in rare and special instances, no oil is found in them.

Terrestrial deposits are little more favorable. They are unlikely sources of oil. Some lacustrine deposits may be favorable; most of them are not. The great source of oil is in marine beds.

The Source of Oil

The sediments carried to the sea are by no means pure mineral matter. In them are leaves, twigs, tree trunks, fish, and carcasses. Bits of all the organic matter of the land and its inland waters are carried out and buried in the mud of the ocean bottom. Marine plants and animals, from seaweed and squids to sturgeon and whales, die and add their carcasses. The billions of microscopic organisms that live in the sea add their skeletons to the muddy ooze and their flesh to its organic content. Most important of all, perhaps, are the countless minute vegetable forms known as algae that live in the water. Invisible or nearly invisible to the naked eye, their number is so great that in the aggregate they add millions of tons to the sediments. Thus, some of the muds and oozes, rarely the sandstones, may come to contain a notable percentage of organic matter.

The pressure and heat that transform the sediments into rocks convert the organic matter, through various intermediate stages, into oil and gas; newly-formed shales and limestones may contain billions of

SOURCES OF OIL

Oil is formed from animal or vegetable materials included in the sediments that in time become sedimentary rocks.

minute particles of oil and gas. These particles are generally too small and too widely scattered to be discernible, much less of any commercial importance. These, nevertheless, are the source beds for which the oil geologist looks, basing his conclusions on what indications he can find as to their original organic content.

Steps in the Formation of Oil and Gas

That organic material may be transformed into oil and gas by long-continued heat and pressure, perhaps aided by radioactivity and bacterial action, is well established. Some chemists have said that the temperatures developed in the course of earth movements are not great enough to distill organic material into oil and gas. The geologist replies that they overlook the time factor; that a little heat through a million years may produce the same result as much heat for a few hours.

How much of the source material may be transformed into oil and how much into gas is another question. The answer is still largely theoretical, though the theory has a number of facts to support it.

It would seem that the first step in the transformation is to a waxy substance or group of substances loosely called *kerogen*.[1] In many parts of the world, and particularly in northwestern Colorado, northeastern Utah, and southwestern Wyoming, are great deposits of kerogen-bearing shales known as *oil shales*. These contain very little actual oil, but if they are subjected to heat, a ton will yield from a pint to a couple of barrels of oil, and some gas. Wherever there has been even a slight amount of folding or faulting of shales, with the consequent development of pressure and heat, some of the kerogen has been changed into heavy, viscous oil, much of it almost pure asphalt and part of it only semi-liquid.

Along the Athabaska River in Northern Alberta, Canada, are the famous *tar sands* containing more oil than all the oil fields so far discovered. The beds have been subjected to comparatively slight heat and pressure; they have been folded scarcely at all and have probably never been weighted down by more than a few thousand feet of over-

---

[1] The term *kerogen* is in some disfavor as being indefinite and as having been commonly used with a precision that is not warranted. It is used here, for want of a better word, to designate the whole indefinite collection of waxlike substances that are intermediate between the original organic material on the one hand, and oil on the other.

lying rocks. The oil is heavy, viscous, and asphaltic, but when heated moderately for a reasonable time—say six hundred and eighty-five degrees for thirty minutes—it becomes a much lighter, more fluid oil comparable to that found in many oil fields.

From these tarlike oils the progression goes upward through the heavy oils of Mexico and the San Joaquin Valley, the lighter oils of the Los Angeles Basin, and the still lighter oils of the Mid-Continent and Appalachian fields, and finally arrives at fields that contain no oil—only gas. In general, though with many exceptions, the rule seems to be that the more the rocks have been folded and faulted, the more they have been subjected to pressure with resultant heat, the lighter is the oil contained in them.

Other factors have their effect, of course. The nature of the source material undoubtedly has much to do with the result, and may explain why one bed consistently carries gas, another oil, in the same area. Contact with underground waters rich in oxygen or sulphur may make an oil asphaltic and heavy. There are probably other factors not yet guessed. Some competent students think these factors, known and unknown, may be more important than pressure and heat. They ask, and with some reason, at what point and how is an asphaltic oil transformed into a paraffin oil?[2]

Nevertheless, with full recognition of the fact that we are in the realm of theory and that our theory is not universally accepted, it would appear that heat and pressure, and perhaps radioactivity and bacterial action, transform the original organic material into kerogen, kerogen into asphalt, asphalt into heavy oil, heavy oil into lighter oils, lighter oils into yet lighter oils, and the lightest oils into wet gas—until the final product is a dry gas no part of which is liquefied under present commercial conditions. Some gas is probably formed in the earlier as well as the later stages of the process, so that gas may be a side product as well as an end product.

It is not likely that all the steps of this transformation take place in the source beds in which the organic material was originally deposited. The first steps, transforming the source material from a solid to a fluid or semifluid state, undoubtedly take place there. Subsequent steps in the transformation may take place in the source beds or in the reservoir bed or even after the oil has come to rest in a *pool*. That's getting ahead

---

[2] Chapter 11 gives more on oil chemistry.

of the story. Let's go back to our minute particles of oil and gas just formed in the source beds.

Reservoir Beds

If these minute particles stayed in the source beds they would be of no value to anyone. Happily, however, the compaction of the rocks squeezes some of the fluids, including particles of oil and gas, out of the source beds into the more permeable beds.

The muds and oozes in which the oil originated are fine-grained, and when transformed into shales and compact limestones they have almost no permeability. In the compaction, they may be compressed to a fraction of their original volume, with a consequent squeezing-out of most of the water which carries with it the newly-formed particles of oil and gas. The fluids thus squeezed out go into the more sandy beds which are not so compressible, and these lose little volume during their compaction into sandstone and sandy limestones; consequently, they are relatively porous and permeable.

A sandstone is a hard rock and looks solid, but from 5 to 30 per cent of its volume is made up of voids or pore-spaces capable of holding fluids. Put a piece under a microscope and see how spongelike it looks. A sandy limestone or a limestone, part of which has been dissolved by water, or a dolomite formed from a limestone, may also be somewhat porous. These are the reservoir rocks for which the oil geologist is ever seeking.[3]

Oil on the Move

When a particle of oil makes its way into sandstone it is not likely to find itself in an oil pool. Instead, it finds itself lost in a great volume of water slowly moving from where it is to somewhere else. This may be sea water entombed in the sediments since their marine deposition. It may be water that has entered along some outcrop and is being forced onward by hydrostatic pressure. The rocks that were laid down in the sea floor have been folded during the period of uplift and now dip down into great troughs or arch up over great uplifts. On the major arches, erosion has been most active; on some, it has cut down to and

---

[3] Shales retain much of their porosity when they are compacted from mud, but they are so fine-grained that they have little permeability—they contain much pore-space, but the fluids contained therein cannot move between spaces.

exposed the edges of the deeper-lying sediments. Into such outcrops rain, melting snow, and running streams feed a never ending supply of water that makes its way slowly into and across the adjacent basins, pushing ahead of it, perhaps, *connate* or *fossil* sea water that the sediments may still contain. Thus, water that enters the Dakota sandstone west of Denver may eventually emerge in springs and wells in central Kansas. In such a sandstone, filled with slowly moving water, the oil particle also begins to move, and here a difference of opinion develops.

Structural Traps

According to the older school of geologic thought, the oil, being lighter than the water, moves uphill up the slope or *dip*, regardless of the direction in which the water may be moving. Eventually, it may come to a minor arch or *anticline* in the rocks; it can go no farther without traveling downhill. There, because of its buoyancy, it remains— deeply buried beneath the surface. Gradually, it is joined by other particles of oil and gas until, on the crest of the anticline, the voids of the sandstone are filled with oil (and probably some gas) and a *commercial field or pool* has formed.

According to a newer school, the particle of oil does not move against the current of the water any more than a chip floats upstream. Instead it travels with the water, whether up the dip or down, until it comes to a place where the water movement is checked or ponded by an anticline. There it remains, to be joined by other droplets until the sandstone pores on the crest of the anticline are filled with oil and gas and a commercial pool has accumulated.

Modern reservoir engineering combines the two schools of thought by combining the factors of each quantitatively with a certain amount of arithmetic. The result helps find petroleum (or helps confuse the finding) and is known to petroleum engineers as *hydrodynamic theory*. Some geologists think the oil travels long distances, and they lay much emphasis on *drainage areas* from which the oil in an anticline may have come. Other geologists think that the amount of migration is slight, and that the oil in an anticline has originated locally, even within the limits of the anticline itself. There is evidence in favor of both views, and each may be correct in specific cases. One field may have derived its oil locally; the oil in another may have come from a distance.

Whichever of these schools of thought you choose will bring you into

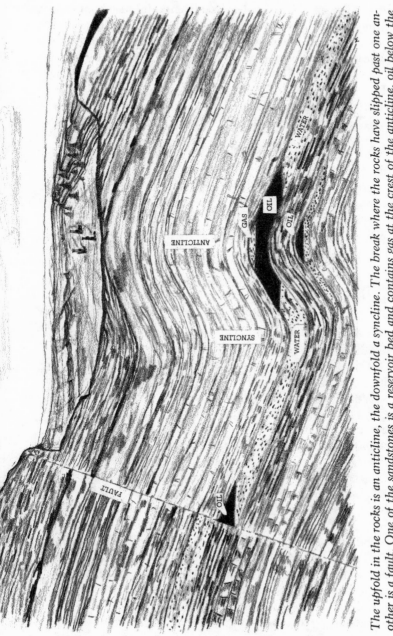

The upfold in the rocks is an anticline, the downfold a syncline. The break where the rocks have slipped past one another is a fault. One of the sandstones is a reservoir bed and contains gas at the crest of the anticline, oil below the gas on the flanks of the anticline, and water elsewhere. It also contains some oil trapped against the fault. The lower limestone contains some oil on the crest of the anticline. Anticlines and faults are called "structural traps."

good geologic company, and all agree that the oil accumulates on the crests of anticlines. When wells are drilled off the anticline, in the intervening troughs or *synclines,* they are likely to produce nothing but water and disappointment.

The travels of our particle of oil may be interrupted before it reaches an anticline. It may come to a *fault,* a place where the strata have broken and slipped past one another, bringing the broken edge of the reservoir bed against an impermeable shale or limestone. There, the particle and its comrades may come to rest, accumulating a commercial pool against the fault.

The attitude of the rocks of an area—the folding or fracturing or displacement or lack thereof, that they exhibit—is called the area's *geologic structure.* The individual folds or fractures or displacements are also called geologic structures. Oil traps that are due to geologic structure, such as those on anticlines or against faults, are therefore known as *structural traps.*

Stratigraphic Traps

Our drop of oil, however, may be caught by a trap that is not due to structure but is due instead to a discontinuity of the stratum in which the drop is traveling. Such a trap is known as a *stratigraphic trap.*

The drop may come, for example, to the ancient shore line of the sea in which the reservoir sand was deposited. There the sandstone ends between beds of shale or limestone. In geologic parlance it *lenses out* or *pinches out.* With no permeable bed along which to travel farther, the oil particle and its comrades are trapped in the edge of the sandstone, and may, if there are enough of them, form a commercial accumulation of oil. East Texas, one of the greatest oil fields yet found, is in such a trap.

We have been assuming that the sandstone in which our drop finds itself is a continuous *sheet sand* underlying a wide area, but this may not be the case. The sandstone may be only a *lens* representing a sand-bar formed on or near the beach of the ancient sea or on some shoal in the sea bottom or a relatively clean turbidity-current sand mass surrounded by silt and mud. Such a lens may be tens of miles long and be completely filled with oil or gas or both. Most of the *shoe-string sand fields* of Kansas and Oklahoma, including such great fields as Glenn Pool and Burbank, and the *Michigan Stray* gas fields of Michigan, are in ancient sandbars; the *Clareton-trend* fields of eastern Wyoming are

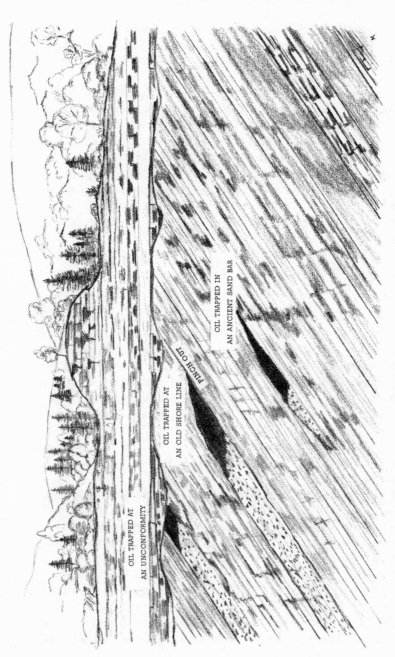

OIL TRAPPED AT
AN UNCONFORMITY

OIL TRAPPED AT
AN OLD SHORE LINE

PINCH OUT

OIL TRAPPED IN
AN ANCIENT SAND BAR

Oil may accumulate wherever its progress through a reservoir bed has been stopped by a discontinuance of the permeability of the bed. These are called "stratigraphic traps."

in sands left by turbidity currents. Another type of lens may represent the sand and gravel deposited in an ancient creek or river bed, in which case it will be narrow and sinuous and completely surrounded by shale. Some of the smaller productive *shoe-strings* of extreme eastern Kansas are of this character.

Traps at Old Erosion Surfaces

Another set of conditions may produce stratigraphic traps. In those we have considered so far the discontinuity of the reservoir bed has been due to discontinuity of deposition. In another class of stratigraphic traps, the bed in which the drop of oil is traveling may originally have been continuous, but a part of its original extent may later have been eroded away, and its truncated edge covered by younger beds.

When the series of sediments, of which the sandstone is one, were elevated and became dry land and were folded in the process, some parts of the series were raised up into arches, as we have seen, and some parts folded down into depressions. The arches may be comparatively narrow like the anticlines we have been considering or they may be broad, low arches or they may be broad, sharper arches such as form mountain ranges. Erosion begins as soon as the area emerges from the sea. On the arches, and particularly on the broader ones, the upper sediments may have been eroded away until the sandstone we are considering was exposed and then eroded—although in the basins it may still be covered by hundreds or thousands of feet of other sediments. Then the area may again be submerged beneath a new series of sediments.

The erosion surface at the top of a series of sediments on which younger sediments have been laid down is called an "unconformity." If the older sediments have been folded or tilted before the younger sediments were deposited, the beddings of the two sets of sediments will not be parallel but will be at different angles or dips from the horizontal. The unconformity is then called an *angular unconformity*. If the area is again uplifted and becomes dry land, the surface will be up in the younger sediments, and the older sediments will be buried somewhere below. During the uplifting movement, new folding may take place which affects both the older and the younger sediments, but the angular unconformity between the two series of sediments will, of course, remain.

Now if our drop of oil comes to such an angular unconformity in traveling along through the sandstone into which it has migrated, it will have come to the truncated edge of its sandstone. If the truncated edge is overlain by impermeable beds, the drop and its companions may be sealed into the sandstone—and may form a commercial accumulation.

The beds above the unconformity may of course contain permeable as well as impermeable beds. Suppose the truncated edge of the sandstone is in contact with a permeable bed in the younger sediments. The oil will then continue its journey in this permeable bed until it emerges somewhere at the surface or is caught in a stratigraphic or structural trap.

A Word About Reefs

Coral reefs and the islands they form in the South Pacific became public knowledge shortly after World War II, about the time petroleum geologists began to recognize them as important petroleum reservoirs. Such reefs or *bioherms* are formed by several small animal types that live in shallow sea water under the right temperature, current, salinity, chemical content, and food conditions. Each builds his own tiny limestone residence on those of his ancestors below. He lives in it and dies in it and when he dies his body dissolves—leaving a porespace. Thus, reefs grow hundreds of feet tall where the sea floor is sinking about as fast as the little animals like to build. Generally, the porosity left by the hosts is reduced considerably by washed-in silt and the remains of other sea creatures that used the reef for hunting and concealment, but many *limestone-trap* petroleum fields, including Redwater and Leduc in Alberta and the Scurry Reef in West Texas, produce their oil from reefs.

Like any other reservoir a reef must be covered by impermeable sediments to be a trap.

Back to Structural Traps

Most oil fields so far found are in structural, rather than in stratigraphic, traps because many structural features such as anticlines and faults are obvious or easy to detect locally, while their stratigraphic counterparts are usually discovered only after thoughtful studies of the rock samples from outcrops and wells over thousands of square miles. The technique of finding anticlines has been highly developed; that of finding stratigraphic traps presents many problems and is only now beginning to catch up. As we learn more about finding strati-

The ridge or rimrock dips away from the valley in all directions, showing that this is a closed dome. The surface of a dome may be a ridge or hill instead of a valley, or the topography may not express or reveal the existence of the dome.

graphic-trap fields, the number of Burbanks, East Texases, and Clare-
tons is increasing.

Because the technique of finding stratigraphic traps is still in its
adolescence, and because most oil geology is still a search for anticlines,
let's push sandbars and shore lines and unconformities into the background for the present and give our attention to anticlines.

Geologic Structures, Good and Bad

In most areas, except those wholly unexplored, the geologist knows
whether source beds are present and what beds are likely to serve as
reservoirs. His main task is the search for anticlines. He calls it looking
for structures. Any folding or displacement of the rocks is a geologic
structure, as we have seen, but the oil geologist has narrowed the term
to mean a structure favorable for oil accumulation. When he says a
well is "on structure" or "off structure" he means it is on or off the
good part of an anticline.

Not every anticline is a good structure, however. It may be too
big; oil is not found in major uplifts such as mountain ranges, though
it may be found in minor folds on or adjoining them. The anticline
may be too small, merely an insignificant wrinkle in the rocks. It may
be so flat on top that no commercial concentration has taken place.
It may be so steep and sharp that its crest has no room for a commercial pool. Erosion may have cut so deeply into it that the reservoir sand
is exposed and has lost its oil through surface seepage. It may be so
faulted and fractured that its oil has escaped along the fractured zones.
The perfect structure is one of moderate size, not too badly fractured
by faulting, toward the crest of which the beds rise from all directions,
making a trap from which the oil has no escape.

Such a structure is known as a *closed structure* and is often called a
dome. It is not necessarily symmetrical or round; it may be many
times as long as it is wide; its shape may be irregular. The essential
thing is that, as you walk in any direction from its crest, you sooner or
later find the rocks dipping downward away from you. When you find
such a structure, in a region of adequate source beds, and with one or
more reservoir beds safely buried beneath it, *drill it, my dear sir, drill it!*

Do All Good Structures Produce?

And when you do drill, you may get a dry hole. Oil geology is not
yet an exact science; probably it will never be. No geologist can tell

you where you will find oil; he can only tell you where the chances of finding it are greatest. Wildcatting is a hazardous business; geology reduces the hazards, but it does not eliminate them.

The odds vary with distance from producing fields, perfectness of structure at the prospective horizon, knowledge of source and reservoir beds and of fluid movements, and other factors. On good structures in the highly developed parts of Oklahoma, for example, the percentage of hits approaches ninety, but westward across the state the odds become longer and longer until they approach the national average, about ten per cent. In 1957's wildcat drilling, New Mexico and West Texas averaged one producer out of every five wildcats; Kansas, one out of every seven; and Illinois, one out of every forty-five. The geologist who gets a hit out of every four or five recommendations is maintaining a notable batting average. The man who tells you that he knows a certain wildcat well will get oil is no geologist; he's a would-be soothsayer.

Salt Domes

One type of anticline, known as a *salt dome*, deserves special mention because of its peculiarities and its productivity. In some regions, such as the Gulf Coast of Texas and Louisiana, the North German Plain stretching south from Hamburg, and the desert along the Colorado–Utah boundary, great masses of salt have been forced up into and through parts of the sedimentary rocks. The source of the salt, no doubt, was in thick salt beds formed by the evaporation of the inland seas in which the sediments were laid down. The alternate drying and filling of such an inland sea, fed by sediment-carrying streams, would give an alternation of salt beds and sediments. The salt flats of Utah, on which the world's land speed record has been made and broken, represent a stage in this process.

Salt under sufficient pressure will move and flow like ice in a glacier. Where thick salt beds have been deeply buried, and earth forces have overloaded them here, and developed points of weakness there, the salt has flowed slowly but inexorably to the points of weakness, where it has rammed its way up to form plugs or domes of salt. The overlying rocks have had to give way; they have been bulged up and tilted and perhaps fractured and shoved aside. The result is a plug or core of salt, surrounded by tilted and perhaps fractured shales and sandstones and limestones. Such salt cores may be of almost any shape and size. Some

SALT DOMES

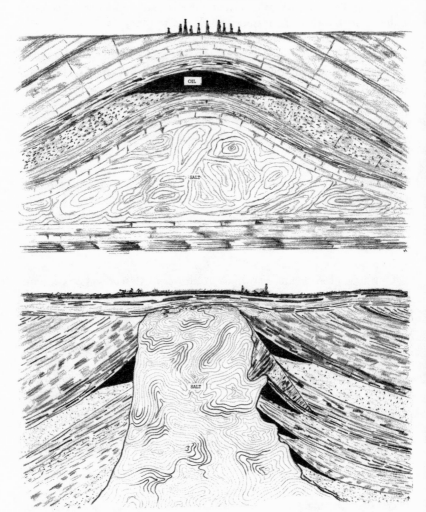

Salt under pressure may flow like ice in a glacier, and may arch up the overlying rocks into an anticline, as in the deep-seated type of salt dome shown in the upper sketch, or may fracture and penetrate the rocks, as in the piercement-type dome shown in the lower sketch. Oil may accumulate in reservoir beds arched over the top of the salt or on the flanks of the salt plug.

Gulf Coast salt domes are a couple of thousand feet across; Paradox Valley, in western Colorado, is thirty miles long and about six miles wide. Salt cores that have been drilled to depths of two miles or more are not uncommon; no one knows how much deeper some of them may be. In some domes, large and small, the salt core extends to, or nearly to, the present surface, having ruptured and upturned the adjacent strata; these are called *piercement domes*. In others, the salt has arched up the overlying strata without puncturing them. Most domes of this type are deeply buried; these are the *deep-seated domes*.

The same causes that collect oil and gas in other anticlines may collect them in the porous beds on the tops or flanks of salt domes. Some of the prolific fields of history, such as Spindletop, Humble, and Goose Creek in Texas, are salt-dome fields. The reservoir beds are usually much fractured and displaced on a piercement structure, impeding free movement of fluids from one part of the top or flanks to another. The flank beds are usually very steep, so that the oil may occupy only a narrow belt around the structure. For these reasons the oil is not easy to locate even after the dome is found. On one relatively small dome eleven dry holes were drilled before a twelfth well located the oil and discovered a productive field.

The Companionship of Oil and Gas

We have been talking much about oil and little about gas, but where there is oil there also will be gas. There are gas fields with no oil, but few if any oil fields with no gas. This is fortunate; as we shall see, gas plays an all-important part in producing the oil.

One of the notable things about the range of chemical compounds that make up the oil–gas succession is that most of the compounds are soluble in the others. When oil and gas are in intimate contact, the oil will dissolve some of the gas, all it can hold. If oil and gas accumulate in a reservoir bed, and if the oil can dissolve and hold all the gas, any well drilled into the trap above the water line will be an oil well. If the pool contains more gas than the oil can dissolve, the excess gas will occupy the top and the oil the flanks of the trap. In that case the structure is said to have a "gas cap." The top wells will be gas wells, those on the flanks will be oil wells and, of course, those far enough down the flanks will be water wells. A well just far enough down the flank to hit the *water line*, where oil and water meet, is an *edge well*.

Oil and Water

In addition to the water that fills the reservoir bed outside of the pool of oil and gas, smaller amounts of water may be associated with the oil and gas themselves. The oil and gas seldom completely displace the water that occupied the reservoir bed before their accumulation in the oil-field part of it. A microscopic film of connate water clings to the sand grains in most fields; it remains after the rest of the original water has given place to the oil and gas. Besides this interstitial water, small bodies of water appear to have been entrapped by the oil and gas accumulations in some fields.

The reduction of pressure as the field is produced may permit water from above, below, or around the oil to filter through permeable channels. For these reasons, oil and gas almost always bring a little moisture to the surface with them. In a good new field, the amount may be so small as almost to defy detection. In other fields the proportion of water to oil is large, and in most fields it increases as the field is depleted of oil and gas. Because most oil sands were laid down in salt water, the water that accompanies the oil is usually salty.

The problems presented by water that is produced with oil are, however, problems for the production engineer rather than for the geologist. We shall see more of the problems and the engineers later. At present, we are interested primarily in geology.

The Age of Oil Geology

Before we follow the geologist further, let's consider the antiquity of his kind. We will find it modern, not ancient, history. Some of the first men to make their living at oil geology are still doing so.

Long after the modern oil business started in Pennsylvania in 1859, wells were located by proximity to oil springs. If there were no oil springs, then salt springs were considered a favorable indication. The direction or *trend* of one field was given much weight in the search for other fields. The trend idea had some merit in Pennsylvania, for many of the trends were anticlines and an anticline might have several domes along its axis. If there were no trends or salt springs, then guesses, hunches, or witch wands might come into play. Many wells were drilled because their locations looked like other places where oil had been found. That wasn't bad business, in the absence of something better; similar geology may (or may not) produce similar topography. Geology was not greatly needed anyhow; the Appalachian country was full of oil fields, and hit-and-miss methods found enough of them to glut the infant market.

In 1861, T. Sterry Hunt, developing suggestions previously made by William Logan and H.D. Rogers, advanced the theory that oil accumulates on anticlines. No one paid much attention, and the location of wells "b' guess and b' gosh" went merrily on. In 1883, Dr. I.C. White, State Geologist of West Virginia, revived or rediscovered the anticlinal theory, and succeeded in convincing most geologists and a few oil men of its correctness. That started the real science of oil geology, and Dr. White has well been called the father of it.

In those days geologists were few, and practically all were teachers or were attached to the various state and federal geological surveys. By 1910, oil geology was taking men from classrooms and government surveys to work for the oil companies or to establish consulting offices, but for years the "practical" oil man looked askance at these newcomers in the business. Then came a time when the demand for oil geologists was greater than the supply, and graduates in geology were snapped up before their diplomas were dry. That day passed, but today the American Association of Petroleum Geologists has some 10,000 active members.

It is not surprising that the practitioner of so new a science should not know the answers to all his problems, but some things he does know, and with these we have become somewhat acquainted. We have seen that he is interested in the presence of source beds, reservoir rocks, and structural or stratigraphic traps; that in most areas he already knows about the first two; and that his most frequent task is to search for the third. Having learned what it is necessary to find, and assuming that we, too, know that source beds and reservoir beds are present, let's go with a geologist to look for an anticline, and afterward for a stratigraphic trap.

The Need for Discretion

Our trustworthiness and discretion will have to be known before we are allowed to go. The geologist does not advertise his field activities; the scouts and geologists of too many rival operators are eager to learn what he is doing and to take advantage of what he may find. The result of a few weeks' work may be worth many thousands or even millions of dollars to the geologist's principals if they alone know them and act on them. They will be worth proportionately less if other operators learn of them and lease some of the promising area. The mere fact that a geologist is known to be working in a locality may lead other geologists to start work there; they may get the same results that he is getting, and their principals may act on them as promptly as his. If you are lunching with a geologist's wife, you may ask, with propriety, "Is your husband in town?" But do *not* ask, "Where is he?" It isn't done.

The geologist with whom we are going to the field knows us, however, and knows that we will not tell where we have been, or what we have seen. He is ready to take us.

CHAPTER 4

Picking the Place
To Drill—In Practice

Hills and Anticlines

IN the search for anticlines, one thing must be borne clearly in mind: They are not necessarily represented on the ground by ridges or hills. An anticline is an upfold of the rocks, true, but the land surface is the product of erosion and may or may not reflect the underlying structure. In some youthful geologic areas like California, topography and structure accord fairly well and the surface over an anticline is more often than not a ridge or range of hills. In Wyoming the most characteristic surface expression of a dome is a valley surrounded by inward-facing cliffs known as *rim-rock*. In most of the Mid-Continent there is little or no consistent relation between structure and topography. So it goes. In some places topography is helpful in finding good structures; in some places it is not, and nowhere can it be relied on entirely. Closed anticlines are not found and mapped merely by looking for hills and ridges.

How To Find an Anticline

How, then, are they found? By diligently hunting, recording, and piecing together every scrap of information available on or below the surface regarding the elevation of the reservoir bed in which oil may be found. A closed anticline, after all, is merely a place where the strata are at a higher elevation than in any part of the area immediately surrounding. If a geologist had the elevation of some stratum or *key bed* at every point in an area he would have complete information as to the geologic structure of all beds "parallel" to the key bed. Such complete information, unfortunately, he never has; all he can do is to assemble and interpret what information is available—and then wish for more.

In areas where rock outcrops are plentiful, and the surface beds are parallel to the prospective reservoir bed, the task is fairly easy. The geologist first studies whatever aerial photographs are available, preferably in pairs with the help of a stereoscope, and plots on them, or on a separate map, the contacts between rocks or soils of different appearances and the dip and *strike* of each outcrop. His next job is to check his photo results on the ground and to add as much as he can to them by direct observation. The horizontal direction or trend of a bed (the strike) and its angle of inclination (the dip) eventually localize the place away from which the rocks dip downward in all directions. There is his structure. If natural exposures are not so plentiful, he may augment them by digging pits or trenches to expose the rocks he wants to observe—anything to get enough dips and strikes to define the structure. Let's cite an example.

Two geologists started south from Cheyenne to look for structures just east of the Rockies. In this area the *normal* or *regional* dip of the strata is eastward away from the mountain uplift. In a lake bank north of Fort Collins they found a sandstone dipping west. This was a *reverse dip*, and it indicated an anticline; their company promptly sent a staff of geologists to work it out. Every natural outcrop in the area was visited and its dip and strike taken. Every irrigating ditch was walked out and every road cut examined in the search for further exposures. Farmers' wells were bailed down and geologists lowered into their dripping depths to take more dips and strikes. Where wells were absent, pits were dug. Gradually a map developed; in a few months it showed a north–south anticline with the beds plunging north at its north end and south at its south end. On it were two *highs*, or domes, with a *sag* or *saddle* in the anticlinal axis between them. On those two domes are now the Wellington and the Fort Collins oil fields.

Unconformities

But what if no adequate outcrops are to be found? Or what if, after being folded, the series of strata of which the reservoir bed is a part has been eroded, and then covered with other strata not parallel to the older ones, so that the structure in the reservoir bed is not reflected in the younger beds at the surface? We have already discussed unconformities briefly. Now we must give them a little further attention.

Through the millions of years of geologic time, there have been in some regions many uplifts into dry land, many subsidences below the sea, and many different periods of folding. In a period when an area was dry land the folded rocks may have been eroded until the surface was far from parallel with the bedding. In a later subsidence below the sea, the folded rocks may have been covered by younger, horizontal sediments so that the new bedding does not parallel either the old surface or the bedding of the folded rocks. During a subsequent period of uplift, further folding may have taken place and the younger rocks may have been eroded to form a new surface. During yet another submergence, still younger rocks may have been laid down. In some regions there have been several such cycles, resulting in several series of beds not parallel to one another.

As we have already learned, younger beds laid down on older folded beds and not conforming to them are said to be "unconformable" on the older beds. Beds above such an unconformity may give little or no indication of the nature or location of folding in the older beds in which the geologist may be interested. Or, even where there is no such unconformity, the rocks may all be covered with soil or sand dunes. What is a long-suffering geologist to do in such a case?

Subsurface Geology

He may do one of several things. He may turn to *subsurface geology*. Water wells or coal-mine shafts or old wells drilled for oil may reach beds conformable with those in which he thinks oil may be found. If he can obtain their records and surface elevations, he may be able to work out the elevation of some *key horizon* or *marker bed* at various points and thus learn what he needs to know. Or, not finding such records, he may use a core-drill or slim-hole drill. By drilling below the soil cover or below the unconformity, he can measure the elevations of the beds conformable with those in which he thinks oil may be found, and he can use the elevations to make a map of the structure, if any.

A coring outfit, which drills by rotating a hollow bit made of hardened special steel or ringed with diamond chips, takes a core of the formations drilled—a core that can be brought to the surface and studied. A slim-hole drill also rotates a bit, but instead of recovering a core, it brings up cuttings of the rock being drilled. With a microscope or a hand lens, the geologist studies the cuttings and makes a *sample log* of the hole. The sample log, perhaps combined with one or more *geophysical logs*, enables him to record the presence (and therefore the elevations) of sandstones, shales, coal beds, and other rocks. This is so especially if there is in the area a logged well or a core-drill hole with which his records can be compared. Both core and slim-hole outfits are usually small, usually mounted on trucks, and usually able to drill holes down to a thousand feet with fair speed. By such means the subsurface geologist may work out a structure.

Subsurface work requires accurate identification of the key horizons or marker beds being used to determine the structure. Such identifications cannot always be made with the naked eye, so the geologist may call in the geochemist who makes chemical analyses of the core or cuttings, the micro-mineralogist who studies them under the microscope for signs of minerals that will identify them, and the micropaleontologist who looks for fossils.

Fossils in particular help the geologist out of many a tight hole. Life in the sea has evolved from lower to higher forms, as has that on land. Some forms of life appeared rather suddenly in the sea; others died out rather suddenly. Years of study have established the geologic time range of many such forms, particularly of certain microscopic marine animals known as *foraminifera*, often irreverently dubbed *forams*. If a distinctive foram appears under the microscope the paleontologist may be able to say, "That rock came from such and such an age or formation or horizon," which may be all that the geologist needs to know.

Geophysics

But if no subsurface information can be found, or the core or slim-hole drill cannot penetrate below the unconformity or reveals no identifiable horizon—then what? Then the geologist turns to his ally, *geophysics*.

Geophysics is the application of certain familiar physical principles —magnetic attraction, the pull of gravity, the speed of sound waves, the behavior of electric currents—to the science of geology. It has brought into the oil business and also into mining a large number of physicists who know geology and geologists who are strong in physics. All are classed together as geophysicists, and many good oil fields stand to their credit.

Measuring Magnetic Forces

Widely used in rapid reconnaissance is a method depending on magnetic attraction. One of the oxides of iron is highly magnetic, and all iron compounds are slightly so. Nearly all rocks contain at least a little iron. Most of the coloring of rocks, particularly the tans, browns, reds, and greens, is due to iron compounds, and of course some rocks contain much more iron than others.

In each of the three great classes of rocks there are wide variations in iron content and hence in magnetic attraction. A red sandstone contains more iron than a white one. But, in general, the igneous and metamorphic rocks are more magnetic than the sedimentaries. It follows that, in a region underlain by sedimentary rocks of uniform magnetic attraction, the area of greatest magnetic intensity is the area where igneous or metamorphic rocks are closest to the surface. If no igneous intrusion is buried beneath the area, then the area of greatest magnetic intensity is probably underlain by an anticline. When sedimentary beds are upfolded into anticlines the underlying *"basement complex"* of igneous and metamorphic rocks is of course upfolded, too. This brings the basement complex to a higher elevation on the crest of an anticline than it is elsewhere. If erosion subsequently levels off the land surface, the igneous and metamorphic rocks with their relatively high magnetic attractions are closer to the surface on the anticline than in the surrounding synclines.

The geophysicist goes to the field with a delicate instrument known as a "magnetometer." It is about the size of a large camera and is mounted on a tripod. It registers the intensity of magnetic attraction with minute accuracy. He levels the magnetometer and records its readings at many points in the territory he is mapping, and calculates these readings into arbitrary magnetic units called "gammas." He plots his gamma records on a map and draws lines or *isograms* through the points of equal magnetic intensity. The area enclosed by the highest isograms is a *positive anomaly*; it is the area in which the most magnetic rocks lie closest to the surface and is, therefore, if there are no confusing factors, underlain by an anticline or dome.

The method is particularly effective in locating buried granite ridges such as the Amarillo Mountains of the Texas Panhandle. These mountains, once a granite range higher than the Adirondacks, have been submerged and buried beneath two thousand to four thousand feet of sediments. The surface above them is today an area of monotonous plains. In the tilted sediments on their flanks are the prolific oil and gas fields of the Panhandle. In the mapping of this buried range, and thus in helping to locate the fields associated with it, the magnetometer has done yeoman service.

That its use is attended with confusing (sometimes confounding) factors and results, every geophysicist will admit. Atmospheric variations in magnetic intensity, including the occasional wide variations known as magnetic storms, may be credited to variations in the rocks instead of in the air and may lead to erroneous conclusions. Not all igneous and metamorphic rocks have the same magnetic attraction; basic rock like basalt and diabase contain more iron than the more acidic granites and gneisses. A diabase dike or a diorite intrusion, in a granite series that forms the floor for the sedimentary series, may give the same magnetometer results as a pronounced anticline.

An outgrowth of the rapid technological advances of World War II is the *airborne magnetometer*. Mounted in a torpedo-shaped *bird* and towed behind an airplane it records the continuous magnetic *profile* of its flight path. Several hundred miles of regularly spaced profiles are commonly recorded in one day, so that a week's flying may give the geophysicist the basic data for a magnetic map covering several thousand square miles. High tension power lines and buried metal such as pipelines bother the airborne magnetometer much less than they do its surface counterpart, and surveys at two or more altitudes, properly correlated, may come close to distinguishing between igneous intrusions and anticlines.

Airborne or surface, the magnetometer must be operated with care by men of skill and its results must be interpreted intelligently in the light of all other available geologic data. Intelligently used, it is highly useful in many regions, though by no means in all.

Measuring Gravity

Another geophysical method, using instruments known as *torsion balances* and *gravity meters*, depends on measuring minute differences in the gravitational attraction of the earth at different places. Gravity is the pull that the mass of the earth exerts on an object. Its force is mainly due to the weight or mass of the earth as a whole, but a slight, a very slight, amount is due to the mass of the immediate part of the earth on which the object is. Thus a man weighs a microscopic amount more when standing on an outcrop of granite, which is a heavy rock, than on an outcrop of shale or sandstone, which is lighter. These differences in gravity from place to place, due to differences in the specific gravities of the underlying rocks, are so small as to be almost, but not quite, incapable of measurement.

The Hungarian physicist Roland von Eotvos invented an instrument, the torsion balance, that not only measures differences in gravity, but points out the direction from which the strongest gravitational pull comes. The calculations of torsion-balance observations and their conversion into geologic deductions are so complex and involved as to stagger an ordinary mathematician; unavoidable errors of observation and calculations are often greater than the anomalies on which conclusions are based!

A newer instrument, the *gravity meter* or *gravimeter*, has almost supplanted the torsion balance. It is an extremely sensitive spring scale that measures the vertical force or pull of gravity with minute accuracy. It serves the same purpose as the torsion balance, but its mathematics is less complex, it is much faster, and by its use an area can be covered more quickly and at less cost. Gravity meters are usually housed in heat-insulated bases or tanks and used on tripods. With appropriate subsidiary equipment they can be lowered to the ground through a hole cut for the purpose in the bottom of a truck or jeep, or over the side of a ship, or packed on a man or a horse.

Fundamentally, though by a different method, the gravity meter crew is seeking the same information as the magnetometer crew. In general, the rocks that are most magnetic, such as the igneous and metamorphic rocks, weigh the most and have the strongest gravitational pull. Where they come closest to the surface, as under the axis of an anticline in level country, the force of gravity is strongest. By measuring the slight gravitational anomalies (very like, and commonly coincident with, their magnetic counterparts) the gravity meter detects the presence of structures not discernible by surface or even by subsurface geology. It is particularly effective in locating salt domes, for salt is a light rock even when compared to the poorly consolidated sands, clays, and shales in the midst of which it so often occurs. The gravity meter and torsion balance have been extremely valuable, especially in the detection of salt domes and buried granite ridges; they have many oil fields to their credit.

Making and Measuring Earthquakes

We come now, however, to a method more successful to date than any of the others described. This one depends on making small earthquakes and then measuring the speed and intensity with which their shocks are transmitted through the rocks. Observatories in various parts of the world have long had large, sensitive, ponderous instruments mounted on stone or concrete bases, known as *seismographs*, for recording the tremors of the earth's crust, even though the shocks that create the tremors may occur thousands of miles away. The geophysicist helping in the search for oil has reduced these delicate monsters to a light and portable instrument that, with all its attendant paraphernalia, can be mounted in a truck and whisked across the countryside. Such seismographs measure earth tremors (in reality, sound waves) that the geophysicist himself creates by well-placed explosions of dynamite.

The principle is simple. The tremors travel through some kinds of rock faster than through others. Some rocks transmit them readily and reflect back only a small part; others, and particularly limestones, transmit a much smaller and reflect back a much larger proportion.

When a shock is set up at the surface the resulting waves go traveling down through the underlying rocks and the waves that are reflected back are recorded. So long as the reflected waves are small, the downward-traveling waves can be assumed to be passing through good transmitters such as shales and sandstones; when strong waves come back to the surface, they indicate that the down-moving waves have encountered a reflecting surface such as the top of a limestone. Knowing the speed at which the waves travel, the geologist can then calculate the depth of the limestone, which, with the elevation of the surface, may be all that he needs to know.

Suppose a geophysicist is asked to locate an anticline. From some well formerly drilled, or from the nearest exposure of the whole series of rocks underlying the area, he learns that the surface is underlain by several thousand feet of shales, these by a bed of limestone, this by a few hundred feet of shale, and these by a sandstone that may act as an oil reservoir. By a few experimental shots near the old well or the convenient exposure he determines the speed with which waves travel through the shales. He then takes his seismograph into the area to be investigated, digs or drills a hole through the soil down to solid shale, and in the bottom of it sets off a few sticks of dynamite. The shock of the explosion oscillates the finger of the seismograph; so does each tiny tremor reflected back by the layers of the shale as the waves travel downward; so do the larger waves reflected back from the top of the limestone. The oscillations jiggle a tiny ray of light that falls on a rapidly moving strip of sensitive paper, and thus all the waves that have

GEOPHYSICS AT WORK

MAGNETOMETER

SEISMOGRAPH

MARINE SEISMOGRAPH

Several geophysical methods help the geologist determine the attitude, character, and content of rocks beneath the surface. The magnetometer and seismograph are useful oil-finding tools in many areas.

been created by the explosion and reflected back to the surface are photographically recorded. The strength or intensity of each reflected wave appears on the strip, with the time of its arrival to the thousandth of a second. Such accuracy is necessary, for the waves travel five thousand to fifteen thousand feet per second.

A simple computation then gives the depth from the seismograph to the limestone. When the elevation of the seismograph is determined, another calculation gives the elevation of the limestone at that point. By *shooting* enough points the geophysicist can determine the geologic structure beneath the whole area, and if an anticline is there he will probably find and define it.

The method just described is known as *reflection shooting*. An earlier method, now little used, is known as *refraction shooting*. The shot is set off at some distance from the seismograph, and the critical waves that reach the seismograph are those that have traveled through a single high velocity bed in the shortest path. The method has an advantage in its ability, under favorable circumstances, to determine the dip and strike of deep-lying beds, but in general it lacks both the speed and the accuracy of reflection shooting.

In regions where a limestone or other good reflecting bed is not present the seismograph method is valueless; in regions where the reflecting beds are lenticular (lens-shaped) or irregular, it is uncertain; in regions where too many beds reflect about the same proportion of waves, the results are likely to be confusing; and in regions like the Gulf Coast where the sediments are poorly consolidated, the results are often fuzzy and not too reliable. Serious difficulties are encountered where a thick mantle of glacial drift or of relatively young sediments lies in irregular thicknesses on the beds whose structure is sought. The method does not work equally well in all areas and in some it does not work at all. But in certain regions where the search for oil is going on vigorously, such as most of the Mid-Continent area, reflection shooting gives results of surprising accuracy. In such areas shooting has found and mapped many structures. So popular has it become, indeed, that many companies hesitate to drill a Mid-Continent structure without having a seismograph report on it. Some of them even demand seismograph reports in areas where the method is not dependable.

You may wonder why, if the method works so well, all of the territory in which it can be used is not immediately covered. The answer is dollars. A seismograph outfit in the United States costs $10,000 to

$20,000 a month; it will cover—on detailed work—from twenty-five to fifty square miles. Consider what would be the cost of shooting the twenty thousand square miles in eastern Colorado that everybody would like to see worked. The cost tends to confine seismograph work to local areas that some other form of geologic evidence indicates to be especially promising.

Electrical Methods

In *electrical prospecting* a number of electrodes are placed in the earth at intervals of a few feet to a few thousand feet, and a current is passed through the surface soil. Theoretically the current passes through all the rocks in a hemisphere the diameter of which is equal to the distance between the electrodes. The relative resistances encountered between different points, or the drop in electrical potential between the electrodes, or the time required for certain electrical "transient" factors to build up, give an indication of the character of the intervening rocks. From this it is often possible to work out the geologic structure.

Electrical prospecting is little used in oil finding nowadays, although in the early years of geophysics it enjoyed a somewhat more prominent role. Its effectiveness is limited to depths of a few hundred to perhaps a few thousand feet, and it seems to be best known for its inaccuracies. In mineral exploration it is in better repute, and with increasingly effective measuring instruments it may come back into petroleum exploration.

A reservoir bed filled with oil has an electrical resistivity markedly different from that of the same bed filled with salt water. So along with soil analysis and radioactivity prospecting, electrical prospecting shares a unique claim: Under certain conditions it may record the presence of oil beneath the surface. All that the other established methods attempt to do is locate and map structure; the drill must determine whether or not the structure contains oil.

Geophysical Well Logs

Geophysical well logs have already been mentioned; they are graphic records of electric and physical characteristics of the rocks a well penetrates. Where holes have been drilled, the geophysical logs are extremely valuable to the subsurface geologist; they comprise much of the data he uses to hunt the structural and stratigraphic combinations that he hopes will lead to new oil fields. We shall learn more of geo-physical logs in the electric logging section in Chapter 7.

Geochemical Methods—Soil Analysis

Another method, which differs greatly from any of those we have discussed, is gradually assuming a place in petroleum exploration. It is based on *analyzing the soil for minute quantities of hydrocarbons* and is geochemical rather than geophysical.

The theory is that every commercial deposit of oil and gas is overlain by the tight strata that prevent the oil and gas from migrating upward through the beds. No such seal is perfect, however, and small quantities of gas escape and find their way to the surface. On their way minute quantities are adsorbed on the surfaces of the sand and clay grains of the rocks through which they pass.

The geochemist collects samples of soil from depths of eight to twelve feet and analyzes them for methane, ethane, and heavier hydrocarbons. The methane is considered separately from the heavier gases; it may be present in the soil as a result of bacterial action on organic matter. The source of the ethane and heavier hydrocarbons, on the other hand, must be petroleum. Analyses made of many thousands of samples have shown that the soil in some localities contains more ethane and heavier hydrocarbons than in others, suggesting that such areas of hydrocarbon concentration are related to the presence of deep-seated oil and gas deposits.

Advocates of the method show two types of concentration patterns. In one, the high concentrations are found above—or nearly above—the accumulation from which the hydrocarbons have come. In the other, relatively low concentrations are found in the soil above the accumulation and high concentrations are around the edges. One geochemist explains this *halo pattern* as follows: The crest of an anticline is usually the place where the rocks have been most sharply folded and therefore are most permeable. In consequence this is where the greatest upward emanation of gases first takes place. These gases carry small amounts of moisture containing mineral matter. As they ascend, the mineral matter is deposited, forming an impervious plug above the crest of the anticline. The soil above the plug is low in hydrocarbons, but the soil at its edge has a high hydrocarbon content, forming the high-concentration halo around the anticline. The place to drill is in the low-concentration area inside the halo.

Skeptics suggest that gases migrating upward from an oil and gas accumulation are likely to follow whatever pervious beds or fault zones they encounter rather than to continue a vertical migration through less-pervious overlying beds. The surface traces are, therefore, more likely to be at the outcrop of some pervious bed or fault zone than to be above the accumulation whence they came. The skeptics point out that such traces are nothing more than minute seepages of oil and gas, akin to the oil and gas seepages that are so plentiful in some oil regions. Such seepages are excellent evidences both of the existence of source beds and of the conversion of source material into oil and gas, but comparatively few of them mark the exact locations of favorable structures or of commercial accumulations.

The proponents of soil analysis reply that seepages in visible amounts take place only where the seal of the accumulation is imperfect because of faulting or contact with a pervious bed, and that the minute emanations that cause soil concentrations take place under different conditions.

Another geochemical method is based upon the belief that when the volatile hydrocarbons reach the surface, they react with oxygen and sunlight and are converted to heavier materials such as waxes and liquids. In surveying by this method, samples are collected from the top few inches of soil and analyzed for these materials. On plotting the data, it is often found that concentration distribution patterns are obtained which resemble those resulting from the analyses of the deeper samples for ethane and heavier hydrocarbons.

Still another method is based on the observation that the growth of certain bacteria is accelerated in the presence of hydrocarbons.

One thing should be noted: The important geophysical methods we have seen locate favorable structures only; they leave the actual oil finding to the drill. Geochemistry attempts to find the oil and gas deposit itself.

Radioactivity Prospecting

All known rocks and soils contain at least traces of radioactive minerals, hence they are at least slightly radioactive. Experimenters in the oil-prolific Gulf Coast region have found that a halo of greater-than-normal radioactivity exists around many oil fields; the extra radioactivity apparently comes from the hydrocarbon traces sought by the soil analysts. The radioactivity surveyor, afoot or in trucks or airplanes, uses a sensitive Geiger counter or scintillator to record gamma-ray anomalies. When he has gathered enough readings he can show a radioactivity map of an area.

Radioactivity prospecting has its proponents and its skeptics. With more history behind it, it may come into its own as a tool for inexpensive, rapid reconnaissance and for checking other reconnaissance methods.

Radioactivity, soil analysis, and electrical prospecting may be worth watching. They are as near as man has come in his long pursuit of results from dowsers and "doodle bugs."

The Structure-Contour Map

Well and good! We have now discovered, perhaps at tiresome length, how the geologist, whether in his own person or disguised as a geophysicist, finds a structure. Having found it, how does he describe it to his superiors or his backers? Pages of description would not convey a clear idea of the details of a closed anticline to a vice-president in charge of development or to a man with money to invest. The geologist has a simpler solution: he makes a *structure-contour* map.

A contour is a line on which every point is at the same elevation above or below sea level. If it is drawn on the surface of the ground it is a *topographic* contour; if on the surface of some rock stratum, it is a structure contour. Suppose you wanted to map a promising structure somewhere south of Oklahoma City and knew that the top of the Viola lime[1] would make a good, regular surface to map. Suppose that you could shovel off all of the thousands of feet of overlying rocks, leaving the top of the Viola bare. Then suppose you found it arched up into an elongate dome, just the thing you would like to drill. Imagine then that your levelman determined that, on the very top of the dome, the top of the limestone was 5,315 feet below sea level. If you then walked down the side of the dome until your feet were at

---

[1] Every well-defined geologic formation has a name, usually given from the locality at which it was first identified or described. In fact the same formation may have one name in one locality and another name in another locality. This arises from no desire to confuse or to cause the formation to masquerade under an alias, but is due to the fact that when the two names were given, the identity of the formation in the two localities had not been established, or, to put it in geologic parlance, "the correlation was not complete."

minus 5,320 feet[2] and started painting a line at that elevation along the surface of the limestone you would presently find you had painted clear around the dome and were back where you started to paint. The painted line would be the minus-5,320-foot contour on the Viola lime. Drop down another ten feet and paint in the minus-5,330-foot contour; you will have a longer walk and use more paint this time. Never-

STRUCTURE-CONTOUR MAP

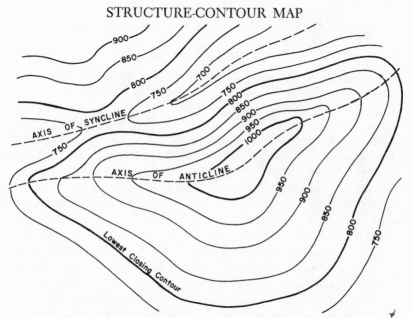

If you could strip bare the surface of some buried bed and on that surface draw a series of lines, each line staying always at the same elevation above or below sea level on the surface of the bed, these lines would be structure contours. This is a structure-contour map of a dome similar to that in the sketch on page 36.

theless, though each successively lower contour lies outside of and is longer than the one above it, keep on dropping down and painting a line at each ten-foot elevation. When you have contoured the whole area in which you are interested, get up in the air a few thousand feet and take a photograph looking straight down. That photograph of the

---

[2] The minus shows that the elevations are below sea level.

lines you have painted will be a structure-contour map, showing the structure on top of the Viola lime.

Obviously and unfortunately, the geologist cannot use such a heroic method as this, but he manages. By one or more of the methods already described he determines the Viola-top elevations at many points. These elevations he plots on a map, and through and between them he plots the structure contours. He may not be quite as accurate as you could be with your paintbrush on the exposed surface but if he has enough elevations, he can show the essentials of the structure.

That Lowest Closing Contour

Perhaps before you quit painting those lines you had an experience that puzzled you. Painting along, well down the side of the dome, perhaps at minus 5,390 feet, you found that instead of going around the dome again you had diverged and started northeast toward Tulsa. Shaking your head, you went back to where you had started that particular contour, and started painting in the opposite direction. When you got nearly to the place whence you had diverged before, you found your contour swinging around and heading southwest for Fort Worth. You had come to the saddle or gap that separates your dome from some neighboring anticline or from the general structure of the region. Your minus-5,390-foot contour is a few feet too low to pass through it to "close" around the dome. Move up a few feet, say to minus 5,385, and paint in the lowest contour that will pass through the saddle and go continuously around the dome. That will be the *lowest closing contour*; remember it, for it marks the limit of your interest in the structure. Outside it (below it) you'll find little if any oil.

That shallowest point in the saddle, through which you have drawn the minus-5,385-foot contour, is the *point of closure* or *spill point*. The vertical distance between it and the minus-5,315-foot top of the dome is the closure. On the hypothetical figures we have taken, our hypothetical dome has seventy feet of closure.

Pay some attention to this closure business, for it is important. You can see that as the reservoir sand on your dome was slowly collecting particles of oil out of the water passing by, it would stop collecting when it was full of oil down to the level of the point of closure; any more oil would simply continue past with the water. What does this mean? Just that the lowest closing contour is usually the outer limit of possible production. Drill outside it and you'll probably get a nice dry hole, full of water.

This doesn't mean that every well inside the lowest closing contour will produce oil; far from it! The structure may contain a lot of oil; it may contain a little; it may contain none. The *water line*—the line within and above which is oil and outside and below which is water—may be at the lowest closing contour or at any level above it. It will be below it only where the dome is a minor one atop a bigger, closed anticline. A closed anticline, like any other vessel, may be clear full, half full, one-fourth full, or empty, but you can be reasonably sure of one thing: It is not likely to be more than clear full. Don't count on its carrying oil below the lowest closing contour.

Yes, there are exceptions. Permeability inhomogeneities and flowing reservoir water have disrupted the water level in places, so the oil extends a little below the lowest closing contour opposite the spill point; but such exceptions are not the rule. More often the structure is not completely full, and the water line all around it is above the elevation of the point of closure. It's a safe rule to stay inside the lowest closing contour.

The next time a man wants you to help finance a well because it is close to a big producer, find out through some good geologist which side of the lowest closing contour the well is on. If it is outside, skip it. It may be only a hundred yards from a 50,000-barrel well; it might better be a hundred miles, for then it might be on some other structure instead of being just off this one.

Searching for Stratigraphic Traps

So far we have been trying to find and map structural traps. How about stratigraphic traps?

The main concentration of the search is on locating ancient shore lines and the sand edges and sandbars associated with them. Less attention has been given to the more difficult problem of locating traps below old erosion surfaces and in turbidity-current deposits. By studying the geologic formations over a wide area, using both surface information and the information revealed by well logs, the geologist may be able to map the area as it was at some former geologic time, plotting with fair accuracy the seas, the mountain ranges, and the lowland areas as they existed. This is *paleogeography*. Not often, however, can the paleogeographer locate with even fair accuracy the ancient shore lines, let alone the beach bars and offshore bars that may have formed along and near it. Still less often can he locate the truncated edges of reservoir beds below an unconformity. The best that a paleogeographic map can do, as a rule, is to narrow the area within which stratigraphic traps can be expected. Paleogeography is not a new science, but the finding of stratigraphic traps within the areas that it delimits is still young.

If electrical resistivity, soil analysis, and radioactive surveys prove able to locate oil-filled sands in structural traps, they may do the same for stratigraphic traps. Under exceptionally favorable conditions, the seismograph may help to locate the edges of sand bodies.

The only other method now being seriously considered is the careful search for slight folds in the surface rocks due to differential compaction of the underlying beds. Sand shrinks only a little when it is compacted into sandstone. Mud, on the contrary, packs down considerably when it is compacted into shale. The shales overlying a thick sandbar are likely, therefore, to exhibit a slight anticlinal wrinkle, and the wrinkle may extend to the rocks at the surface. Careful searching for slight folds succeeds in locating some stratigraphic traps, though apparently no surface evidence exists of most of them.

Aside from these methods, the only method now in sight is exploratory drilling, including the drilling of small holes with portable outfits (at a fraction of the expense of drilling full-scale wells).

What Geology Cannot Do

From all this discussion of oil geology, one thing should stand out clearly: Leaving aside the possibilities of electrical resistivity, soil analysis, and radioactivity surveying, which still must prove their worth, no amount of geologic knowledge can tell whether or not the structure or stratigraphic trap contains oil. If the trap meets all the geologic requirements it is a good place to drill. If, in addition, it lies in an oil-producing region there is a good chance—though by no means a certainty—that it will be productive. If a projected well is on a producing structure and within the closure, but on a lower contour than any well yet drilled, it may or may not produce. If there is a producing well on a lower contour the chances approach certainty, though they never quite reach it, since no reservoir sand is everywhere of equal permeability and the projected well may hit a tight spot where the sand is not permeable enough to yield oil. All that the best and most competent oil geologist can do is to evaluate the chances and advise where to and where not to drill. He can never guarantee the results.

What are the wildcat well's chances of success? Here are some figures: In 1959, 7,031 new-field wildcat wells were drilled in the United States. Of these, 558 were drilled for reasons not known to the compilers of the information, 5,947 were based on geology—including geophysics and geochemistry—and 526 were drilled for other reasons, such as lease requirements, showings in old wells, promotion schemes, and hunches. Because we do not know the reasons for drilling the 558 we will have to eliminate them from our computations and consider only the other 6,473 drilled that year.

Of the 5,947 based on geology, 5,242 were dry and 705 were producers. Of the 526 drilled for other known reasons, 494 were dry and 32 were producers. Geology scored 11.9 per cent hits; other reasons scored only 6.1 per cent. Thus geology nearly doubled the chances for success, but—and here is the point—the odds were still about eight to one against getting a producer.

Doodle Bugs

There are men and women who claim to "locate oil." Some of them depend on "psychic powers," but most of them have instruments called by high-sounding names. All of them and their users the oil man lumps under the inclusive name, *doodle bugs*. The instruments, though some of them pretend to operate on geophysical principles, are the descendants and heirs of the witching sticks, divining rods, and dowsing rods that have preyed on the credulity of man since the days of Babylon. If you believe in witchcraft or clairvoyance, spend your money on doodle-bug locations; otherwise you'd best leave the doodle bugs alone.

The Law of Averages

You may wonder how the oil companies keep going if they can never be certain that a wildcat will produce oil. They survive and even profit because they depend on the law of averages. If a company drills ten wildcat wells a year, in a region and on geologic advice that nets one producer out of every five wildcats, it should get two producers for its year's work. The two producers have to pay for themselves and for the eight dry holes drilled, if the company is to make money. If they are good wells and the price of crude is right they will do much more.

If you should ever consider putting money into wildcatting for oil, give a thought to the law of averages. If the money goes into one well in good territory the chances are four or five to one that you will lose, but if the well comes in, you stand to make a lot of money. If the money goes into a ten-well program in good territory the odds are in your favor, and one successful well out of ten may give you a handsome return on the full commitment. The same law governs wildcat royalties; one such royalty is a long-shot gamble, but an interest in ten wildcat royalties, in good territory and bought at a reasonable price, comes close to being an investment. All of which assumes that you would not put your money in an oil venture without knowing the management to be honest and capable, and the geology good.

The Place Is Picked

So much for picking the place to drill. When you have rested, your interest may revive enough so that you would like to know what happens next, which obviously is to get the right to drill where the geologist recommends. After a hundred-million years with a geologist, a few weeks with a lease man should be refreshing.

CHAPTER 5

Acquiring the Right To Drill

The Lease Man

THE lease man's job is to acquire the right to drill wells and produce oil on the lands that the geologist recommends. I wouldn't want to mislead you; there are still a few companies so benighted that they do not employ geologists; there are a few promoters so hardened or so fatuous that they can't be bothered with geology; there are regions in which excitement is so intense that the safe course is to lease first and geologize afterward. Then there are areas where geologic work is so slow and difficult that lands are leased by the wisest companies without regard to known geology, usually on a *checkerboard* basis, a quarter section out of each section or a few sections out of each township, hoping that future geologic work or drilling will find that a fair proportion of the leases are *on structure*. Thus, the lease man does not always follow the geologist, but as a rule he does. His efforts are usually circumscribed by the geologist's lowest closing contour, plus enough margin to allow for the well-known fact that geologists usually must make interpretations from incomplete data.

Suppose, then, that a geologist has found the substance of his waking dreams, a closed anticline, underlain within reasonable drilling depth by a good reservoir sand and by shales that were once rich in organic material. What happens next?

If the geologist is working for a company he submits his map and recommendations to the office, and, if the proper officials concur, they go to the land department, which assigns a lease man to the job. If the geologist is working for himself he may do his own leasing or take an independent lease man into partnership. If he is working for a promoter, either the promoter or the geologist or an independent (or associated) lease man may do the job. Whoever does it is a lease man for the time being.

69

The Lease Man's Job

Let's assume that the structure covers fifteen thousand acres. In most settled regions that would mean from one hundred to four hundred landowners. Farmers, widows, estates, banks, realtors, absentee landlords, and still more farmers—from each a lease should be obtained and the lease terms should be as nearly uniform as possible.

Imagine that all the treasure of Captain Kidd, Morgan, and Jean Lafitte were reported to be buried deep in the back yards of a Nantucket village, and that you were given the job of getting from each of two hundred villagers the right to explore for it and dig it up. Imagine further that a dozen rivals might learn of the treasure at any time and try to obtain the same rights. Think of the bickering and dickering you would have to do, and the speed and discretion with which you would have to do it. No villager would have the least idea of the methods and costs of such an operation. Everyone would be hoping to make a better deal than his neighbor and fearful his neighbor would make a better deal than he. Each would be sure most of the treasure was in his yard and afraid he might not get his full share. The average landowner knows as little about finding and producing oil as the villagers would know about digging for treasure, and for shrewd trading ability the average farmer or small-town banker is equal to any citizen of Nantucket. Ponder such a situation, and then think of the lease man, to whom such an assignment is routine business!

What business? To get from every landowner, so far as possible, and before a rival lease man gets to him, an oil and gas lease giving the right to enter upon his land, drill thereon, and produce therefrom all the oil and gas that may be found.

The Term of the Lease

The lease may be for any term, usually five or ten years, but don't overlook the words that follow: "and for as long thereafter as oil or gas in commercial quantities are produced from said lands." This means that the *operator* has five years, or ten, or whatever the fixed term may be, within which to find oil or gas on the leased premises; if he fails to do so within the fixed term, the lease is at an end. If he succeeds within the fixed term, the lease will remain in effect as long as oil or gas is produced. Landowners seldom object to this clause, nor should they; no one can expect an operator to spend large sums developing a property and have his tenure end while there is still oil in the ground—oil that can be produced, with practically no additional investment, through the use of facilities representing the investment he has already made.

To the rule just given—that the lease lapses unless oil or gas is found within its fixed term—there may be one exception. An increasing number of leases provide that the lease will not lapse while a well is being drilled, if the well was started within the fixed term. The Supreme Court of Oklahoma has held that such a provision is implicit in the lease, whether expressed or not, unless the contrary is expressly stated. With such a provision in the lease, either express or implied, the operator may start a well on the last day of the fixed term, and so long as he drills it diligently the lease is in force. If he gets oil or gas in commercial quantities the lease will remain in effect as long as he continues to produce them, just as though the production had been obtained before the expiration of the fixed term.

Gas Storage Leases

In recent years still another kind of lease has become important in some areas. Geologists and engineers have found that natural gas can be pumped into certain types of rocks, deep beneath the surface, and can be kept there until it is needed to meet the demands of the consumers. The gas supplier would like to hold excess gas produced in the summer months in such an underground storage close to the market area until the next winter. So the geologists and engineers must find a promising reservoir-caprock-trap combination, near town, where the production process can be reversed and gas pumped back into the ground. The lease man must then lease the ground to get the right to put the gas into storage under the landowner's properties. Storage leases may run for 20 years or more, or be perpetual as long as the rock reservoir is used for gas storage.

Drill or Pay Rental

Much argument usually arises over the drilling clause with its attendant rental provision. This provides that drilling will commence within a certain time, failing which, a fixed rental per acre per year will be paid until drilling does commence.

The annual rental varies with the leasing intensity of the area. In regions where there has been no successful development, it may be as low as ten cents an acre; in more active areas it is usually fifty cents or a dollar; and may be two dollars or more in particularly "hot" territory. Payments are made in advance, either quarterly, semi-annually, or annually depending on the wishes of the lessor and lessee as expressed in the lease.

You may think so small a rental does not amount to much, but remember two things: (1) In drought and depression, the lease rental has been the only real cash that many a farmer has seen from year to year; it has prevented many a foreclosure. (2) The rentals add up to a big aggregate for the lessee. To a small operator, the $1,500 to $15,000 per year required to hold our hypothetical structure will be important money. If he is big enough to be playing the law of averages, he has rentals to pay on several other structures as well as this one. The rental payments of some of the big companies run into millions of dollars a year. Thus even a small rental per acre serves its intended purpose: it furnishes an incentive to the operator to drill rather than pay rentals.

When and Where Must Drilling Start?

More troublesome questions are: How long may the operator have before he must drill or pay rentals, and where must the first well be drilled? In regions that already contain producing fields the answer is pretty well standardized: The operator must commence a well on each lease within a year or pay rentals at the end of the year. In areas remote from production much more latitude is given. For example, in a geologic province in which no oil has yet been found the operator may be permitted to avoid the payment of rentals for a fairly long time by drilling a well anywhere on the structure. He may even satisfy all the rental requirements of a large area containing several structures by drilling a well anywhere within the area. Such a lease may provide that rental shall be paid unless a well is started within a stated time on "the lease block of which this leasehold is a part, defined as follows, . . ." or "within blank miles of the land covered by this lease." The lease may further provide that a well must be commenced on the leased premises within a certain time after the completion of the test well in an adjacent area, or else payment of rentals must begin.

Landowners and Liberality

These provisions do not represent undue liberality on the part of the landowner; on the contrary, they are to his ultimate advantage. They make it easier for the lessee to finance the drilling of a well, never an easy task in new territory. Thus they bring nearer the day when oil may be found in the area and the subsequent day when wells may be producing on the lessor's land.

It should be noted that none of these provisions extends the fixed term of the lease. The operator is given his choice of complying with the drilling requirements or paying rental, but rental payments will not extend the lease beyond its fixed term. If oil or gas is not being produced at the end of the term the lease lapses, except that a well drilling on the leasehold may keep the lease alive until the well is completed, as we have seen.

It is scarcely necessary to say that reasonable provisions with regard to drilling are not always easily obtained. Every landowner naturally wants the first well drilled on his land. Now and then one insists on it, and thereby shows lack of wisdom. The first well should be drilled at the place where oil is most likely to be found, which is usually on or near the crest of the structure. Nothing can be worse for all concerned than to have the first well drilled in the wrong place and prove a dud; a long time may elapse before another well is drilled in the area. Consequently the lease man is usually adamant on this point.

Being adamant on important points is habitual with the lease man, for if he gives in to one landowner he will probably have to give in to all, and presently the deal is loaded with unbearable terms and conditions, and in the end is abandoned. More drilling deals than you would think are sunk without trace through failure to get reasonable lease terms.

What Is "Commencing a Well"?

Before we leave this matter of drilling requirements, note one little matter of wording: There is a lot of difference between *commencing a well* and *commencing the drilling of a well*. Hauling in materials, digging mud pits, developing a water supply, even building a road to the location—any of these may constitute commencing a well. Only "spudding in," the commencement of *actual drilling*, commences the drilling of a well. The lessor and the lessee should be in clear agreement as to which they mean, and see that the lease is worded accordingly.

The Landowner's Royalty

Now we come to a most important consideration. The landowner's royalty[1] is the proportion he is to receive of all oil and gas produced and sold from his land. This is his major stake in the affair. If his land is highly productive the royalty will pay him a lot of money, for the operator has to pay all the expenses of drilling and equipping the wells, building tanks, installing pumps, and operating the property. The landowner pays out nothing and receives, delivered free of cost into the tanks or pipeline, his percentage of all oil and gas produced and sold, or its value at the prevailing market price.[2]

No wonder royalty interests have a ready sale value and royalty companies are formed to buy, hold, and deal in royalties. Such companies, if ably managed and guided by good geologic advice, may be most profitable. It is no wonder, either, that such an investment field should be invaded by high-pressure promoters, and that thousands of people have made ill-informed royalty purchases and have been stuck unmercifully. In buying royalties or stock in royalty companies, as in other things, it pays to know through whom and with whom you are investing.

What proportion of the production does the landowner get as royalty? The standard figure is one-eighth, or twelve and a half per cent. Under exceptional circumstances it may go up to fifteen per cent or even higher, or down to ten per cent. For a time, one-sixth prevailed in some of the California fields. The federal government has demanded one-sixth on certain Indian lands, and demands even higher royalties on leases of public lands that lie on producing structures. These, however, are exceptions.

Is an Eighth Enough?

The landowner sometimes thinks a royalty of one-eighth is a pretty small proportion. "After all," he says, "I own all the oil under my place.

---

[1] Not all royalties are landowner's royalties. The lessee may assign his lease to someone else, reserving to himself a proportion of the oil and gas that may be produced. This reserved proportion, over and above the royalty payable to the landowner, is an *overriding royalty*.

[2] Usually the royalty owner takes his royalty in cash, but under most leases he may take it in kind—as oil or gas—if he wishes.

Why should I let someone else have seven-eighths and I get only one-eighth?" A little consideration, however, usually convinces him that his one-eighth without risk or expense is about as good as the operator's seven-eighths with all the risk and all the cost. In this he is likely to be right. The operator's working interest may prove to be more or less that the royalty; the relationship depends on the productiveness of the field, the drilling and operating costs, and the price of oil, none of which can be foretold. In any case the operator takes the risk. A man given his choice between all the oil royalties and all the oil leases in the United States would do well to take the royalties.

Landowners as a class are as intelligent and reasonable as other folks, and once the relative risks and rewards are explained, a royalty agreement is usually readily reached. There are exceptions, of course, and now and then a landowner insists that an eighth is not enough and that he should have a tenth. Ask any experienced lease man whether he has not had such a demand at least once. Don't embarrass him by asking whether he granted it; most lease men are honest but all are human, and such a demand is a temptation. To their credit, be it said that not many have yielded, but a few have been soft-hearted enough to "raise" the royalty to a sixteenth or even a twentieth!

Such incidents aside, note that the royalty is not a proportion of all the oil and gas produced but a proportion of the oil and gas produced *and sold*. Oil or gas used for fuel on the lease, oil that escapes and runs down the creek, which seldom happens nowadays, and gas that blows into the air—these don't count. The landowner gets his cut only on the part of the production that the operator can market.

Minor Provisions of the Lease

What else will the lease contain? A number of minor but nevertheless important provisions. The operator has the right, of course, to drill wells, erect tanks and pumphouses, lay pipelines, and do everything else he needs to do to "operate the property." He must bury all pipelines below plow depth if the landowner so requests and usually he may not drill within two hundred feet of a house or other building. If gas is obtained, the landowner may have, without charge, enough of it to heat and light the principal dwelling on the premises. The operator has the right to pay off any mortgage or other lien and thereby acquire the rights of the mortgagee thus paid off. The operator may surrender part of the lease and retain the rest, paying rental on only the part retained.

One or more of these provisions may be omitted, but nearly all of them are in most leases.

Provisions by Law or Custom

The lease also contains a couple of important provisions that may not be stated in it; they are there by force of law, or by custom. One is that the leasehold must be so developed and operated as to protect the landowner's interest in the oil and gas beneath it. Oil and gas are fluid materials, capable of moving about in their reservoir bed. In their undisturbed state they are usually under high pressure and they tend to move toward the nearest point at which the pressure is most rapidly reduced—the bottom of the nearest producing well. They have no conscience about crossing property lines. If you and I hold adjoining tracts, and if I sit idle while you drill your tract and produce from it as rapidly as you can, and if I finally bestir myself and drill my tract I will not get as much oil as if I had acted promptly. A lot of my oil will have gone to fill your tanks and pocketbook, and you will have depleted the gas pressure of *reservoir energy* that could otherwise drive the oil out of the reservoir into my wells. It follows, therefore, that to protect the landowner and his royalty rights, the development and operation of each lease must keep step with that on adjoining leases.

A rigid part of this requirement is that all *offset wells* must be drilled promptly. An offset well is one located opposite, and as near the property line as the well on the adjoining property. If the well is located across a corner of the adjoining property instead of across a side line it is a *diagonal offset*. The terms "offset" and "diagonal offset" are frequently used to apply to adjoining and cornering tracts, not only to wells.

The second requirement is a counteractant to the first: No well may be drilled less than two hundred feet (sometimes three hundred feet) from the boundary of the lease. This prevents an operator from attempting to get his neighbor's oil by *crowding the line* with a row of wells along the boundary. This requirement is now covered by statute in many states.

OFFSET WELLS

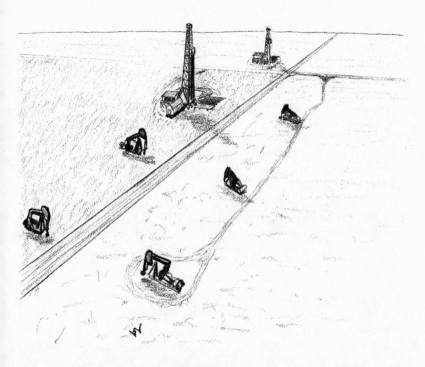

Every well on a tract in producing territory should be "offset" by a well on the adjoining tract at the same distance from the common boundary. This is intended to prevent drainage of oil from one tract to the other.

Cash for the Landowner

So much for the express and implied provisions of the lease. You will notice that all of them are "if and when" in nature. A rental and a royalty will be paid, drilling will be started, pipelines will be buried, and this and that will be done if and when this or that happens or does not happen. Is there anything certain in this transaction, anything tangible that the landowner gets? There is: he gets a cash bonus when he signs the lease. In undeveloped regions the bonus may be only one dollar, it is true, and often he is asked to acknowledge receipt of the dollar without actually getting it; but in more active areas the bonus is likely to be from twenty-five cents to five dollars an acre, and in really "hot" territory it may run into hundreds of dollars an acre. The bonus is money in the bank today, regardless of what happens tomorrow or next year.

The canny landowner, moreover, is likely to convert a part of his royalty into cash. A royalty buyer may pay him from fifty cents to fifty dollars an acre for half his royalty,[3] the price depending on the geology and activity in the area. This is cash money! It is good for a new television set or a payment on the mortgage. If the land proves dry, the landowner is ahead of the game by that amount. If the land produces, on the other hand, the half of his royalty that he keeps may yield him more money than he ever expected to have.

Mutual Advantages of the Lease

Analyzing the advantages of the lease to the two parties we see that the landowner may get a cash bonus for giving the lease; he may get a test of the area in which his land lies; he gets an annual rental until drilling starts on his land. If the area is productive he gets his land developed and the oil and gas under it produced, and he gets as royalty a percentage of the oil and gas—all without spending a cent or risking a dollar. The operator gets the right to find and produce all the oil and gas that may lie under the land, and to keep the greater part of it—all at his own risk and expense.

As in all good business transactions, the benefits are mutual, inter-

---

[3] Royalties are usually bought and sold by the *royalty acre*, which is the equivalent of a one-eighth royalty on one acre. Thus if the owner of 160 acres has leased at one-eighth royalty, half of his royalty is 80 royalty acres. Undrilled royalties commonly sell for $1 to $100 an acre; the price depends on geology, location, and drilling activity. Producing royalties have sold at $20,000 an acre and even higher.

dependent, and rather evenly balanced. That is why the lease man accomplishes his seemingly impossible task and does it within a reasonable time.

The balance of benefits has not come about by chance. The operators learned long ago that a dissatisfied lessor is a poor associate and that the best lease is one that is fair to both lessee and lessor. Back of the lease man is a hundred years of experience as to the form of lease that will be most satisfactory to both parties. More than 500,000 wells have been drilled and are producing under the general form of lease he offers. It has proved an agreeable working arrangement between thousands of operators and tens of thousands of landowners. The fairness of its various clauses has been studied by the courts in hundreds of cases. With such an instrument to work with, the lease man's customary success is not so surprising after all.

Producers Form 88

The form on which most leases are written is known as *Producers Form 88*. It varies in wording and minor details with different companies and in different territories, but its essential features are almost universal in the United States. An example is given among the appendices.

The lease man carries a stock of these printed forms, fills in the blanks as he discusses the terms with the landowner, and then takes the landowner and his spouse, if any, to the nearest notary public—unless, as he often does, he has a notary with him. The lease is signed and acknowledged before the notary, and the lease man gives the landowner a fifteen-day or thirty-day draft on the lessee for the amount of the bonus. The landowner gives the draft to his bank, which sends it to the lessee's bank. The lessee's lawyer or legal department checks the property title, and if this is good the lessee pays the draft. He then receives the lease, which may have been held meanwhile by either the lease man or the landowner's bank. The lessee's bank remits the amount of the draft to the landowner's bank, which pays it to the landowner, and the transaction is complete. When enough such transactions have been completed the leasing campaign is done. In oil man's language, the area has been *blocked*.

Rival Lessors

This does not always mean that our lease man has obtained a lease on every tract in the area. Some tracts may be so tied up in court or in estates that they cannot be leased. Some may be owned by absentees whom it is impossible to locate. There may be one or two landowners who simply refuse to lease to anyone. These are all part of the lease man's regular worries; he does his best with them and when he can do no more he quits.

His greatest worry is rival lease men. Now and then an area is *geologized* and leased before competing operators learn about it, but such cases are rare. In wildcat territory remote from production, the rival operators who learn about the "play" may stay out of it, knowing that in such an area only a solid block of leases will induce anyone to drill. The discoverer then can lease the area, drill a well, and thus advance everyone's knowledge of the geology. Wisely the competitors prefer this course to breaking into the play and reducing the chance of the well's being drilled. The lease man then has the field to himself, with nothing to worry about but absentees, estates, and stubborn landowners. When he has finished he may have all of the area leased, in which case he will say that the lease block is "solid."

In or near active territory, however, rival lease men are nearly always on the job by the time the leasing campaign is well under way. If they represent responsible operators, the terms offered by all of the lease men will be substantially the same. If the play is hot they may start bidding against one another on the size of the cash bonus, and a campaign that started at fifty cents an acre bonus may finish at two dollars. In general, though, the number of leases that each man gets is determined by the merits of his company, its intentions with respect to early drilling, and most of all by the personality and persuasiveness of the lease man himself. The first man in the area has the first chance to establish himself in the confidence of the landowners; by the time rival lease men appear he may be on friendly terms with everyone in the neighborhood. He is likely to emerge from the campaign with a substantial part of the area blocked up, under lease to his principal.

The lease man may be an employee of the company taking the leases; therefore this is his principal work. On the other hand, he may be an independent or free-lance lease man who *hires out* or *works on a ticket* for a company that wants leasing done in a particular area. On a big lease play, he, in turn, may hire a whole staff of lease men to work under him in order to block up the country as fast as possible before his competition has a chance to "get in." He will probably pay his *lease hounds* a modest salary plus an acreage bonus for each acre they can assemble. He, himself, will also be paid a per-acre bonus for everything that he and his staff can get in the designated area.

When Will Drilling Start?

The landowner should not expect the drilling crew to appear the day or even the month after his lease is signed. If the lessee is the geologist who found the structure, or is an independent promoter, he may still have to find someone who will drill the well. For this he may give a substantial interest in the lease block, keeping a minor part of the leased acreage *checkerboarded* through the block, or he may keep a minor, undivided interest in all the leases in the block.[4] He may sell part of the leases to raise enough money to drill the well,[5] or he may organize a company and sell shares for the same purpose. If the lessee is an established company it may have more important wells to drill before it gets around to this one. Possibly it may want to wait, either until other development indicates the promise of the area, or until oil is scarcer and its price higher. The leasing of an area is no assurance that drilling will start in the near future. Many a block has been leased and then abandoned, either by nonpayment of rentals or by expiration of the term of the leases. Many a block so abandoned has been leased again and either drilled or again abandoned.

In time, however, if the geology is right, the time comes for a test well to be drilled in the area. If you are interested we shall see how it is done.

---

[4] A complete checkerboard, or alternating patterns of lease parcels, retained by the lessee is known as a *divided half interest*. An *undivided interest* belongs to the lessee who retains a given percentage of his interest in each lease of the entire block.

[5] This is known as *promoting a well*. Sometimes enough leases are sold that the promoter makes money even though the well is dry; sometimes he fails to sell enough to *clear the well* and is stuck for the balance of the cost. Promoting wells is always exciting and sometimes profitable.

CHAPTER 6

Drilling the Well

Naming It

THE location[1] having been made, whatever road necessary having been built, and an adequate water supply having been provided, we are now ready to do the one thing and the only thing that will tell us whether or not oil is to be had; we are ready to drill a well. Let's begin by christening it.

The naming of a well follows an established practice. First comes the name of the operator drilling the well, then the name of the landowner from whom the land has been leased, and finally the number of the well on the lease. Thus if our well is the first to be drilled by the Anonymous Petroleum Corporation on a lease from John Doe its name is "Anonymous Petroleum Corporation John Doe No. 1," which in everyday use is shortened to *Anonymous Doe No. 1*. It may from time to time be called the Jonesville well from some nearby locality or the Brown Creek well from the structure on which it is located, but these are nicknames. Its official name, by which it will appear in the trade journals and on the scout tickets, is Anonymous Doe No. 1.

We Decide on Cable Tools

Now we have an important decision to make; shall we drill with cable tools or by rotary? You may remember from Chapter 2 that in

---

[1] "Location" is a word of wide and varied use in the oil business. It means the determination that a well has been decided upon at a certain point, as in the phrase, "Amerada has announced the location of a well in such and such a place." It means the point at which a well is to be drilled, as in the phrase, "Amerada is moving tools to the location of such and such a well." In fields where wells are spaced according to a certain fixed pattern, such as 330 feet apart, it means the conventional distance between wells, as in the phrase, "The Amerada well is two locations north of such and such a well." All of which antedates Hollywood and its phraseology.

cable-tool drilling[2] the hole is punched down into the earth by raising and lowering a heavy bit suspended from a cable, and that in rotary drilling the hole is made by rotating a cutting tool on the end of a pipe.

The first cost of a cable-tool outfit is much less than that of a rotary outfit. A complete string of cable tools may cost $10,000 to $65,000; a heavy-duty rotary outfit may cost $90,000 to a half-million dollars. Rotary drilling costs more per day than cable-tool drilling. In most formations, however, rotary drilling is faster than cable-tool drilling, so that the total drilling cost may be less with rotary, despite the greater cost per day. Less casing is usually required in rotary drilling than in cable-tool drilling, and casing is one of the major items in the cost of a well. Large gas flows under high pressure can be controlled in rotary drilling more safely than with cable tools.

Cable-tool drilling gives a more accurate record of the formations drilled than does rotary drilling, and an accurate *log* of the formations is important, especially in wildcatting. Coupled with this is the rotary's tendency to "mud up" and seal off producing formations. More than one good producing sand has been "passed up" in a rotary hole without its presence being suspected. These rotary disadvantages can be overcome by coring, electric logging, and drillstem testing, as we shall see, giving a record even more reliable than a cable-tool log and avoiding the danger of passing up productive horizons.

The rotary was formerly handicapped by its inability to drill through hard rocks and by a tendency for rotary holes to be crooked; but these handicaps have been overcome almost completely.

The choice of methods often depends on the individual preference or prejudice of the operator, or on which type of equipment is on hand, but the trend is toward rotary drilling. Cable tools are being relegated, in the main, to shallow wells, to wildcat wells where the operator wants to avoid the expense of coring, and to small drilling programs where the initial cost of a rotary outfit would not be warranted. Truck or trailer-mounted cable tools may be used to start a hole to be continued by rotary and are therefore called "spudders" (although many holes are spudded by rotary). Nine-tenths of today's drilling in the United States is rotary drilling.

We won't drill Anonymous Doe No. 1 by both methods; yet we should watch both methods at work. Let's drill Anonymous Doe No. 1

---

[2] Some of the old books refer to *churn drilling* or *percussion drilling*, and cable tools are often called "standard tools."

with cable tools and then drill the offset with a rotary. We shall find out very quickly that most of the procedures and the names of the pieces of equipment around a well being drilled with *standard tools* come from the early days of wooden derricks, boilers, and bull wheels. Wells are still being drilled in some parts of the country with the antiquated machinery that is reminiscent of the days of Colonel Drake's first discovery in 1859.

We might do well to look over the old standard rig, now modernized in many essential parts, for it was the very foundation of the oil business. This then will be a cable-tool well.

The Cellar and Foundations

The first step is to dig the cellar, a hole in the ground eight to ten feet square and six to twenty feet deep. If the ground is *cavy*[3] or if we expect high gas pressure we may be wise to have a heavy anchor for casing and to line the *cellar* with concrete. The cellar is to provide space below the derrick floor for jointing and unjointing pipe, and for the valves and fittings that may be necessary.

Next comes the foundation for the rig; heavy timbers or concrete blocks beneath each corner and on these *floor sills*, *mud sills*, and *main sills*. On the sills rests the derrick floor, usually three to eight feet above the ground. Running off to one side, on lighter foundations and usually at a little lower elevation, is the flooring for the engine house, dog house, and walkway, at which we will presently look. Extending out from one side of the walkway, near the derrick and on a level with the derrick floor, are two or three heavy steel or timber frames. Together they are the pipe rack on which casing is stacked preparatory to being lowered into the well.

The Derrick

The cable-tool derrick is a square open tower, seventy to a hundred feet high and twenty to twenty-two feet square at the bottom, tapering to five to seven feet square at the top. The four heavy uprights that form its corners are called "legs," and are connected by horizontal

---

[3] A formation that caves or slumps or sloughs into the well is called "cavy" or "cavey." A driller will tell you that in such ground "the hole does not stand up." One of the most cavy materials is bentonite, a clay that may swell up to six or eight times its original volume when soaked in fresh water, though it swells much less in salt water.

members called "girts," and diagonal members called "sway braces." All the members are in the outside faces of the derrick, none across the inside. The inside is clear and unobstructed, kept so for the handling of drilling tools and casing.

The derrick, once made of lumber, is now usually of steel. Wooden derricks are becoming antiques. It looks spindly and frail, especially if it is steel, but it will withstand a stout gale, and many tons may be hung from its top without collapsing it.

Forming the top is the *crown block*, a steel frame in which are mounted vertical, grooved pulley-wheels, known as *sheaves*. Over the sheaves will run at least three steel cables: the *casing line* for raising and lowering the casing, the *sand line* for bailing the well, and the *drilling line* for raising and lowering the drilling tools. Sticking up from the crown block is usually an inclined beam or a vertical framework known as the *gin pole*, used for hoisting the crown block into position.

The sides of the derrick may be enclosed with canvas or iron sheeting up to a height of ten or fifteen feet above the ground, but are open to the dog house and the walkway. The room thus enclosed is a little bigger than an ordinary two-car garage.

The Rest of the Rig

About twenty-five feet from the derrick, and connected with it by both an outside walkway and an enclosed passageway, the *belt house*, is the *engine house*, about the size of a small hen house. It houses the drilling engine, which once was steam, but now is diesel, gas, or electric.

When the moving trucks depart, the drilling crew will move in and *rigging-up* will begin. The fuel lines will be connected to the engine, a forge will be installed, and everything will be made shipshape. Let's take a look at what we have.

The Completed Rig

Step up on the walkway near the engine house, stroll to the derrick, and look in. In front and slightly to your left you will see a hole in the exact middle of the derrick floor. That's where the well will be. At your left is a huge upright beam, with another big beam balanced across its top, half in and half out of the derrick. It is so fastened that it can teeter up and down. The upright is the *Samson post* and the balanced beam is the *walking beam*. Notice that when the walking beam is horizontal, its derrick end is over the well hole. Its outer end, beside the walkway, back toward the engine house, is connected to a vertical rod or beam or bracket called the "pitman"; the lower end of the pitman in turn connects to a steel arm known as the "crank."

In front of the Samson post, and directly under the walking beam, you may notice a smaller but nevertheless sturdy upright. This is the *headache post* and its purpose is to save the driller's skull if the walking beam should pull loose from the pitman or the crank. Projecting in front of the headache post, about shoulder high, is a small pulley with a wire round it that reaches back to the engine. These are the *telegraph wheel* and *telegraph cord*; by turning the wheel the driller starts, stops, and controls the speed of the engine.

In front of you, at the opposite side of the derrick from the Samson post, is a big horizontal steel drum, wound with steel cable three-fourths of an inch to an inch in diameter. The drum is the *bull wheel* and the cable is the *drilling line* by which the tools are suspended in the hole when drilling is being done. *Running* the tools, which means lowering them in the well, and *pulling* them, which means pulling them out of the well, are done with the bull wheel. At your left, beyond the Samson post, half in the derrick and half in the belt house, is another horizontal drum with steel cable on it. This drum is the *calf wheel* and the cable is the *casing line*. Casing is run or pulled with the calf wheel.

Back in the belt house is a huge wooden or iron wheel called the *band wheel*. It is turned by a belt from the flywheel of the engine or by a transmission. On its axle are a grooved pulley from which the bull wheel is run by a rope drive, a sprocket from which the calf wheel is run by a chain drive, and the crank that moves the pitman that moves the walking beam that moves the tools up and down in the well.

Back of the band wheel and run from it by a friction or chain drive is a smaller drum, also carrying a cable. This drum is the *sand reel* and the cable is the sand line that lowers and raises the *bailer* with which the well is bailed to remove the cuttings from time to time as drilling progresses.

Hanging from the front or derrick end of the walking beam is a slender vertical frame, five to seven feet long, in which is a screw almost as long, with a set of clamps at the bottom of the screw. This arrangement is the *temper screw*; its function is to lower the drilling tools gradually in the hole while they are "making hole." The driller lowers the tools by letting out the screw by hand so that the tools keep pace with the deepening of the hole.

When drilling starts, the tools are fastened to the end of the drilling line and lowered into the hole by unreeling drilling line off the bull wheel. The drilling line is then clamped at the bottom of the temper screw, so that the tools are suspended just above the bottom of the hole and attached through the drilling line and temper screw to the walking beam. The engine, through the band wheel, drives the crank 'round and 'round; this moves the walking beam up and down, and this in turn jerks the drilling line and the tools up and down in the hole, which is the way cable-tool drilling is done.

Drilling has not yet started, however. The small building at one side of the derrick houses the crew lockers, some of the tools, and a long bench which is the top of a storage box. That's the *lazy bench*, where the scouts and tool pushers and visitors sit to watch the drilling crew work. On it a driller has been known to nap while the tool dresser handled the tools; that's one way in which toolies get the experience to become drillers. Let's make ourselves comfortable on it and watch what's going on.

Spudding In

Rigging up is completed, motors are running, and the magic moment has arrived. The well is about to be *spudded in*. The first few feet will not be drilled with the regular tools and the walking beam; some hole has to be drilled before the long *string* of regular tools can be suspended from the temper screw. That first thirty to three hundred feet of hole are made by attaching a heavy manila rope to the wrist pin on the crank, running the rope over the crown block and down to the middle of the derrick, and there attaching to it a heavy bit known as a *spudding bit*. The revolutions of the crank, imparted through the rope, raise the spudding bit and drop it. The driller stands by the hole, holds the tools steady, and sees that the bit revolves slowly as it is raised and dropped. The impact of the bit pulverizes the ground and gradually pounds or "drills" a hole into it. If no subterranean water runs into the hole, water is run in with a hose, to keep the pulverized materials or *cuttings* in suspension while the drilling goes on.

Spudding in bears the same relationship to a well that turning the first shovelful of earth does to a new post office. The drilling of a well commences with spudding in.[4]

When enough hole has been made with the spudding bit, so that the bit is *buried* below the level of the derrick floor, both the bit and the spudding line are removed and the regular drilling tools are lowered into the hole with the drilling line and the bull wheel. Look them over before they are in.

The Drilling Tools

At the bottom of the string of tools is the bit. Suppose you started with a big round steel bar four or five feet long and cut about a fourth of its diameter off each side to leave it flat on two sides and round on the other two. Then cut wide trenches into its flat sides for the lower four-fifths of its length. At the bottom, dress the rounded sides out slightly and shape the bottom end to a broad blunt wedge. If you did all this you would have the semblance of a bit. The wedge-shaped bottom looks too blunt and dull to cut anything; as a matter of fact, it pounds and breaks the rock formations rather than cutting them.

The top end of the bit is tapered and terminates in a short heavy screw known as the *pin*. Onto the pin is screwed a long steel bar known as the *drill stem* or *auger stem*. This adds weight and stiffness to the bit. Screwed on above this is a set of *jars*, which look as though a drill stem had been refashioned into a pair of long interlocking links of chain. Their purpose is to impart a jarring impact to the tools, especially if they should become stuck in trying to pull them from the hole. Above the jars will be another drill stem called the "sinker." Into the top of the sinker is screwed a *rope socket*, and into the top of this the drilling line is fastened. The rope socket contains a *swivel* that permits the tools to rotate without twisting the cable.

When the bit, stems, jars, and rope socket have been screwed together, or *made up*, which the driller and tool dresser do with jacks and heavy wrenches (usually manipulated by hand but sometimes by the use of power from the engine), they constitute the *drilling string*. The string may be from twenty to forty feet long and may weigh from two

---

[4] Remember the distinction we drew in Chapter 5 between *commencing a well* and *commencing the drilling of a well*? You can now understand the difference. Anonymous Doe No. 1 was commenced a couple of weeks ago or more, but drilling has just now commenced.

CABLE-DRILLING TOOLS

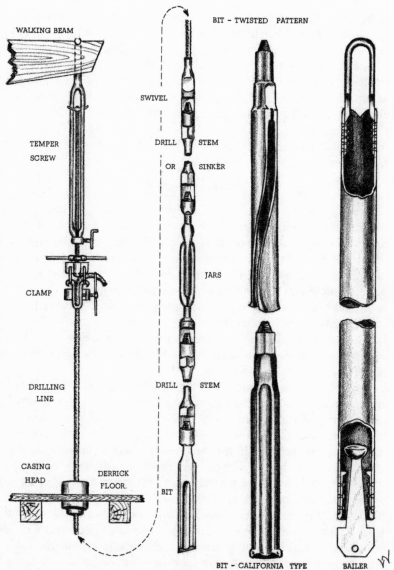

WALKING BEAM

BIT – TWISTED PATTERN

TEMPER
SCREW

SWIVEL

DRILL STEM

OR SINKER

JARS

CLAMP

DRILLING
LINE

DRILL STEM

CASING
HEAD

DERRICK
FLOOR

BIT

BIT – CALIFORNIA TYPE

BAILER

The rope socket, sinkers, jars, drill stem, and bit that make up a string of cable tools weigh a couple of tons or so. The bailer is used to remove cuttings or water from the hole, and sometimes, temporarily, to bring up oil.

to five tons. You can see why the successive impacts of this dainty outfit, dropped from a height of a few feet every second or two, will pulverize almost any rock formation.

The driller lowers the tools nearly, but not quite, to the bottom of the hole, controlling their descent by a brake on the bull wheel. Then, as already described, he clamps the drilling line securely into the clamps at the bottom of the temper screw and slacks off the line above that point, so that the tools are suspended from the walking beam instead of from the crown block and bull wheel. Now we and the hole are ready for regular drilling.

The Portable Rig

Now that we have watched the old standard outfit rig up and get ready to drill and have seen where many of the terms originated, let's look at a modern cable-tool rig *stacked* in the corner of the pasture.

On this one we see almost the same items of equipment, but they have been condensed into a tight package of motors, cable spools, pulleys, and frames that can travel from location to location all on one or two big trucks or trailers. The old rigs lack mobility. With the keen competition for drilling jobs that has grown over the years and with the increased costs of drilling and producing oil, time-saving and labor-saving mean money in the bank; hence the demand for and common use of easily portable drilling rigs.

We note first the slender steel mast of the portable outfit. It can either be broken down into pieces and moved by truck or the derrick can be folded like a jackknife. The bull wheel, the calf wheel, and the band wheel are all mounted together; the bull wheel and the calf wheel are two separate drums on the same axle, controlled by separate clutches and brakes. The diesel or gas engines drive them through sprockets and chains and drive shafts or even through fluid-drive transmissions like the one in the family bus. The crank, pitman, and walking beam are all here, but we hardly recognize them because there is no separate band wheel. The crank is now a crankshaft driven from the transmission, which lifts a pitman located directly behind the cable drums. On the old rig the pitman raises and lowers the far end of the walking beam. Here the walking beam is pivoted at the far end and the pitman pushes it up and down near the middle. No part of the walking beam is over the hole; indeed the derrick end is engineward from the legs of the mast and it carries a sheave instead of a temper screw and clamps.

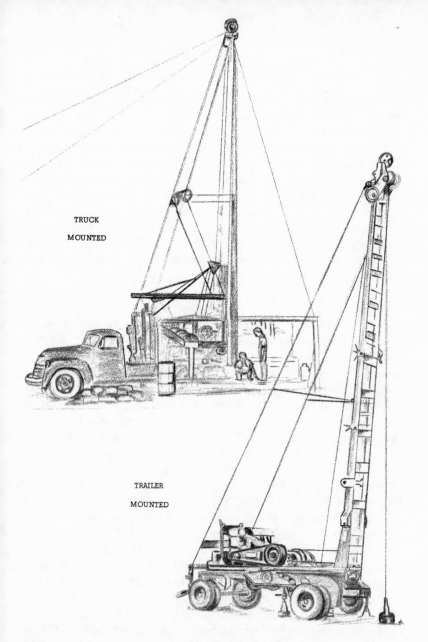

TRUCK
MOUNTED

TRAILER
MOUNTED

Cable-tool drilling is done by raising and dropping a string of tools on the end of a cable. The up-and-down motion is imparted by the walking beam. The tools are pulled from or lowered into the hole by winding or unwinding the drilling cable on the cable drum. Casing is raised or lowered by the casing line and calf wheel. The cuttings are removed from the hole by the bailer, which is raised and lowered by the sand line and reel.

The drilling line spools off the bull wheel, passes under the sheave, up over one of the sheaves in the crown block, and down to the tools in the hole. The drilling action is the same but the walking beam pulls down on the line to lift the tools and then lets them fall sharply on the return stroke. An automatic brake release has replaced the temper screw; it permits only a set amount of line—determined by the drillers —to spool off the drum. Instead of working inside the legs of the derrick, with the hole being drilled through the center of the derrick floor, the mast leans out over the hole so that there is open working space all around the hole. The driller stands back near the derrick leg at the drilling station where he can see all the machinery and the collar of the well.

Drilling

The driller starts the engine and the walking beam starts its seesaw motion. Its oscillations raise the tools in the hole, then drop them. As the hole is deepened, the driller—by means of the temper screw— lowers the tools just fast enough so that they will strike the bottom with precisely the right amount of impact. If he lowers them too slowly they strike the bottom too gently and then tend to miss every other stroke or "peg-leg," as the driller calls it, and he does not *make hole* as rapidly as he should. If he lowers them too rapidly—drills with a *loose line*—the lift will be insufficient for a full impact and, what is worse, the tools may get out of the vertical line and drill a *crooked hole*. A good driller can tell from the feel of the cable and the behavior of his engine just how hard the tools are striking; most of the time he can also tell the character of the rock being drilled.

After a few hundred feet of depth, the stretch of the cable acts like a spring, lifting the tools several feet higher than the stroke of the beam, then on the return they *reach-out* and smack the bottom with a crisp blow and retreat. The driller likes this action since it gets away from tamping the bottom of the hole and sets up a more violent turbulence of the drilling water to clear away the cuttings.

You may imagine that while the bit is pulverizing the rock, the rock may also have some effect on the bit. It does. A bit may drill many feet of soft shale without much wear, but a few feet of hard sandstone or limestone dull it and wear down its edges so that it drills a smaller hole than intended. It must then be *dressed*: heated in the forge, hammered out to its original size and shape, and retempered. On a large lease this may be done in the central shop; some rigs have power hammers; but on most wells this is a job for the driller, the tool dresser, and a couple of sixteen-pound sledge hammers.

We are ahead of ourselves, however; on this well we have only started drilling. Let's get back to our lazy bench.

Bailing

Presently—in half an hour to several hours, depending on the formation being penetrated—a *screw* has been drilled, that is, the temper screw has been let out to its full length. By this time the *cuttings*, the rock that has been pounded up in making the hole, are beginning to cushion the impact of the bit on the bottom of the hole. It's time to bail. The drilling line is unclamped from the temper screw, the bull wheel is started, and the tools are pulled out of the hole and tied to the side of the derrick out of the way. The bailer, which has been standing inconspicuously in a corner, is lowered into the hole and brought up full of water and cuttings, which are dumped through a hole in the derrick floor into a "slush pit" beside the derrick. Samples of the cuttings are saved to help identify the formation being drilled. The bailer is run until the hole is relatively free from cuttings. Then it goes back into its corner and drilling is resumed.

Have a look at the bailer. It's a cylinder, small enough to fit comfortably into the hole and eight to forty feet long. It is suspended from the sand line, which runs over a sheave in the crown block and down to the sand reel. At the bottom of the bailer is a ball-and-seat valve. The ball is inside the bailer where it rests on the ring-shaped seat. From the bottom of the ball a prong or "dart" projects through the seat and from four to eight inches beyond the bottom of the bailer. When the bailer lands on the bottom of the hole the dart pushes the ball, thus opening the hole so that the water and cuttings can rush in. When the bailer is raised the ball settles into place and closes the hole and the bailer carries its load to the top. There, when the bailer is lowered into the *dump box* or *slush-pit launder*, the dart pushes the ball up and the water and cuttings rush out and run off into the slush pit.

The Log

Notice the driller dipping his hand or a kitchen sieve into the bailings as they escape from the bailer? He's grabbing a sample of the cuttings, to tell him the nature of the formation being penetrated. He already has a good idea from the way it drills, but he is supposed to keep an accurate record. Watch him making an entry in his log book. That log is expected to tell everything essential about the well: when it was spudded, with what size hole, when and at what depth the drilling tools were hooked on, how many feet are drilled each day, the character, color, and description of each rock stratum penetrated, the point at which each string of casing is landed, and everything else of importance.

Nowadays, in addition to this driller's log, another log is kept on most wells. The driller saves samples of the cuttings, putting them in bottles or envelopes or sacks, each marked as to the depth from which the sample comes. These go to the geologist and to the offices of the other companies that are scouting the well, and also perhaps to the state geological survey. There they are studied under the microscope and sometimes subjected to chemical tests, and a *microscope log* or *sample log*—giving detailed information as to the character, mineral constituents, fossils, and the like—is prepared. The geologist calls this procedure "running samples."

From these logs the geologists tell at what geologic horizon the well is drilling, with a view to predicting what lies ahead of the drill. If you are about to drill into a gas sand or a cavy zone, it's helpful to know it and prepare for it. This is the job of the geologist who is *sitting on the well*.

The logs later serve another important purpose. As a field is *drilled up*, the accumulating logs give an increasingly accurate picture of the subsurface geology of the field. New contour maps of the structure are prepared as the logs arrive at geological offices, and the new maps are used in locating additional wells and in planning the further development of the field.

Drillers have been known, too frequently, alas, to neglect their logs or even to fake them, and not all drillers' logs are reliable, but the geologist has to do the best he can to insure that accurate records are kept. No driller is a good driller who does not keep a careful log and collect good samples.

Not too rare, unfortunately, are the drillers who have been caught keeping footage of hole "in the kitty" as a reserve to draw on "the morning after the night before" when they may be late getting started. Perhaps they have hoped that the mental arithmetic and manipulation involved in doctoring the log and trying to fool the geologist trained them for future superintendency.

Keeping water in the bottom of the hole is essential; the water keeps the cuttings in suspension. Otherwise the hole would have to be cleaned out every few inches. If nature does not supply it, water is put in the bailer or is run in from a hose if the formations are not cavy. Sometimes nature does even better, however; during *drilling-in* into an oil or gas reservoir, the oil or gas flood may clear the hole of cuttings or at least keep them from packing in the bottom of the hole. No cuttings on the bottom mean there is nothing to bail, too.

Casing

Sometimes nature supplies too much water. The drill may penetrate a porous formation containing a lot of water under enough pressure that it will rise and flow out at the surface. If so, something must be done about it or it will interfere with the drilling and probably cause the well to cave. The solution is to run a string of casing.

Unwelcome water is not the only reason for casing. A gas sand or even an oil sand may be encountered and be *cased off* in order to reach a more important producing sand below. So much *open hole*, which means uncased hole, may have been drilled that there is danger that the wall may cave in even without incoming water to aggravate the danger. In each case the answer is to run casing.

Casing is heavy steel pipe that comes in sections or "joints"[5] usually thirty to thirty-two feet long, which is a little higher than an ordinary telephone pole, though joints up to forty feet long are being used on more and more wells. Each joint has threads on the outside at one end and a collar with threads on the inside at the other. A heavy ring known as a *casing shoe* is screwed onto the threaded end of a joint, the joint is picked up by the casing line which runs over the crown block and down to the calf wheel, and is lowered into the hole, shoe-end down, until the collar-end is just above the derrick floor. There it is held by heavy clamps while another joint is screwed onto it, the threads on the

---

[5] In the oil business a *joint* may be a place where two things are joined together, but more often it is a length of pipe or tubing. If a man used the word in both senses at once he might correctly say that a joint of pipe is a length of pipe between two joints, just as the "second joint" of a chicken is the part between the knee joint and the hip joint. A piece of pipe materially shorter than the length of a standard joint is called a "nipple," a "pup joint," or a "sub."

second joint screwing into those in the collar of the first. The two joints are then lowered until the collar of the second is just above the derrick floor and a third joint is added. This is continued until the *string* of casing reaches the bottom of the hole or some higher point where it is to be *landed* or *set*. It remains suspended from the clamps on the derrick floor until it is set on the bottom or on a shoulder farther up the hole or until it is cemented, as is described later.

If something goes wrong, or if a string of casing is to be recovered after the well is completed, the process is reversed, and the casing is raised and unscrewed joint by joint until it is all out and on the pipe rack.

If there is reasonable certainty that a string of casing is to remain in the hole and will not need to be *pulled*, which means recovered, the joints may be beveled instead of threaded and welded together instead of screwed. This is not a common practice, but it is used where conditions favor it. If plans go awry and the string has to be pulled, it is cut into joints as it is pulled, either by electric or oxy-acetylene torches or by steel tools known as *casing cutters*.

Another practice sometimes used with screw casing is to weld the collar after each connection. Both welding and collar welding are giving way to cementing the threads with epoxy resin. Epoxy-cemented connections are claimed to be as tight and strong as welded ones, and a cemented connection can be unscrewed if it is heated to soften the cement.

Floating the Pipe

With the deep drilling being done nowadays, the weight of a string of casing may be greater than the tensile strength of the pipe or than the strength of the threads in the collar. No one wants a string of pipe to pull in two, for the part that drops is usually badly damaged and hard to get out of the hole and such an incident may multiply the cost of the well or force its abandonment. If the driller thinks there is danger he may *float in* his pipe; all he does is screw on a water-tight drillable casing shoe, fill the hole with water and then run the casing. He is merely taking advantage of the principle that makes a steel ship float. If this doesn't give him all the buoyancy he wants, he fills the hole with mud just thick enough to be reasonably fluid, instead of water. When the string reaches the bottom of the hole he can easily drill out the shoe and the casing is safely *landed*.

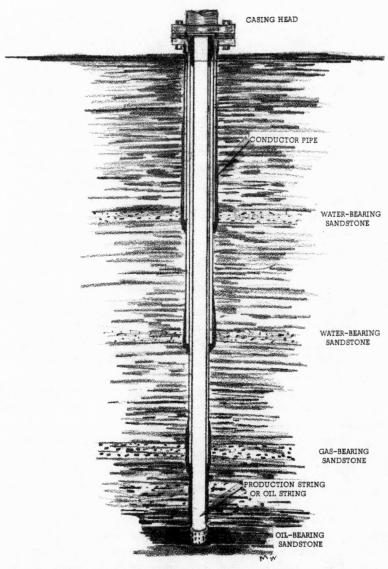

CASING HEAD

CONDUCTOR PIPE

WATER-BEARING SANDSTONE

WATER-BEARING SANDSTONE

GAS-BEARING SANDSTONE

PRODUCTION STRING OR OIL STRING

OIL-BEARING SANDSTONE

Each "string" of casing extends from the surface down to the point where it is needed to exclude water or gas or to support the sides of the hole. The first and largest but shortest is called the "surface string" or "conductor pipe." The last, smallest, and longest is called the "oil string" or "production string."

Casing Within Casing

This is the first casing job we have watched. When you watch the next one, notice that the second string goes down inside the first string, that it extends on down below the first string to wherever the second casing point may be, and that it extends from the top of the hole down. Outsiders are likely to think that each string of casing starts at the bottom of the preceding string and extends downward. Not so; each string runs from the surface down. When several strings are in, the top of the well looks like the concentric rings on a target, with the last and smallest string constituting an inordinately big bull's eye.

The one exception is the occasional use of a *liner*, which is a short string of casing, generally only a joint or two, that may in special circumstances be set from the bottom of the inside string of casing to the bottom of the hole.

Setting and Cementing Casing

How is the casing set firmly enough so that fluids cannot enter the well around the bottom of it? When circumstances are favorable the weight of the casing will seat the casing shoe so firmly on the rock where it is *landed* that no water or gas or oil will leak past the shoe and get into the casing. This is known as a *formation shut-off*. Where the formations are not stable, or where the gas pressure or water pressure back of the pipe is high, a formation shut-off is not reliable and the casing must be cemented.

Cementing a string of casing consists in filling the small annular space between the wall of the hole and the casing with cement and letting the cement set. Only a few feet or a few joints at the bottom may be cemented, or the outside of the whole string may be cemented clear to the top of the hole.

How would you go about it? Would you pour cement down the space, an inch wide or less, between the casing and the wall of the hole? If you did you would be lucky to get the cement as far down as the third joint; the space is too small and irregular to be filled unless the cement is forced in under pressure. Oil men—usually a crew supplied by a company that makes a business of cementing wells—go about it in the opposite way.

First they calculate how much cement it will take to do the job.

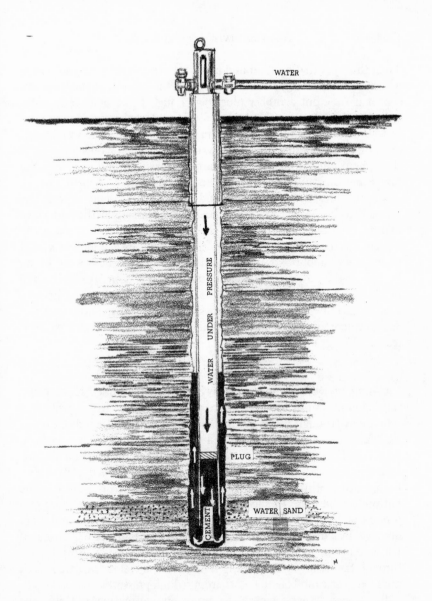

The object of cementing casing is to exclude fluids from the hole by forming a solid seal of cement between the casing and the wall of the hole. Enough cement for the purpose is put into the casing, a plug is put on top of it, and water is forced in under pressure until the plug, driving the cement before it, reaches the bottom, by which time the cement has been forced back up outside the casing, where it is allowed to set.

Then they mix that amount and put it into the well, inside the casing, which has been raised a short distance off the bottom of the hole. On top of it they put a rubber plug, big enough to be practically water tight, inside the casing. Then they fill the rest of the casing with mud or water and keep pumping more in under pressure. The weight and pressure of the column of mud or water drive the plug down and force the cement out through the bottom of the casing and up between the casing and the wall of the hole. When the plug hits the bottom of the casing the pressure rises sharply and the cement is where the driller wants it to be. Valves are then closed on the casing at the surface to maintain a balanced pressure at the bottom of the hole. When the cement has set, which with the special quick-setting cements now used is a matter of about twenty-four hours, the plug and the cement in the bottom of the hole are drilled out and the job is done. A test is made by bailing the hole dry and letting it stand awhile to see whether any fluid is entering. If the test is satisfactory the driller logs, "9 in. set 3237; cemented with 500 sacks." This is often more abbreviated, so it reads, "9″ csg @3237′ w/500 sx."

Perforating and Notching

Casing, as we have seen, is heavy steel pipe intended to keep unwanted fluids out of the well and to prevent the hole from caving, but sometimes (though this is getting ahead of our story) it is desirable to let fluids in. For example, if the oil sand itself is cavy the only way to keep it out of the hole is to case past it. But the oil must be allowed access to the well. Or the operator may want to produce from a sand he has drilled through and cased off, in addition to the sand at the bottom of the hole. In such cases, a few joints of the casing are *perforated* or *notched* with the desired number of holes. The holes are punched by a device lowered into the hole to the joints to be perforated (or notched). It works by shooting explosive jets or steel bullets through the casing wall, burning notches like a welder's cutting torch might do, or sand-blasting slots from a high-pressure nozzle.

In some instances, where the producing formation is thick, the casing will be landed above the producing zone. Then a joint or two of slotted or perforated "liner" will be hung inside and from the bottom end of the casing with special clamps. The liner helps to support the hole but offers plenty of openings for the oil to enter.

How Big Is the Hole?

Well and good! We've set and cemented our string of pipe, we've gotten rid of the water and cavings that were bothering us, and we have a clean, dry hole. What next?

The first thing we'll have to do will be to change tools; the ones we were using before won't go down through the casing. The casing is inside the hole that they drilled, which means that the casing is smaller than they are.[6] So we set up a string of smaller tools, which means that from now on we will be drilling a smaller hole. Every time we set a string of pipe we reduce the size of the hole that will be drilled thereafter. Which leads to the question, how large is a well?

The size of casing for cable-tool holes was formerly designated by inside diameter, but present practice is to name inches of *outside diameter*. The standard sizes run from 4¾ to 24½ inches. The inside diameter is three-eighths to an inch smaller than the outside diameter. The threaded collars are, of course, larger than the pipe, and the hole drilled to take a string of pipe must be slightly larger than the outside diameter of the collar. The bit that drills a hole for a certain size casing must therefore be an inch to two inches larger than the nominal casing size. As a rule, though not always, the tools and the hole are referred to in terms of the casing that is to be run, and when the driller says he is drilling a seven-inch hole or using seven-inch tools he means that he is using a string of tools that will make a hole big enough for casing measuring seven inches outside diameter.

The oil man likes to finish his well with a hole big enough to permit setting at least a 5½-inch string of casing to the producing formation. This last and smallest string of casing is called the "production string" or "oil string." To finish with a 5½-inch string the driller must start with a considerably larger hole to allow for the successive decreases in diameter of the hole as successive strings of casing are set. If he starts too small and encounters a couple of unexpected water sands or cavy streaks he may *run out of hole*, that is, have to reduce his hole to a size too small for practicable drilling. Cable-tool holes have been

---

[6] Strictly speaking, this may be true only of the bit, and sometimes only the bit is changed. As a rule, the whole string is changed to maintain the proper balance of weight and length.

finished as small as three inches, but such small tools have little weight and impact and the man who finds himself with a hole less than five inches will usually save money and time by *skidding the rig* and starting a new hole.

No one, on the other hand, wants to start his well larger than necessary, for big casing costs more than smaller casing, and casing is a major item of the cost in a cable-tool well. All of which means that the operator should, and usually does, take into account all the geologic and other knowledge he can get before deciding how big a hole to start.

If the well is to be shallow and only one horizon needs to be cased off he may start a 9-inch hole. If he expects much "surface water" he may spud a 10¾-inch hole, set a 10¾-inch *conductor pipe*, lighter than ordinary casing, and then drill ahead with an 8⅝-inch hole. A well that is to be deeper, and likely to encounter more water or gas sands or cavy horizons, may be started with 13-inch, 13⅜-inch, 16-inch, or 18⅝-inch tools. Few wells are started at 24 inches in the United States, though a few 50-inch rotary holes are drilled each year to serve as access shafts to underground petroleum storages and to aid nuclear explosives research.

Under-Reaming

"But," you say, "suppose a driller sets his 7-inch casing at 2,800 feet and drills a 5-inch hole from there down, anticipating no trouble until he gets the productive sand at 3,400, and then at 2,950 hits an unexpected water sand; is there nothing he can do to avoid setting the 5-inch pipe at 2,950?"

Yes, there is; he can under-ream.

An under-reamer looks like a bit, but at the bottom, in place of the usual dressed-out bit-face and edges, it has lugs that, when free, give the under-reamer much the same shape and effectiveness as a bit. The lugs can be pressed downward and inward, so that the under-reamer, with stems, jars, and the rest of the string, can be lowered through the casing. When the under-reamer gets below the casing, strong springs force the lugs out and the thing will drill a hole big enough to take the casing through which it has been lowered.

In the case you have suggested, the driller, if he has not cemented his 7-inch, can under-ream down through the water sand at 2,950, lower the 7-inch, and case off the water. If the water makes the hole cavy he can let the casing *follow down;* that is, keep lowering the casing as he drills, so that the bottom of it is always just above the drilling tools. When he has set the 7-inch below the water sand, he can drill a 5-inch hole on down to the oil sand.

Under-reamers would be used more if the lugs hadn't a bad habit of coming off in the hole and thus causing *fishing jobs.*

Fishing

Fishing in the oil business may be sport, if you want to call it that, but it's never a pleasure. It's the all-too-frequent job of getting out of the hole something that doesn't belong down there. It is largely responsible for the driller's extensive vocabulary.

As we have seen, an under-reamer lug may come off in the hole. That may not be so bad; as a rule they are fished out rather easily; if not, they can usually be drilled up. But now and then a string of tools or various parts of one come unscrewed or the drilling line or the sand line breaks. Then there are a few pounds to a ton of junk in the hole, to be fished out if possible.

It wouldn't be so hard if the pesky things would stand up straight in the hole where any one of a dozen ingenious fishing tools could take hold of or screw on to them, but when they bury their heads in the wall or hide them under the side of the casing shoe, then successful fishing depends on one or all of a striking array of *spears, bull-dogs, baskets, magnets,* and *sockets,* each with its special use for special circumstances. When none of these works, the driller will fashion a still more special fishing tool, sometimes *taking a picture* of the situation by lowering an *impression block* into the hole, pounding it down on top of the object for which he is fishing, and using the impression thus made on the block to help him decide on the character and shape of the fishing tool he needs. The tool selected is lowered into the hole and manipulated with the drilling line, and by the feel of that line the driller must tell what is happening in the bottom of the hole.

Remembering that the hole may be a half mile or a mile deep, it's a wonder any fishing job is successful, yet a surprising percentage of them are. Distressingly often, however, they run into weeks and months, and into thousands of dollars before success is attained. It's no fun to have "fishing" as the only entry on the log day after day, when every day is costing hundreds of dollars.

Drilling-In

Here on Anonymous Doe No. 1, thank fortune, we have had no fishing jobs. You and I have been watching it for weeks, until the lazy bench is polished like a night-club bar. At last the well is down to within a few feet of the point where the objective sand, the hoped-for *pay* horizon, is expected. The tool pusher sticks around, the superintendent and probably the operator are on the way, and the place is all cluttered up with scouts and geologists. There is tension in the air. We are about to drill-in.

A *control head* is put on, a heavy fitting that screws on to the top of the innermost string of casing. It can be opened to let the tools in or out of the hole, and closed to fit snugly around the drilling line. Out of it leads a pipe—or several—fitted with heavy valves to conduct the oil to the field tank that has been built close by. If heavy pressures are expected, the thing has many pipes and valves and is called a *Christmas tree*.

The tools are lowered into the hole and drilling resumed, cautiously, the driller noting every change in the pace of the engine or the tension on the drilling line that might indicate an inflow of fluid into the hole.

A Dry Hole

A screw is drilled; nothing happens. The tools are pulled and the bailer run. Before it reaches bottom the sand line slackens; it has encountered fluid rising in the hole. It is pulled out and is full of water. Oh—oh; we've drilled a dry hole!

Such is wildcatting! Let's go home.

The crew will pull and salvage all the casing they can, strip the rig, load up the tools, and move to the next job. In a few days all that will be left to show for so much money and such high hopes will be a bare spot and a piece of conductor pipe protruding a foot or two above the ground. The party is over.

That's one thing that can happen, and too often does. Thank heaven, there may be another conclusion. We dreamed that one; now we are awake and watching.

A Well

The tools are dropping regularly; the screw is not quite drilled. Suddenly the line slackens and a spray of gas and oil begins to sizzle around it at the control head. Glory be! We've got a well! The control head begins to vibrate; must be a big one! Slowly a valve is opened and a stream of oil shoots into the tank. Careful now; if the well is opened up too quickly it may shoot the tools up against the control head and wreck the works. Gently, easy does it, until all the valves are open. Man, man; she's filled that hundred-barrel tank in less than twenty minutes! We've got a 7,000-barrel well! Shut her in 'til we get more tankage.

Feeling delirious? If so you're not lonesome. Look at the rest of the crowd, all trying to appear calm and none of them succeeding. Even the driller, to whom this well has just been another job, is grinning like a busy ape. That roar you hear is the starting of the scouts' cars headed for the nearest telephone. Look at the president of Anonymous Petroleum Corporation; he poured his last dollar of credit down that hole; now he's trying to tear himself away from the tank to go wire his wife and his office. It's a great day for a lot of people. If you feel able, we'll go somewhere and relax.

Making Big Ones From Little Ones

"Suppose," you say as we drive into town, "that had been a little well instead of a big one? What if it had had a good thick producing sand, but had come in for seven barrels instead of seven thousand? Would they have let it go at that?"

Probably not. The size of a well depends on several things, among them the porosity and permeability of the sand at the spot where the well penetrates it. If the well had been a *teaser* they would have studied samples of the sand bailed out of it. If it had been mostly shale instead of clean sandstone, they probably would recommend a *frac job* to open up fractures in the sand. But if the samples showed a sandstone clean of shale, but unusually fine-grained, or a sandstone made up of reasonably coarse grains cemented together with silica or lime, the operator might send for a well-shooter. If the samples had shown a limestone with some porosity but not enough to let the oil escape readily, then the operator would have sent for an acidizer.

Few reservoir rocks are of uniform porosity and permeability[7]

---

[7] Porosity is the amount of pore space in the sand or other reservoir rock, often referred to as the *percentage of voids*; it determines the amount of oil per cubic foot the sand can hold. Permeability is the measure of the ease with which the

throughout; most of them have porous areas and *tight* areas. The reservoir rock may be tight where the well hits and be more porous only a few feet away. If the permeability can be increased between the well and the porous area, the production of the well may be greatly increased. *Opening-up* the formation near the well bore increases the effective permeability.

Acidizing

If the imperviousness or tightness is due to calcium carbonate—limestone—cementing the reservoir bed, it is often possible to dissolve out part of the lime with hydrochloric acid, thus increasing the porosity and permeability. Many a small well has been converted into a big one by acidation. The job is not one for an amateur. A good acidizing concern studies the rock to be treated, adapts the strength and character and amount of acid to the job to be done, and takes proper precautions to protect the casing and adjacent formations that may be keeping water out of the well. If the first dose is beneficial subsequent doses may be given, either at once or at considerable intervals, depending on the patient's behavior.

Shooting

If the reservoir rock is a sandstone, and especially if the tightness is due to siliceous cement or to fineness of grain or clay, acid is out and the only hopes are in the well-shooter or the fracing operation. The well-shooter shatters the sand by exploding nitroglycerine in it, and for ninety years he has been making big ones out of little ones. The nitroglycerine or *soup* (usually in solid form) is placed in tin torpedoes about half the size of the casing; each torpedo is lowered into the hole (need I say carefully?) by a special line wound round a spool attached to the sand-reel shaft. The bottom of each torpedo nests into the top of the one below it. Any amount, from two quarts to two hundred, may be used, depending on the thickness and hardness of the sand.

---

oil or other fluids can move through the sand. Porosity depends on the total volume of the pore spaces rather than on their size, for many small spaces may be equal to a few large ones. Permeability depends on the size and connection of the pore spaces. Thus a sand may have high porosity and low permeability, though the reverse is not likely to happen.

The *effective porosity*, which is the important thing, is the volume of connected pore space per unit of volume of the sand.

When all the torpedoes are in, the hole is filled or *tamped* with mud or water, and a detonating mechanism is lowered in. Then the last string of casing may be lifted far enough to be safe from injury, and at the appointed time the deed is done.

Formation Fracturing

Even though the shooting of a well with "nitro" is a tried and true stimulation technique, it has been largely replaced with fracing (it rhymes with cracking), a relatively new and very ingenious method. Thousands of gallons of oil or treated water carrying thousands of pounds of sand are injected into the nearly-impermeable reservoir as fast as possible and at very high pressures. This takes big, high-pressure pumping equipment, supplied by specialists. Massive high-powered pump trucks drive up to the rig, hook up a maze of heavy duty plumbing to the wellhead and then start their pumps. From one of the tall tanks near the rig, the pumps draw crude oil, diesel oil, or water and mix it with chemicals or acid as prescribed by the engineer. Then, as the frac fluid is pumped down the hole, a special grade of sand is mixed with it at a controlled rate.

Such tremendous pressures are applied to the oil sand by the pumps that the formation splits and shatters, allowing the "frac fluid" to enter and travel far back through the resultant cracks. As long as the pressure is maintained, the cracks will probably be held open, but as the pressure is released they would close again except for the sand grains that prop them open. Thus the sand holds the walls of the fractures apart and permits the oil to flow to the well bore.

Fracing is like walking into your loaded clothes closet; you part the clothes, step in between them, push them apart with your elbows and bring out the hunted item. If you just step back out, the clothes flop back in place and you cannot find a trace of where you entered. Just prop those clothes apart with a coat hanger placed crosswise and you can thereafter move in and out through the same opening.

Acid is used in the frac fluid where the reservoir is cemented or partly cemented with limestone or dolomite. Then it helps to enlarge the fractures chemically and the job is properly termed an *acid frac*.

By 1960, about 80 per cent of all wells completed as producers were fraced to stimulate production, and fracing is even credited with changing a long-standing trend toward deeper drilling in the search for new oil fields. Many an old field, once depleted to the stripper stage, has been given a new lease on life, and many tight sands known to be saturated with oil have been re-opened or redrilled and fraced for new production.

Now We Have a Well

And so at last Anonymous Doe No. 1 has taken its place as a producer. When they had enough tankage to give it a real test, it did 6,800 barrels *natural*, which means without acidizing, shooting, or fracing. You have recovered your sanity, and are ready to watch a rotary in operation. There will be one spudding across the road in a few days.

CHAPTER 7

Drilling Another Well

A *Rotary Well*

"**G**OOD morning, Mr. Johnson. Richard Roe speaking. Anonymous Doe No. 1 has been turned into the line[1] and is making sixty-five hundred.[2] Your lease on my land offsets it. When are you going to start my offset? The surveyor has staked the location."

"Right away, Mr. Roe; I've already let the drilling contract and the rig is on the way."

Thus Johnson Roe No. 1 is born.

Moving in the Rig

Mr. Johnson is the operator or the oil company executive who has committed his company to the complete development of the oil pool which partially underlies Mr. Roe's property. Mr. Johnson does not own a rotary drilling rig; he has arranged for a drilling contractor to drill the hole for him. Mr. Johnson likes to know exactly how much the Johnson Roe No. 1 well will cost, before it is drilled. Therefore, he has signed a *turn-key contract* to have the hole drilled for twelve dollars and fifty cents a foot. Now we can begin to see why oil well drilling is a costly business. Our contractor has had to include in the bid his costs of moving the rig from another hole, which he has just completed, to the location stake of the Johnson Roe No. 1.

---

[1] *Turned into the line* means that a pipeline has begun to run the oil from the field tanks into which the well's oil goes first, so that the well's production is being marketed.

[2] You remember that Anonymous Doe No. 1 was originally estimated as a 7,000-barrel well. When given its first test it flowed at the rate of 6,800 barrels a day, and when it was turned into the line it did only 6,500 a day. Such reductions, and in fact much larger ones, are common, though now and then a well increases during its first few days on production.

Let's drive over and watch the rig being made ready to move from the site of its previous job. The derrick must be let down like closing a jackknife and disconnected into two or more pieces. Each of these pieces of the derrick must be hoisted or pulled up onto the bed of a large diesel oil-field truck and driven to Mr. Roe's property, usually behind an escort. Other trucks, probably fifteen of them, back up to each remaining piece of machinery and winch it gently on to the body of the truck. Everything is lashed down with chains and the trucks move out with hundreds of tons of equipment. Now, back at the location of the Johnson Roe No. 1, we see the whole process is reversed. The derrick is reassembled and unfolded and all of the motors, tanks, pipe, and equipment are re-set alongside several deep trenches which have just been dug in Mr. Roe's pasture. A bulldozer is still pushing dirt up into a dike around a pond which has suddenly appeared next to the well location. Several water tank trucks are shuttling back and forth from Mr. Roe's water well to fill the trenches and the pond. These are the mud pits and the sump; they will be an important part of the drilling operation. This is all part of moving in. It will be another day or two of hard work before all the equipment is hooked up so that the drillers can start to drill. While the crews are setting up the rig, let's look over the equipment which has just been unloaded.

The Rig and Equipment

You need not be an oil man to notice the difference between this well and Anonymous Doe No. 1. The derrick is taller—it may be as much as a hundred and ninety feet high, about the height of a seventeen-story building—and at its top and near its middle are little railed platforms seldom seen on cable-tool rigs. It has only two legs, from all appearances, and seems to be ready to fall away from the engines. There is no walking beam, no pitman, no temper screw, no Samson post, no headache post, no band wheel, no bull wheel, no calf wheel, no sand reel. You will look in vain for the familiar bits, stems, jars, sinkers, or bailer.

For that matter, you may look in vain on one rotary rig for things grown familiar on another. Rotary practice is being improved so rapidly that some new item of equipment, stronger or simpler or more automatic than that previously in use, appears every few months. Steam power, so long familiar around the cable-tool rigs and later with rotary, has largely given way to diesel, gas, and electric power. The equipment you will see on Johnson Roe No. 1 may look primitive by the time the field is fully developed.

The Substructure and the Blowout Preventor

Notice, even the cellar is missing? Modern methods, bowing to the need for speedy and efficient moves, place the entire derrick and attendant machinery on an elevated platform six or eight or ten feet high, so that the crew has easy access to equipment under the rig. The *substructure*, as it is called, consists of box-like sections of heavy steel girders and braces. Each section can be moved by truck and bolted to its neighbor quickly at the next location. Most of the plumbing for the well is connected up through the substructure.

As soon as the first string of casing, the *surface pipe*, is set on the Johnson Roe No. 1, the crew will weld the *blowout preventor* to it in the substructure, just above ground level. The blowout preventor is a big valve in a massive frame with hydraulic lines and control handles attached for both remote and direct control; it will be fixed to the surface casing all during the drilling operation. During drilling, if a heavy flow of water, oil, or natural gas is encountered, the driller can close the preventor around the drill pipe and forestall a blowout and its accompanying danger, fire.

The Table

In the middle of the derrick floor, where you rightly judge the well to be, you will see a heavy steel turntable, ten to thirty-six or more inches in diameter, going round and round. It rests on a steel base level with or just above the derrick floor and rotates on roller bearings running in oil. This is the rotary or rotary table or turntable or table.

The rotary table has a twelve- to eighteen-inch square or hexagonal open hole in the center which is shaped to receive the *kelly bushing*.

The Kelly Bushing

In order for the rotary table to turn the drill pipe, yet permit the drill pipe to move down as the hole gets deeper, a heavy block of steel is fashioned to match the hole in the rotary table. In its center is a square or hexagonal or ribbed hole, four to six inches in diameter, shaped to fit the sliding *kelly* that turns the drill pipe. More about the kelly and the drill pipe in a moment. When the rig is drilling, the

MOVING IN

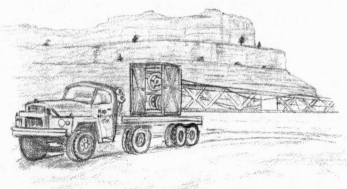

DRILLING RIG

SUBSTRUCTURE

kelly bushing is fixed firmly in the rotary table and the greased kelly moves freely up and down through it as it turns.

Since the top of the kelly bushing is the last place we can see a mark on the kelly as it follows the drill pipe into the hole, most measurements of depth are made from the kelly bushing. The top of the kelly bushing may be a few inches above the rotary table, and the rotary table slightly higher than the derrick floor, and the derrick floor may be ten feet above the ground level. So, with all of these points to measure from, we still find elevations and depths labeled K.B. or R.T. or D.F. or G.L., whichever they were measured from, although K.B. is preferred in present practice.

The Hoist or Draw Works

On some of the older rigs the table is driven through a chain drive from the *hoist* or *draw works*, a heavy case full of gears, the brakes, and the drum for the drilling line. It is about halfway between the table and the engine, which is about where you are used to seeing the band wheel. This machine is the center of the driller's attention on a rotary job. It is the power unit that raises and lowers the drill pipe, lowers and removes casing, and on most rigs it turns the table that rotates the kelly that turns the drill pipe that carries the bit that drills the hole. On the newer rigs the table is rotated by a separate driving mechanism from the main engine. The draw works weighs as much as several Ford cars.

The Grief Stem or Kelly

We have already found the kelly bushing and noted that it had a square or hexagonal or ribbed hole through its middle. Through this hole is a similar-shaped length of very thick-walled pipe called the *grief stem* or *kelly*. It fits loosely enough that it can slide up and down freely. If it is round it has slots or grooves in its sides to permit a firm grip by the ribbed bushing and is a *slotted kelly*. If it is some other shape, the bushing grips it as a "square peg in a square hole." It is four to six inches through, thirty to forty feet long, and is open like a water pipe from end to end; it has couplings to attach it to the swivel above and the drill pipe below. A stopper block is attached to the kelly below the kelly bushing, and when the kelly is picked up out of the hole, the bushing comes with it and is lifted out of the rotary table. Both the kelly and the bushing can then be set aside by lowering the kelly into the *mouse hole* at one corner of the derrick floor. Mouse holes are drilled off to one side especially to have a place to park the kelly when it is not in use.

The Drilling Line or Rope

The Anonymous Doe No. 1 was drilled with cable tools where great lengths of woven wire cable were required. As a matter of fact, that rig had to have more footage of cable than the maximum depth expected to drill. The cable had to reach all the way to bottom, and still have some left on the drum. That was more than three-fourths of a mile of *rope!*[3]

On our rotary rig here at the Johnson Roe No. 1, none of the drilling line ever gets into the hole. One end is spooled on the draw works drum and is then wrapped around and around between the pulley wheels or *crown sheaves* in the top of the derrick and the sheaves in the traveling block and finally back down to its attachment on the leg of the derrick. These parts constitute a tremendous block and tackle rigged to lift the weight of the drill pipe or the casing. Instead of needing nearly a mile of rope, our rig probably has less than a thousand feet.

The *weight indicator* is mounted on the dead end of the drilling line near where it is attached to the leg of the derrick. The pull on the drilling line is measured by this device and reads in pounds of weight hanging from the traveling block.

The Swivel, Its Support, and the Mud Hose

Above the kelly is a ball-and-roller-bearing affair, of which the upper part is stationary, the lower part rotates with the kelly. This is the swivel. It hangs by a hook from the traveling block. By raising or lowering the traveling block the swivel, kelly, drill pipe, and bit are raised and lowered.

Sticking out at one side of the upper, non-rotating part of the swivel is a heavy hose, bigger and heavier than a fire hose. Mud, good fluid mud, is an essential factor in rotary drilling; the mud hose carries it to the drill pipe.

If you let your eye follow the mud hose you can see that it hangs in

---

[3] *Rope* means wire rope in rotary driller's parlance. Rope of hemp or jute or sisal or nylon is *soft rope.*

a deep loop from the swivel, then back up near one leg of the derrick to its connection with the riser or standpipe. The standpipe transmits the mud from the mud pump to the mud hose for its trip down the drill pipe.

The Drilling Mud

Are you still looking for the bailer? You won't find it; rotary wells are not bailed; the cuttings are flushed out with the mud.

The mud pump or slush pump (there may be two or even more of them) is built to pump mud under several hundred pounds of pressure. Out back of the derrick are the slush pond and mud pits, which we saw being dug, where mud is kept at just the right consistency to do the job at hand. If the hole is standing up well and there is no gas or water to be overcome, the mud will be thin, about the consistency of good cream; but if the well has encountered a high-pressure gas sand that has not been cased off, the mud will be as thick as can be put through the pump. In most cases special material is hauled in in bags, and in extreme cases the mud is made with barite or some other heavy mineral.

The slush pumps draw the mud from the slush pit and force it through the riser, the hose, the swivel, the kelly, the drill pipe, and out through small holes in the bit, against the bottom and sides of the hole, and back between the drill pipe and the walls of the hole to the surface. With it come the cuttings.

Now, if the mud flushes out the cuttings, what happens if the pump is stopped for repairs or so the crew can put a length of drill pipe in the drill string? The chips of rock that are coming up the hole in the mud stream start falling back to the bottom, there to pack so tightly around the bit that we may not be able to move it again. We can see that the mud has to be thick enough to hold the cuttings in suspension during any shut-down period when the driller stops the pump.

Now we have a problem, because the mud is used over and over again, and we don't want to pump the cuttings back down the hole. This is where those deep pits back of the rig do their job. The crew has set up an ingenious shaking-screen device at the edge of the first pit. When the mud comes out of the hole it gushes out onto the *shale shaker*; the big cuttings slide off into a pile and the mud goes through into the first mud pit. In the pits the mud moves lazily toward the intake of the pump at the other end, all the time losing the

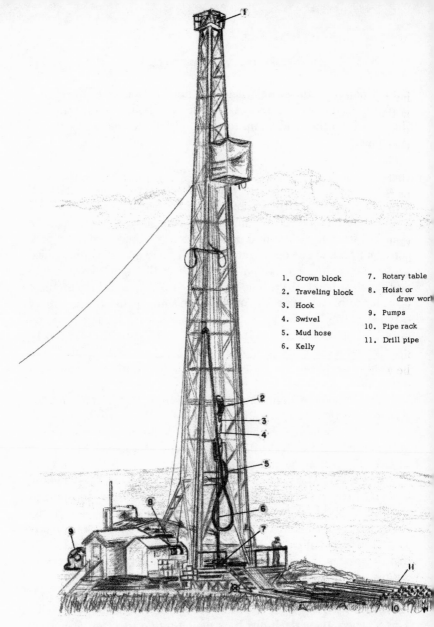

1. Crown block
2. Traveling block
3. Hook
4. Swivel
5. Mud hose
6. Kelly
7. Rotary table
8. Hoist or draw work
9. Pumps
10. Pipe rack
11. Drill pipe

Rotary drilling bores a hole by rotating a bit on the end of a string of pipe, called drill pipe, some of which is shown stacked ready to be "run-in" the hole. The draw works at the left is the mechanism that runs most of the job. The pumps force mud down through the drill pipe and back up between the drill pipe and the wall of the hole, slushing out the cuttings and plastering the wall of the hole so that it will not cave.

fine particles of the cuttings and sand grains which were not caught by the shale shaker. We can call this the settling pit. The mud has to be just the right consistency to let the small particles fall out here where we want them to.

The geologist on the Johnson Roe No. 1 wants to know exactly what kind of rock the bit is drilling in at all times, so that he can keep his formation log of the hole. This is easy, for in the pile of cuttings at the end of the shale shaker are sample chips of everything that has been brought up by the mud. He can collect samples as often as he wants them, wash off the mud, look at the chips under a microscope and write up his log. But for the geologist to write up an accurate sample log, the mud must carry the cuttings all the way up from bottom without grinding them up and without losing the heavier particles.

The mud, as already suggested, plays other roles besides flushing out the cuttings. It keeps the bit cool. Coming up between the drill pipe and the hole wall—under heavy pressure, remember—it plasters the wall with a good stiff layer. The result is that the walls of a rotary hole stand up when the walls of a cable-tool hole would be caving. It is seldom necessary to case a rotary hole to prevent caving, and a rotary may safely carry a footage of open hole that would scare a cable-tool driller to death.

That's not all. In addition to plastering the walls the mud penetrates slightly any porous strata that have been drilled. By so doing it seals out of the hole the fluids they contain. Only unexpected, high water pressure or high gas pressure can put water or gas into a rotary hole against the driller's wishes. Within reasonable limits of quantity and pressure a rotary can drill through water sands and gas sands and go merrily on its way with no necessity for casing. Thus Johnson Roe No. 1 may use only two strings of casing where Anonymous Doe No. 1 had to use five.

The mud provides flotation to support some of the weight of those two strings of nearly a mile of casing. Thus the derrick doesn't have to hold up the whole weight while the crew runs the casing in the hole. We can also be certain that thick mud around the drill pipe helps to keep the drill string from whipping in the hole and knocking off chunks of the wall.

If we think that rotary machinery has been improved by leaps and bounds in the last few decades we should look at the mud recommendations up on the bulletin board in the dog house. Let's go see. Here is a printed form that has been filled out by a specialist in mud chemistry, after he tested our mud and diagnosed our problems. He is the *mud engineer*. His suggestions read like a doctor's prescription, but instead of writing so many pills per day he prescribes his chemicals in terms of adding five sacks per hour or three pounds per ton or fifteen sacks per tour.

We must give a lot of credit for record-breaking deep holes and drilling speed records and the solution of knotty drilling problems to the vast improvements made by mud chemists and mud engineers.

Special Muds

If the well is drilled into the producing sand with heavy mud in the hole the mud may so plaster the sand that little or no oil will be produced. The well is then washed by pumping water into the hole and if necessary churning it with a bailer or cable tools. Experience has shown, however, that the mud may so penetrate the pores of the sand that washing will not remove it and the well may take hours or days or even months to expel the mud particles thus deposited. Some wells never overcome the restriction of the mud.

Such sealing of the pay horizon may be avoided in several ways. Oil may be used instead of mud, if the pressures are not great enough to require a heavier fluid. If we want to use straight oil for a drilling fluid we have to remember that the cuttings may fall back to bottom when the pump stops. The tool pusher may order some heavy crude oil for the drilling fluid. More likely he will call upon the mud engineer to prescribe chemical additives and a small amount of water which will thicken the oil to make it more like a mud. Then he will have an *oil-base mud* because oil is substituted for water as the main fluid.

If oil doesn't do the job required of a drilling fluid or possibly the fire hazard of a crude oil is considered too great, then the mud engineer may prescribe something quite different, but still use some oil along with lots of water. He will call this an *oil-in-water emulsion mud*. Actually, what he is doing with this combination of fluids is to gain some advantages of each. The oil does help to save on drilling time and since it is a lubricant the bit lasts longer. A slippery fluid around the drill pipe and drill collars makes it easier for the engines to turn the drilling string. This saves power and wear-and-tear on the string. Then, too, if the bit is cutting in a sticky or gummy shale, the oil in

THE MUD PUMP

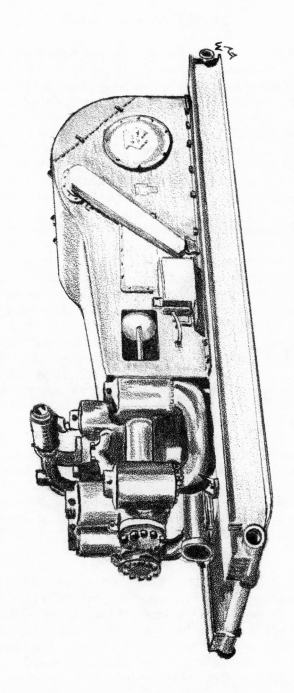

The mud pump circulates the drilling mud that cleans the hole, cools the bit, and plasters loose sediments against caving. The mud pumps on a big rig may use more than a thousand horsepower.

the mud helps to keep the cuttings from packing in around the teeth of the bit. Our driller calls this *bit balling*.

Not only may a water mud seal off a pay horizon but the water may also soak into certain kinds of shales and cause them to swell. The driller says that this kind of a *heaving shale* can push out into the hole and pinch the drill pipe so hard that he could be *stuck*. Oil in the mud helps to keep the water from soaking up those shales.

One of the biggest headaches to our rotary driller is to find that he is pumping mud down the hole at a good clip, but none of it is coming back to the mud pits. This is expensive, because he may have lost several hundred barrels of mud. He has drilled into a *lost circulation zone*. This can happen wherever the connected open spaces in a formation are more than three times larger than the biggest particles in the mud. The weight of the column of mud at the bottom of the hole pushes outward against the formation with more and more pressure as the hole goes deeper and deeper. So, if the bit cuts into rock which has less formation pressure than the pressure of the mud, then the mud runs back into the formation away from the hole. What can we do about this costly situation? Call the mud engineer and order a truck load of *lost circulation material*, mix up more mud and get ready to plug the holes in the formation with larger particles. Lost circulation material might be strips of cellophane, cottonseed hulls, plant fibers, chips of limestone, nut shells, mica flakes, straw, burlap sacks, or willow branches. Anything that can be circulated in the mud, to clog up the holes in the formation, will probably be satisfactory. Let's not forget that pressure is important, too. Maybe the mud can be made lighter in weight, to reduce the pressure, to stop the flow. A new prescription by the mud engineer might do this. Oil mud is lighter than water mud so he might prescribe the addition of oil. And so an oil emulsion or an oil base mud is again a special kind for curing or preventing lost circulation.

Drilling With Air

About 1951, drillers beset with severe lost circulation problems started experimenting with compressed air to take the place of drilling mud. There were lots of problems, but they reasoned air should cure the problem of sealing a formation with mud. Shows of oil and gas should be easier for the geologist to see in the cuttings because they wouldn't be concealed or washed out with mud. Since there would be little weight to a column of air in a hole, the bit should be able to go right on through open fractures and other open spaces in a formation without the costly loss of liquid drilling fluids. They soon found out that air could drill the hole faster and bits lasted longer and the total cost of drilling could be reduced. Only a few real problems remained to be solved.

After a decade of experimentation, and only sixty years after the first rotary rig drilled a hole at Spindletop, air drilling has become a commonplace technique. Imagine the problem that is solved by air drilling in the desert, miles and miles from a water supply and possibly hundreds of miles from a distributor of mud materials!

The Drill Pipe

Screwed onto the bottom of the kelly or grief stem is the tough, heavy drill pipe, joint after joint of it, screwed together after the manner of casing and extending almost to the bottom of the hole. It is 2½ to 6 inches in diameter, with 3½ and 4½ inches being the most popular sizes. The joints or pipe sections are all about thirty feet long and have a threaded cone or "pin" on one end and an enlarged threaded fitting, called a "box," on the other.

The Drill Collars

We said that the drill pipe went almost to the bottom of the hole, so what else should we find on the lower end of this drill string? Our driller has found out that a mile or so of drill pipe is about as flexible as a piece of dry spaghetti, so he puts several joints of very heavy, thick-walled, slick steel pipe on the bottom end of his drill string. This stiffens the string right above the bit and gives some extra weight to help the bit chew its way through hard rock. His drill pipe can always be used in tension, so the drill collars' weight helps to keep the drill string straight, like a weight on the end of a thread.

In order that all of the joints of drill pipe and the drill collars can be kept handy to the derrick floor, before they are put into the string they are laid down in layers on the pipe rack next to the raised walk. The walk is the narrow platform we see sticking out about fifty feet straight away from the open side of the derrick.

Bits

Screwed on to the bottom of the collars is a bit, and rotary bits are of many kinds. The kind formerly used almost exclusively for shale and other soft formations is known as a *fish-tail*. Take a steel wedge

ROTARY TOOLS

OVERSHOT

CASING
RELEASING
SPEAR

JUNK BASKET

WITH MILL SHOE

THREE CONE BIT

INTERNAL
PIPE
CUTTER

about two feet long and about three inches thick at the thick end, with a threaded pin in the center of the thick end to screw into the bottom of the drill collars. Cut a six-inch slit running up from the center of the bottom or thin end. Bend one side of the thin edge toward you, the other side away from you. When you have done it you will have a pretty fair representation of a fish-tail bit. As the bit is rotated the edges of the fish-tails shave and gouge away the rock and thus the hole is drilled. Modern bits for soft formations have three or more "tails" and are otherwise modified from the original fish-tail form. These may be used mainly for cutting the surface hole.

In harder formations one of the many forms of *rock bit* is used. Instead of shaving and gouging the rock they crush and grind it away through the action of revolving toothed cones set in the bottom of the bit. The *cone bit*, which is the one in most common use today, is a short, heavy steel cylinder with two or more cones set at slight angles from the horizontal near the periphery at the bottom. Some have smaller additional cones or cylinders on vertical or nearly vertical axes in the sides of the cylinder near the bottom. The outer surface of each cone is serrated with wedge-shaped or truncated wedge-shaped teeth. As the bit is rotated the cones roll on the bottom of the hole and on the side near the bottom and gouge and scrape and tear chips out, thus eating their way down through the rock.

Dozens of styles of cone bits are on the market, each designed for a particular job and variety of drilling conditions. If we expect to cut mostly soft shale the driller puts on a long-tooth bit. Maybe the Johnson Roe No. 1 will find some hard *digging*, in a tough limestone where the rock would just break off long teeth; then he changes to a stubby-tooth bit. Down below the limestone he may hit even harder rock, such as quartzite, which grinds the bit teeth off like an emery wheel. What then? He has still another type bit with armoured pellets which take the place of teeth molded into the cones. These tough little knots, protruding from the cone, crush the hardest rock into powder. Cone bits come in many standard sizes from 3¾ to 26 inches in diameter, but we are using 8⅝-inch bits to drill the Johnson Roe No. 1.

We almost forgot to notice that our driller has not been screwing the bit directly into the drill collar. He has a short sub screwed in first; then he puts the bit on the end of the *bit sub*. The bit sub is short and far less costly than the big drill collar, and the driller uses it to take

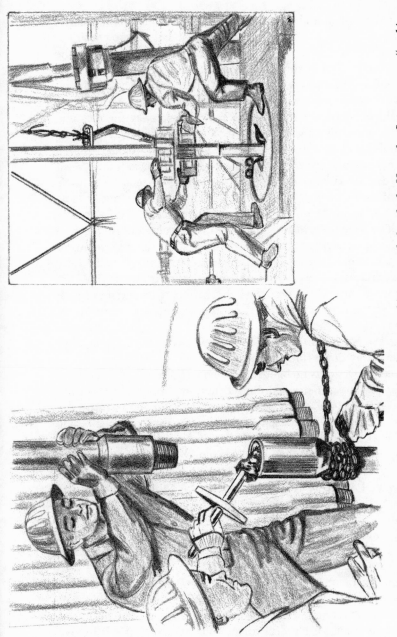

A rotary rig uses three "tours" (towers) or shifts of men working around the clock. Here the floor men are "making a connection"—screwing a new joint of drill pipe into the top of the string already in the hole. Floor men are called roughnecks; another roughneck, high out of sight, is the derrick man.

the wear and tear of changing bits rather than wear out the threads in the collar.

Drilling

Now let's watch a rotary work. The hole is down a thousand feet, the drill string has been pulled out to put on a fresh bit, the bushing has been taken out of the table on the kelly, the drill pipe is standing in one side of the derrick in lengths of two, three, or four joints each, and the crew is ready for *running in*. The bit is held in a special clamp in the hole in the rotary table called a "bit breaker." A stand of drill collars is picked up by the traveling block, swung over the bit, and screwed on. Bit and collar are lowered into the hole until the top of the collar is just above the top of the table. There another drill collar is screwed on; and the string is lowered again. After collars, then drill pipe—length after length—is added until the bit is within ten feet or so of bottom. Then the kelly with its bushing is screwed on to the uppermost joint of pipe and the whole is lowered until the bit is resting on bottom. The mud pumps are started and mud is forced to the bottom of the hole.

Then the driller throws a couple of levers and opens the engine throttle; drilling has begun. The engine turns the rotary table drive mechanism which rotates the table, the table rotates the kelly bushing and the kelly, the drill pipe and bit rotate with the kelly, and the bit eats its way down through the rock. It's as simple as that.

When the bit is worn and needs replacing the table is stopped, the kelly and kelly bushing are hoisted clear of the table and swung to one side and stuck in the "mouse hole." Then the pipe is raised, unscrewed, and stacked in lengths in one side of the derrick. On the last drill collar and sub is the bit, which is taken off with the bit breaker and a new bit is screwed on.

Did you catch the crew's language while they were pulling pipe— the printable part of it, I mean? They may have been "breaking" the pipe, which means unscrewing it, in lengths of two, three, or four joints each. A two-joint length is a "double," which is commonplace enough, but a three-joint length is a "thribble" and a four-joint length is a "fourble."

Logging Samples

The log of a rotary well is just as important as that of a cable-tool well but is not so easy to obtain. The cuttings from a cable-tool well are brought up in a bailer of water. They come, with little contamination, from the bottom of the hole and represent pretty accurately the formation being drilled there. The cuttings from a rotary hole are picked up by the mud, while drilling is in progress, and eventually, ten minutes to an hour or so later, reach the surface. They pour out with the mud on the shale shaker, where they collect for the geologist. He or a roughneck washes and bags the samples at regular intervals. How is the driller to know with any accuracy from what depth they come? By constant watchfulness and sampling, by careful measurement of volume of mud pumped per minute, and by calculation of the velocity of the mud compared to the rate of drilling, an astute geologist (the driller hasn't time) makes a shrewd, scientific guess. Improvement in pumps and hydraulics, and use of a drill-time logger have made the geologist's job far easier than it used to be. Using all of this information the geologist draws a sample strip log of the well as it is being drilled, with colors to indicate the different rock types being cut by the bit.

Reliable knowledge of the formations penetrated or of the depth at which any particular formation is reached may be sought by other methods. The method which furnishes the most complete and accurate information is coring.

Coring

The coring of a formation is the taking of a cylindrical *core* of it as it is drilled. The device for doing it is called a "core barrel." Core barrels are of many types and are constantly being improved. In its simplest form the barrel is a long steel tube about the size of the drill pipe, that goes on the string above the bit. On the bottom of the core barrel a special type of bit is screwed in, called a core head or coring bit. It is open in the center, and around the bottom edge are diamond chips, carborundum, or teeth of special steel. As the bit is rotated it cuts a ring, leaving a core of rock in the center. When the barrel has drilled its length the core is inside the barrel, where it is held by clips or clamps. When the drill pipe is pulled the core barrel takes the core to the surrace with it.

Sometimes it is desirable to take a core without removing the drill pipe from the hole. It can be done with a different type of core barrel, lowered to the bottom of the hole by a cable and rotated by a ratchet actuated by the jerking of the cable. This type is particularly adapted for use with cable tools, but it can be used on a rotary rig, with the drill pipe in the hole, if a sand reel, separate from the hoist, has been provided.

A good set of cores beats all the drillers' logs ever written, for there is the rock itself, to be seen, studied, and analyzed. The only trouble is that it slows down drilling, which, since drilling time costs several hundred dollars a day, is another way of saying that it is expensive. It is done, therefore, when accurate information is important. The unfortunate thing is that so many operators, including some of the biggest, seldom realize how important accurate information is to the future management of a new oil field.

Drilling-Time Logging

Other methods besides coring increase the accuracy of the sample log. One of the most important of these is the logging of drilling time. One of the pieces of standard equipment on a rotary rig is a *drilling-time logger* which measures and records continuously on a chart the pressure on the bit on the bottom of the hole; it also records drilling speed, measured by distance drilled per unit of time. Shale drills faster than sandstone; hard limestone takes longer than sandstone. But recall that the type of bit being used is important when formations of different hardness are met in quick succession. A long-tooth bit may dig very well in soft shale because it gouges large chunks, but a harder rock may break these teeth or tear them to pieces. In just the opposite manner a stubby-tooth or crushing bit may just pack down a shale and spin like a tire in a muddy road. By correlating bit-type and condition, and weight, the pump pressure and the drilling-time log, the geologist can gain a good idea of the character of rock being drilled. This, with microscopic examination of the cuttings, may enable him to compile a fairly accurate log of the well.

Electric Logging

It is now almost universally accepted practice to supplement and confirm the drilling and sample logs with *electric logs*. This method of logging was pioneered in the '30's by Marcel and Conrad Schlumberger. Schlumberger is now a company name, looks like German and is pronounced in French, so that oil men came to call electric logging the "Slumber Jay" method.

Today this early logging method has grown into a whole family of services done on a well by a number of different companies besides Schlumberger. More than 30 different tools can be lowered into the well one at a time on a cable, controlled and operated from a truck driven to the well for this specific job. Because of the cable, this whole field might better be identified as *wire-line services* and be further defined by the particular uses of the various tools.

THE LOGGING TRUCK

The logging truck is equipped with probes and sondes, miles of cable, and several electronic recorders. It records the electrical properties of the rocks around the well, and may also record their sound-wave velocities and radio-active-particle emission and take the well's temperature.

Still the leader in usage is the electrical survey. Although its present form is different from that used during the '30's, the basic principle is the same. A measured electrical current is forced to flow through the strata close to the logging device. The differences in their electrical properties are measured and relayed to the recorder in the truck. The recording instrument draws a series of wavy lines on a long strip of photographic film. These lines on the log tell the geologist what happened to the current and in turn the type of rock, and whether or not it can be a reservoir for oil or gas, and whether or not oil or gas is in the open spaces between the mineral particles.

Modern electric logging tools come in many shapes and sizes. Some of them log a large part of each formation; some a very small part. Some tools need no connection to the formation, others are pressed against the wall of the hole. Each is designed to contribute to accurate interpretation of what is down in this hole we are drilling.

Another class of wire-line logging devices are those which record radioactivity in the ground. These use a special Geiger counter or scintillator in the hole, somewhat similar to the uranium prospector's hand instrument. Their contribution lies particularly in determining the type of rock and its capability of being a reservoir for fluids. Our geologist may call for a *gamma ray-neutron log* to give him such information.

Other wire-line tools log physical characteristics of rocks rather than electrical. For example, the speed of sound through a formation can be correlated with the formation's porosity. And this is our basic interest; the fluid in the porosity may be oil. Because we need porosity measurements in the Johnson Roe No. 1, we have asked the logging engineer to bring his *sonic tool* along when he comes out to run the electric log.

Occasionally the temperatures at various depths in the hole can tell us something about the condition of the casing or help to calibrate some other logging tool, so we may run a temperature survey and record that alongside one of the other logs.

The space age is coming in logging tools also. Research is digging into the very heart of the molecules in oil productive horizons, measuring, for example, how the atoms in oil behave as contrasted to the atoms in water. Doubtless research will make today's tools look primitive to oil explorers a few years hence.

The Rotary Crew

When you first came to the Johnson Roe No. 1, you doubtless wondered why so many men. After watching it awhile you knew. On Anonymous Doe No. 1 the driller and the tool dresser handled it all, though if the engines or light plant needed service or there were other extra work to be done they had a roustabout to do it.

On the Johnson Roe No. 1 there are three crews each working an eight-hour *tour*, drilling around the clock. Each crew is headed by its foreman, the *driller*, and he in turn takes his orders from the *tool pusher* who supervises the whole operation. Each driller hires his own crew of three or four *roughnecks* who serve as mechanics, *floor men* and *derrick men*. They often hire on a job as a team that travels and works together. The tool pusher is a more experienced man than the drillers, so he not only handles all of the records, orders supplies, and passes out the pay checks, but is called upon as the chief trouble-shooter whenever the drillers need advice. The tool pusher or *push* works under the direction of the *drilling superintendent* of Mr. Johnson's company.

In a big drilling company the superintendent may have several rigs to attend to, so he hires a tool pusher for each rig, who reports to him each day. A rotary rig uses several engines and tons of equipment. There is more machinery than on a cable-tool rig, and the machinery needs oiling and other attention. The real work comes, however, when the drilling string is being pulled and run in, which is as often as a bit is dulled. Breaking and stacking drill pipe in lengths of thirty, sixty, or ninety feet, where minutes cost money, calls for a sizeable crew of fast, well-trained men.

Crooked Holes and Directional Drilling

You have doubtless been assuming that the well you are watching goes straight down in an undeviating vertical line. That's what the oil men used to assume until they learned that many if not most wells are crooked. Not so many years ago when methods of surveying wells were devised, drillers and geologists alike were astounded to learn that a well in Texas had deviated from the vertical by about 47 degrees; it was going sidewise faster than it was going down. Hundreds of wells that had been considered straight because tools and casing had run into them smoothly were found to be many degrees out of plumb. This was particularly true of rotary holes, which are harder to keep straight, but many cable-tool holes were found to be crooked also.

One reason for crooked holes is the deflection of the drilling tools by steeply inclined hard strata; faults or slips in the beds may also deflect the tools; but the most common cause is a desire to make hole too rapidly. Within limits, the more weight that is put on a rotary bit the faster it drills, but if too much weight is carried, so that the drill pipe is slack instead of taut, a crooked hole is almost certain to result. Drilling a cable-tool well with a slack line is likely to give the same result. In drilling wells, as in other things, haste makes waste.

Once the evil was recognized, prompt steps were taken to insure straightness, and straight-hole clauses were written into the drilling contracts. Today an unintentional crooked hole occurs only when the well has had extreme hard luck or a careless drilling crew, or the tool pusher has let himself get talked into planned imprudence near a property line, hence near a neighbor's oil.

The discovery that they had drilled crooked holes unintentionally naturally led some oil men to the idea of drilling them intentionally. The initial idea, it must be admitted, was for a man to start a well on his own lease and finish it under some neighbor's more prolific lease, but that phase quickly passed and *directional drilling*, as it is known, found a legitimate field of usefulness. It is being used extensively in California to tap, from derricks on the shore, reservoir beds lying off-shore under several hundred feet of ocean. In some of the salt-dome fields it is easier to reach the producing zone at an angle than vertically. In still other areas where geologists have found oil pools miles and miles from land, out under the ocean, it is possible to drill to many widely separated spots from the same drilling platform. The techniques and problems and costs of *off-shore* drilling are unique in the oil industry.

The methods are simple. In one a *whipstock*, which is merely a beveled bar of steel, is put in the bottom of the hole at the point where deviation is desired. This diverts the tools in the desired direction. Another uses a *knuckle joint*, which is a drilling tool so hinged that it can be deflected at an angle. By such methods holes have been drilled at angles as great as sixty-two degrees from the vertical, under such complete control that the bottoms of the holes are practically at the points originally intended.

Testing a Rotary Well

In time our well drills into the producing sand and everyone is agog to know what it will do. The cores show a good saturation of the sand with oil, and a little gas comes bubbling up through the drilling mud. We have a producer, all right; the question is, how big a one?

If there is no danger of the hole's blowing out we shall pull the drill pipe and watch what happens, as we would with a cable-tool well. There may, however, be no casing in the last few hundred feet of hole and we may not want to set casing before the test. In such a situation we will *drillstem test* to evaluate the sand.

In a drillstem test the drill pipe is run into the hole, equipped with a drillstem test tool which isolates the mud pressure from the produc-

ROTARY SPECIAL EQUIPMENT

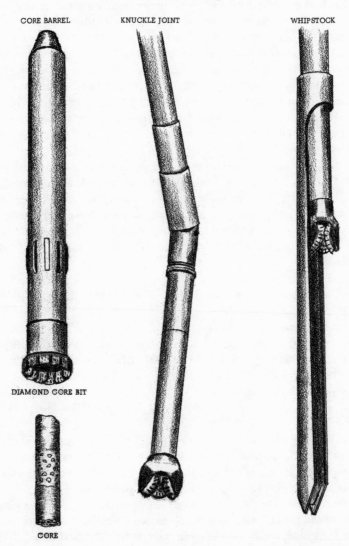

CORE BARREL   KNUCKLE JOINT   WHIPSTOCK

DIAMOND CORE BIT

CORE

When the operator wants to know the exact character of the rocks he is drilling, he uses a core barrel and takes a core which his geologist or micropaleontologist or micromineralogist or chemist can study. When he wants to divert his hole from the vertical he uses a whipstock or knuckle joint.

ing sand and permits the formation fluids to flow into the drill pipe. The production rate is estimated from the height to which the fluid rises in the drill pipe in a given time, the kind and duration of the *blow* during the test, the pressures in the sand before, during and after the test, and how much oil and water are recovered from the drill pipe when the test is concluded.

The pressure-recording instruments are down in the hole in the testing tool. The first blow is the air from inside the drill pipe being driven out by the oil and water entering from the bottom. If there is gas with the oil it, too, escapes during the test, drives the air ahead of it, and finally reaches the surface. This may increase the blow. The gas may be sampled and pressures recorded at the end of a vent pipe, called the "blooie line," during the test so that analyses can be made in the laboratory and the amount of gas determined. After sampling the blow from the blooie line the "testing engineer" may set fire to the gas so he can observe the length and color of the *flare*, and so that the heavy constituents of the raw gas do not settle and create a fire hazard around the rig. The Johnson Roe No. 1 could be both an oil well and a gas well if the gas is of good quality and there is enough of it. It might even turn out to be a flowing oil well. All of these possibilities make a lot of difference to Mr. Johnson as to how he will complete the well and what equipment he will have to buy.

How Deep Is a Well?

Before we leave this drilling business let's try to answer three questions often asked.

First, how deep is a well? The well that started the modern oil business was drilled near Titusville, Pennsylvania, in 1859, by Col. E. L. Drake. It was sixty-nine feet deep. The deepest well drilled to date is the Phillips Petroleum Company 1-EE University, in West Texas, drilled to 25,340 feet. The bottom was about 5,000 feet farther below the surface than the top of Mt. McKinley is above sea level. The well was dry. The world's deepest producing well at the moment is Richardson and Bass, et al, No. 1 Humble LL & E in southern Louisiana, producing from 21,443 feet. Some other well may hold the record tomorrow.

Not many years ago anything more than 2,000 feet was considered deep drilling, and only the boldest and best financed operator would consider going below 5,000. Now wells more than 2 miles deep are numerous and wells more than 3 miles deep get only casual notice in the trade journals.

It's anyone's guess how deep wells will eventually be drilled. The limiting factors are probably the tensile strength of the drill pipe and the strengths of the draw works, table, swivel, and other equipment. When the pipe is not quite on bottom, the weight of the entire string is on the joint at the top, and 25,000 feet of drill pipe weighs more than 160 tons, with the allowance for buoyancy. Comparable weights, of course, are on the draw works, table and swivel at different stages of the operation. A few years ago no string of drill pipe would have supported such a weight, nor would any draw works, table, or swivel have stood up under such stresses. Steady improvement in the tensile strength of special steels and in the ruggedness of surface equipment has made present-day deep drilling possible. How much further this can be carried remains to be seen, but oil men are speculating on the possibility of some day drilling deeper than 30,000 feet.

At any rate, the 15,000-foot well is still the exception. The great majority of wells are less than 5,000 feet deep, and the average of all wells drills in the United States in 1960 was 4,079 feet.

Drilling Speed

The second question is, how fast is a well drilled? It's almost like asking how long it takes to cross the ocean; the answer depends on which ocean and what boat. Like the ocean question, the well question can be answered after a fashion.

In general, and with exceptions, cable-tool drilling is not so fast as rotary drilling. In general also, the deeper a well is, the more slowly drilling will progress. On a cable-tool well it takes longer to run and pull both tools and bailer in a deep well than in a shallow one, just as it takes longer to run a mile than to run a block. This element is much greater in a rotary well, where, in addition to the longer distance to be traveled between the surface and the bottom, so many more joints of pipe have to be broken and stacked and then picked up, screwed on, and lowered.

The fastest cable-tool drilling is in shale soft enough to drill easily but hard enough to stand up. In such a formation a cable-tool well may make from forty to one hundred feet of hole a day; some have made considerably more. In hard limestone ten feet a day may be good progress. Now and then a hard streak is encountered in which daily progress is measured in inches and the crew spends most of its time changing and dressing bits.

Rotary wells are not immune to such troubles. Now and then, though not quite so often, one of them may measure progress in inches for a few days. But when the drilling is good a rotary will really make hole, and more than one has drilled better than a thousand feet in a day. Here, however, is a prize story that held the world's record for a time.

A company let a rotary contract for a well in northeastern Colorado, and agreed to pay part of the contract price at each thousand feet of depth. The company treasurer watched the well spud[4] and drove back to Denver, thinking he had a week or two to get together the first payment. Next day the contractors called him to say that the first two payments were due. They had drilled 2,360 feet in twenty-four hours! World's records do not last long in the oil business, however. More recently a well in the Tepetate field in Louisiana drilled 2,375 feet in eight hours. Changing trends toward air drilling may double this old speed record before long.

The Cost of a Well

That brings us to the third question, how much does a well cost? Again the answer depends on what well, how deep it is, and where it is located. The range is from one thousand to several million dollars.

Depth is an important factor; the same causes that make drilling speed decrease with depth make costs mount rapidly with depth. A deep well, moreover, uses more casing than a shallow one, and casing is an important part of the cost of a well. For these reasons the cost of a well is not directly proportional to the depth; the increase is more nearly proportional to the square of the depth.

Depth is not the only factor, however. The formations are soft in some fields, hard in others. In some fields caving slows down drilling and requires additional casing; other fields are free from such troubles. It follows that the cost per foot in one locality may be twice that in another.

Location is of course another factor, for it involves freight, haulage,

---

[4] Strictly speaking, a rotary well does not spud-in, that is it does not use anything corresponding to spudding tools. The term has been carried over from cable-tool practice, however, to designate the starting of actual drilling.

OFFSHORE DRILLING

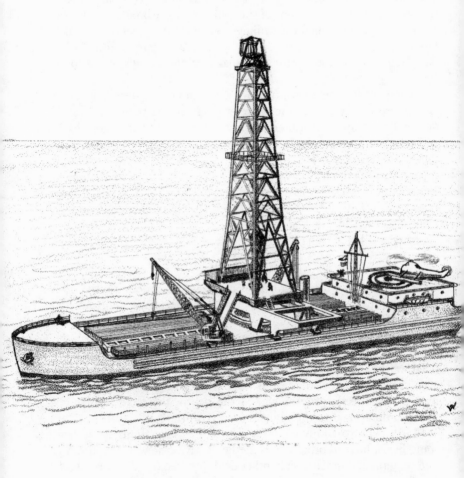

*An offshore drilling rig may be mounted on a relatively small platform—
or on a conventional-hull-design ship as shown here. A modern offshore rig
will operate efficiently in water depths of 600 feet in the worst winds and
seas. Mooring systems provide stability during drilling, and powerful thrust-
ers in the bow and stern rotate the ship to maintain the most favorable
heading at all times. The rig may be capable of both surface and underwater well completions.*

and delays. Equipment costs a lot less in Pennsylvania than in Montana, for example, and a well close to a good machine shop is less likely to be shut down for repairs than one fifty miles from a highway.

The last factor, but not always the least, appears in the estimates as *contingent and unforeseen*. A single fishing job may double the cost of the well, or an unexpected gas pocket may cause a disastrous and expensive fire.

You can readily see, therefore, why drilling cost varies through so wide a range. A four-thousand-foot well in one field may cost more than an eight-thousand-foot well somewhere else. The average drilling cost in the United States in 1937 was $22,000. In California, it was $45,000. By 1960 the average drilling cost per well had risen to just over $55,000 and about 90 per cent of these wells were drilled with the rotary method. Back in 1937 about half of the holes were drilled with cable tools. With the rapid developments in deep offshore drilling, in recent years, national averages of costs will mean less and less as drilling becomes more costly and more complicated. The oil man usually figures his costs on the basis of dollars and cents per foot of hole drilled. For instance, in 1937 that figure was $7.50 but in 1960 the cost was up to about $13.50 per foot.

New drilling equipment and techniques such as air drilling, revolutionary types of bits, and sweeping changes in the design of all the tools give promise of stemming the increases in costs. These drilling costs, by the way, do not include the costs of equipping wells for production, usually several tens of thousands of additional dollars per well.

Now We Have a Field

An undrilled structure is wildcat territory. When the first well comes in it's a *proven* structure and the well is the *discovery well*. When two wells are in, it's a *field*. Subsequent wells will be drilled farther and farther from the discovery well; eventually one of them will be drilled so far down a flank of the structure that it will get water instead of oil; the field is then *delimited* in that direction. When dry holes have delimited the field in all directions, it is *fully defined*. Mr. Johnson will still have to drill a *confirmation well* before "Roe Field" is recognized as a field.

It's a far cry from the completion of Johnson Roe No. 1 to the full definition of the field, and a lot of oil will be produced meanwhile as well as thereafter. Let's take the opportunity to watch awhile.

CHAPTER 8

Producing the Oil—The Principles of It

The Production Crew's Job

THE job of the production crew, from the field superintendent down to the last man in the bull gang, is to produce the most oil at the least cost. This does not usually mean the most oil in the least time. The old-time practice was to drain a field as rapidly as possible, but the industry has learned that more oil can be recovered usually by less hasty production methods. Even with the best production practice an operator does well to recover 40 per cent of the oil in the productive sands; more often, in all probability, he recovers 15 to 30 per cent.[1] The rest stays in the sand, adhering to the sand grains, stripped of most of its gas content and thus robbed of the force necessary to move it to the wells. Part of such oil may be produced by repressuring, water flooding, or some other *secondary recovery* method, of which more anon, but such methods are costly. When applied late in the life of the field, their cost per barrel produced may be unprofitably high. The wise procedure is to get as much of the oil as possible by applying such methods early or by avoiding the necessity of applying them at all. To understand why rapid production may defeat this end, let's consider what makes a well produce.

What Makes a Well Produce

You remember that the oil is contained in the minute voids or pore spaces of a reservoir bed, usually a sandstone or porous limestone. These voids are about like the spaces you would see between a double

---

[1] Oil men use "recoverable" to mean "producible," and the "ultimate recovery" of a field is the amount it has produced or will ultimately produce.

rack of billiard balls if they were piled in a pyramid on the pool table. If the grains of the reservoir bed are irregular in size and shape, as they usually are, some of the voids are larger and some are smaller, but to travel from one point in the reservoir bed to another the oil must squeeze through spaces too small to be seen without a microscope. It must do this in spite of its viscosity and of the forces of adhesion and adsorption that tend to make it cling to the grains. A considerable amount of force of some sort is obviously necessary to cause such a movement.

Now a well produces only the amout of oil that moves out of the reservoir bed into the well. What makes the oil do it? Gravity may play a part, but in most fields the chief propulsive force is either water pressure or gas pressure or both.

Water Pressure

Practically every oil and gas pool is surrounded by water under pressure. If you don't believe it, drill a well just outside one and see what you get. From the same sand that may be producing oil a few hundred feet away you will get nothing but water, and the water will be under about the same pressure as the nearby oil. The water pressure in the reservoir bed may be due to one or all of several causes: It may represent the *hydrostatic head* of water entering the reservoir bed at a higher elevation, perhaps hundreds of miles away; it may be caused by the weight of water in the overlying rocks; it may be due to past or present pressures built up when the beds were being folded. Geologic opinion is not unanimous as to the causes of water pressure in the reservoir bed, but every oil man knows that the pressure is there.

You remember, of course, that one of the definitive characteristics of fluids is their tendency to move from a point of high pressure to one of lower pressure. You can see, then, what happens when a well is drilled into a sand in which oil under pressure is surrounded by water under approximately the same pressure. The removal of oil from the well lowers the pressure in the reservoir near the bottom of the well; more oil, pushed by the water, moves to the well and is removed; and the process continues until water reaches the well and *drowns it out*. When this happens oil men say the well has "gone to water."

A well produced too fast creates a local area of low pressure in the reservoir. Water may then rush in faster than the oil in a cone protruding from its level beneath. This effect is known as *water coning*. Once the oil-bearing pores between sand grains have become filled with water, oil will probably never again re-enter them. After a well has coned water or gone to water there is little hope that it can be made productive again from the same zone. Our production engineer therefore has to determine just how fast each well should produce so that he can take advantage of the optimum rate of movement of the oil without water slipping in on him from below.

In some fields water is the sole expulsive force moving the oil to the wells. If water moved inward from the edges at a uniform rate through all parts of the sand, driving all the oil ahead of it, practically all of the oil in such fields might be recovered, but such an ideal result is never attained. Sand bodies vary in permeability from point to point; water moves faster through permeable parts than through less permeable ones; a multiplicity of wells sets up complex pressure relationships. The result is that some of the oil is by-passed by the water and left in water-surrounded lenses, and some is left in pores so small that the water fails to drive it out. Nevertheless, the percentage of recovery from a field in which water is the chief propulsive force is probably higher than for any other type field.

Gas Pressure

In most fields, however, the active expulsive force is gas. As already pointed out, the oil in practically all fields contains some gas in solution, in the same way that gas in a bottle of soda water is in solution. In many fields the amount of gas is so great that the oil does not dissolve all of it, so that some of it occupies the top of the structure as free gas, like the free gas in the soda bottle above the liquid. The gas, both free and dissolved, is under pressure, as is the oil, the pressure being due, of course, to the same causes that create the pressure in the water in the reservoir bed surrounding the oil *pool*. The amount of gas that the oil holds in solution is dependent, as we have seen, upon the amount of the pressure. If the pressure is lowered some of the gas comes out of the solution and becomes free gas, leaving the oil heavier, more viscous, and less mobile than when more gas was dissolved in it. If the pressure is lowered enough, practically all of the gas comes out of the oil, leaving the oil *dead*, just as the soda becomes dead when the cap is left off.

When a well is drilled into the oil-bearing part of a pool, oil and gas are released and removed. This lowers the pressure in the reservoir at the bottom of the well, and oil and gas move from the surrounding area to the point of lower pressure. So long as the pressure at the bottom of the well is maintained above that at which gas comes out of solution, the oil retains fluidity and mobility, and an efficient movement of oil and gas to the well takes place.

If the well is allowed to produce too rapidly, on the other hand, the pressure at the bottom of the well drops below the point at which gas begins to escape from solution. In that case some of the gas will escape from the oil and make its way to the well, leaving behind part of the oil in which it had been dissolved. The viscosity of the oil and its tendency to stick to the sand grains are thereby increased, and the oil becomes less mobile. It is increasingly hard to move to the well and the gas pressure to move it is decreasing. More force is needed and less force is available.

If the water surrounding the pool moved in uniformly and fast enough to maintain the pressure no damage would be done; but water moves through microscopic spaces much more slowly than gas, and in many fields moves so slowly that its effect in maintaining the pressure is negligible. With the gas pressure depleted, a large percentage of the oil is left in the reservoir bed with no propulsive force to bring it to the well. In time, the amount of oil reaching the well becomes too small to pay for the cost of operation, and the well is abandoned.

The Decline Curve

From what has been said one thing is clear: The production of all wells in which gas is the chief expulsive force declines rapidly when they are produced to capacity. This decline is especially noticeable in the early stages, from the *initial production* through what is known as the *flush* period.[2] If the well came in at two thousand barrels a day, in six months it may be down to a thousand barrels, and in a year to six hundred. The decline is less noticeable after the *flush production* is gone and the well is on *settled production*, but it continues just the same.

If the well is allowed to produce only a part of its potential production the decline may not be noticeable for a long time; the decline in

---

[2] The initial production is commonly designated by the initials I.P. A well that has been shot or acidized has two I.P.'s, the first called "I.P. natural" and the second "I.P. after shooting" or "I.P. after acidizing."

pressure will be slower, for one thing, and for another, a well allowed to produce only twenty barrels a day will probably behave much the same whether its full productive capacity is two thousand barrels or only two hundred. Sooner or later, however, the well will fail to make the twenty barrels or one hundred or whatever amount it has theretofore been producing, and from that time on its decline will be apparent. Unless it goes to water, the well may produce for twenty or fifty or seventy years, but each year it will produce less than the year before.

Fields and individual wells vary greatly, but in general this year's production from a settled well produced to capacity will be from 10 to 30 per cent less than last year's.

Production engineers have a habit of plotting a curve for each well and each property and each field, showing the daily production in each month. This is known as a *decline curve*, a name that bears testimony to the inevitability of ultimate falling off in production. Valuation engineers, whose business it is to evaluate oil properties, project these decline curves into the future and thus estimate how much more oil a well or a property will produce before its abandonment.

Delayed Decline

The beginning of the decline may be deferred by producing the well below its capacity, either with or without repressuring with gas or air or flooding with water. As we shall see later, a great many of the wells in the United States are now operated under *proration*, producing only a fraction of their potential capacities. So long as the amount produced from a well is materially less than the amount the well can produce, the decline in its potential production will be proportionately small. The effect of restricting the production is to retard the lowering of the pressure; the lowering goes on, but at a retarded rate. Unless the field is repressured the time will come when the productive capacity of the well is down to the amount being taken. From that time on the well's production will decrease. The incidence of the decline curve is deferred, but it cannot be avoided.

No one can understand the oil business who has not grasped the universality of the decline curve, the fact that once the full potential production of the well begins to be taken, next year's production will be materially less than this year's. It affects the whole economics of the industry, as we shall see. For the present we must get back to our gas.

Flowing Wells and Pumpers

Bringing the oil to the well is only a part of the gas' service; it may also bring the oil out of the well. *Flowing wells,* which oil men no longer refer to as *gushers,* are wells with enough pressure to blow or lift the oil to the surface. As long as there is enough pressure to do this job the well flows, but when the pressure at the bottom of the well becomes less than the weight of the well-full of oil, diluted and *aerated* with gas, then the oil must be lifted out of the well by some other means. From that time on the well is *on the pump* and is a *pumper.* We show this in our notes as *P.O.P.*

The existence of pumping wells seems to surprise many people; they appear to think that all oil wells are gushers, and that they all produce huge quantities of oil. The facts are that many wells never flow but must be pumped from the start, and that nearly all the rest become pumpers a few months after being drilled. In 1960, some 89 per cent of all wells were producing with a pump or other artificial lift.

How Much Oil Does a Well Produce?

As to amount of production, wells range all the way from a few pints per day in some Appalachian fields, to one capable of producing 184,920 barrels a day in the Yates field in Texas. Spectacular fields like Kettleman Hills, Oklahoma City, and East Texas contribute their quotas of wells that could produce 10,000 to 90,000 barrels a day, if allowed to do so, but other fields in the same states have plenty of one-barrel and two-barrel wells, many of them pumped only a few hours each week. The average production per well per day for the fields in Pennsylvania is less than one quarter of a barrel a day. That's only ten gallons per day per well. The average daily production, in 1960, of all wells in the United States, big and little, flowing and pumping, was *11.5 barrels.* Let's emphasize that, for it is important to an understanding of oil economics. The average oil well produces less than twelve barrels a day.

A Bit of Review

Before we go on, let's see if we have learned our lesson about wells that flow because of gas under pressure, which means the great majority of wells that flow at all. What makes such wells flow? Gas under pressure. What brings the oil to the well, whether flowing or pumping? Gas under pressure. What should the operator do about his gas pressure? Conserve it. How can he conserve it? By trying to make each cubic foot of gas he takes from the well bring with it the maximum amount of oil, or, to put it conversely, by producing the minimum amount of gas per barrel of oil.

The Gas-Oil Ratio

The number of cubic feet of gas produced with each barrel of oil is known as the *gas-oil ratio*, and the constant effort of the production staff is to keep the gas-oil ratio as low as possible. A high gas-oil ratio means that full use is not being made of the expulsive force of the gas, that the gas pressure will be prematurely depleted, and that the ultimate recovery of oil will be less than it should have been.

The proper gas-oil ratio is different in every field, and may be different for every well in the same field. It can only be determined by study of and experimentation with each well. The outstanding fact is that in at least nine cases out of ten, during the flush production stage, the proper gas-oil ratio will not be attained by producing the well to capacity. A flush well flowing *wide open* nearly always has a high gas-oil ratio, which means that it is depleting the expulsive force on which its maximum ultimate production depends.

The ideal gas-oil ratio is the proportion of gas that is dissolved in the oil at the reservoir pressure. If a sample of oil is taken at the bottom of the well at the reservoir pressure, and if when brought to the surface and reduced to atmospheric pressure the amount of gas that escapes from it is equivalent to six hundred cubic feet for each barrel of the oil, then 600 is the ideal gas-oil ratio. Under ideal operating conditions the well will be allowed to produce only six hundred cubic feet of gas for each barrel of oil.

The Law of Capture

The ideal gas-oil ratio is seldom attainable in actual operation; too many factors, many of them hard to determine and some of them hard to control, must be taken into account.

One of the factors that must be reckoned with is the rate of gas and oil withdrawal from adjoining lands. If your neighbor produces his wells faster than you produce yours, the pressure drop or *gradient* from your land to his wells may become greater than that from your land to your own wells. If so, much of your gas and eventually some of your oil will be produced through his wells. In such cases it is no longer yours; it is his. Under the *law of capture*, which is the prevailing law in the United States and Canada, oil and gas do not belong to the man under whose land they originally were; they belong to the man who reduces them to possession through wells drilled on his own land.

If you want to get the oil that underlies your land, therefore, you must reduce your gas pressure as rapidly as your neighbor does. If you don't he will do it for you, or rather do it to you. You will suffer all the consequences of a depleted gas pressure, will lose a part of your oil to him, and will have much of the rest left in the ground without adequate gas to keep it fluid or to bring it to your wells. Thus, in the absence of regulation agreements, the rate of withdrawal from a field is all too often determined by the field's greediest or most reckless operator. One producer, more interested in immediate returns than in conservation of gas or in ultimate recovery of oil, can force the entire field into wasteful operation, and everybody loses out in the end.

Evils of the Law

The disadvantages of such a situation go beyond the fact that the maximum ultimate recovery of oil from a field is dependent on equal wisdom on the part of every operator; the thing has other serious aspects.

A new field is usually capable of producing a lot of oil in a hurry. During its comparatively brief flush period it may yield more oil, if produced to capacity, than it will produce during the many succeeding years of settled production. If every operator drills as fast as he can, and produces his oil as fast as he can, a large quantity of oil is thrown on the market. Soon more oil is produced than can be sold. The producer who cannot sell his oil must do one of two things: If he has the money to do so, he can build or buy tanks and store his production; if he cannot afford a storage program, he will have to shut-in his wells. Storage not only costs money to build; it costs money to maintain and carry. What is worse, some of the lighter and more volatile parts of the oil will be lost by evaporation. Storage is wasteful of both oil and capital. Shutting-in wells while neighboring wells produce is even worse; the gas pressure of the tract is depleted through the neighboring wells, some of the oil is drained away through them, to the loss of the shut-in owner, and worst of all, much of the oil is left in the ground, irrecoverable because of loss of gas pressure. The shut-in producer loses money and the public loses oil that might have been recovered for its use.

Another usual result of such overproduction is to break the market price of crude. With the price down, thousands of older and smaller wells with higher production costs per barrel are either operated at a loss or are abandoned, leaving unrecovered oil in the ground. Thus to the waste of capital, and to the loss of money by the weaker operator, are added the waste of valuable natural resources in the form of gas dissipated, oil left in the ground, and gasoline lost by evaporation, and to these are added the economic losses that result from the demoralization of a great industry.

The Doctrine of Correlative Rights

Out of a recognition of these evils there has emerged a new doctrine: that every owner of oil and gas in a field has an inherent right to the use of his share of the *reservoir energy*, and that he should be protected in that right against wasteful practices by his neighbors. This belief, which is usually referred to as the *doctrine of correlative rights*, has found wide acceptance among production engineers and operators and has gained increasing support from the courts. The day has come when a suit can be sustained against an operator who wantonly or wastefully depletes his neighbor's reservoir energy.

Proration

Meanwhile, without waiting for the courts to engraft the doctrine of correlative rights upon the law of capture, another method for curbing the evils of unrestricted production has come into general practice. It is based on the principle of *ratable taking* and is known as *proration*. It means that each operator in a field produces only a stipulated proportion of the oil his wells are capable of producing. In some states the production allowed a well is a straight percentage of the well's capacity. In other states the acreage drained by the well is taken into account. The trend is toward considering acreage, reservoir pressure, and other factors in addition to productive capacity.

The first proration was by voluntary agreement among the operators in a field, but the trouble with voluntary proration was that one reckless operator could break it wide open. If he overproduced his property, then his neighbor had to do likewise, and then their neighbors, until the contagion covered the whole field. It was not long, therefore, before state legislators began giving to state regulatory bodies the power to determine the ratable take in each field and to penalize the operator who exceeded his quota. At the present time most of the important producing states have such conservation laws.

A number of producing states, moreover, have entered into a treaty known as the Interstate Oil Compact, authorized by Congress, under which appointed representatives meet periodically to consider their common conservation problems. They may even discuss the amount that each state should produce, though they have no power to allocate production to the various states or to control the action of any one of them.

Enforcement of the conservation laws is not always easy, and in their earlier day ingenious operators found many ways to get away with more than their share. Aside from such mechanical devices as concealed pipelines, they developed a nice technique in shipping excess production, known as "hot oil," across state lines, where the conservation authorities could not get at it. Congress then passed the "Connolly Act" forbidding interstate transportation of hot oil or its products. As the operators have become more used to the idea of proration and have come to recognize its benefits the tendency to violate the conservation laws has decreased and they are now generally observed and enforced.

Some Effects of Proration

When we consider the factors influencing the price of crude oil, and when we later review the oil history of the United States, we will see some of the broader economic effects of proration. At the moment we will consider only the immediate effects on the conservation of oil and the cost of recovering it.

It is obvious that by lessening the waste of gas or reservoir energy, proration increases the total amount of oil that will be recovered. The amount of this increased recovery will not be known for many years for any given field, but experience with a large number of prorated fields now near exhaustion has proven that their recoveries have been materially increased. It can now be said that the increase in recovery— the saving of oil that unrestricted production would have left in the ground—is very large.

Some of the more immediate results of controlled production are apparent and can be estimated with fair accuracy. For one thing, in the states which restrict the drilling of wells as well as the rate of production, the saving in drilling costs is at once apparent. If, for example, only one well is allowed on each forty acres, instead of one well to ten acres as formerly, and if in that field the wells cost an average of $50,000 each, there is a saving of $150,000 on each forty-acre tract. Such restricted drilling also has its effect on the conservation of gas, for each well drilled and produced uses some of the gas in the reservoir. In some prorated Texas fields where the drilling cost is estimated at about four cents a barrel of ultimate recovery, the drilling cost would have been about fifteen cents a barrel without proration. A saving of eleven cents a barrel, 73 per cent of the cost of drilling, is no small matter.

Equally important is the saving in production costs. As we shall see, finding the oil is only one item in the cost of getting the oil to the surface, and one of the major items of cost is pumping. Flowing oil is cheap oil; oil that has to be pumped is expensive. Proration, with its conservation of gas pressures, increases the percentage of the oil that can be produced by flowing, before resorting to pumping. Certain Texas fields that promise to flow 75 per cent of their total production under proration would probably have flowed only 40 per cent under unrestricted production. The great saving in pumping costs is obvious.

Potential Production

The conservation laws have augmented the oil man's vocabulary—sometimes in more ways than one! The amount that a well is capable of producing, in barrels per day, is referred to as its *potential*. The amount that it is considered to be capable of producing, as a result of some arbitrary test, is its *rated potential*. If the potential has been fixed by a state regulatory body it is an *official potential*. The amount that the well is permitted to produce is called its *allowable*.

The one thing to remember about potentials is that they are almost always too high. Few wells, if produced to capacity, would make their potentials for more than a short time; many of them would not hold up for more than a few hours, some would not make their potentials at all.

The reason for this is simple. In a proper and reliable production test, a well should be produced at capacity for several days. To do this simultaneously for all wells in a flush field would result in the very evils that the conservation laws are designed to avoid. The wells are therefore tested for short periods, and the results, perhaps reduced by an arbitrary factor, are taken as the potentials. Now if a well has been prorated at 5 per cent of its capacity for a month or more and is opened up for an hour it may put out a disproportionate amount of oil, an amount far in excess of what it would produce in the twenty-fourth hour, for example. It is not likely that because a well produces a thousand barrels of oil in an hour it will make twenty-four thousand barrels in a day. Would you like to bet that a sprinter who does a hundred yards in ten seconds can do a mile in three minutes? Most states therefore cut the hourly rate in two or reduce it by some other factor; several states require the potential tests to be made by producing the wells through a small aperture called a "choke," seldom more than one-fourth inch and often only one-eighth inch in diameter. These precautions have brought the rated potentials down more nearly to the actual potentials, but most of the rated potentials are still too high. The potentials of all the fields in the United States, added together, would give a high figure of production per day; the amount the fields could actually produce at the end of thirty days if thrown wide open would be very much less.

Too high or not, in the main the rated potentials serve their purpose, which is to provide an equitable basis for determining allowables. It does not matter much whether a well is rated at two thousand barrels and allowed to produce 5 per cent or rated at one thousand barrels and allowed to produce 10 per cent; the allowable is one hundred barrels in either case.

Proration Weaknesses

The state conservation laws, good as most of them are, do not wholly meet the needs of the situation. For one thing, they do not always prevent the drilling of too many wells; in fact, the proration laws of some states encourage excessive drilling. If a field is prorated by allowing each well to produce a certain percentage of its rated potential, the more wells a man drills the more oil he can produce. What one man does his neighbors must do. In consequence the *drilling pattern* of a field may be a well to each five acres, when a well to each twenty acres would give more efficient recovery. Such a pattern means that four times as many wells are drilled as should be drilled—and drilling and equipping wells costs money. Every well, moreover, does its share of reducing the gas pressure and thus reducing the percentage of oil recovered.

Such states as Oklahoma, New Mexico, Louisiana, and Arkansas, which take into account the area from which a well drains in fixing its allowable, are thus curbing the tendency toward excessive drilling. Some of them, moreover, have laws under which they can force adjoining small tracts to be drilled and operated as units, which further prevents excessive drilling.

As bad as excessive drilling, perhaps, is the fact that an arbitrary proration by percentage of rated potentials takes no account of the different conditions and needs of different properties in the field. Reservoir beds vary in permeability and thickness from point to point in a field; one well yields its gas rapidly, another more slowly; the correct gas-oil ratio for one well may be all wrong for neighboring wells. Attempts are being made by some state bodies to adopt proration rules that will make allowance for these differences, but the trouble cannot be cured by broad rules; each well needs individual diagnosis and treatment. Unfortunately, individual treatment cannot be given where a field is cut up into a multitude of leases, operated by competing lessees under rigid proration rules of uniform application. Proration is only a step toward ideal operation, and it by no means reaches the goal.

Unitization

The rest of the distance is easily covered, however, if the field is operated as a unit. Proration then merely fixes an allowable for the field as a whole, leaving to the unit operator the distribution of the allowable among the various wells. The unit operator can study the field and the market, drill only the wells required to get the desired recovery, determine and use the best gas-oil ratio for each well, and in general develop and produce the field as he believes it should be done. Excessive drilling, violation of proration rules, sharp competitive practice, waste of reservoir energy—all these become unnecessary; the field can be handled in accordance with the best known practice.

Is it ever really done? Oh, yes. In foreign fields particularly, unit operation is common and highly efficient. Colombia, Venezuela, Russia, Iran, Iraq, Kuwait, Saudi Arabi, Sumatra—call the roll of the great producing regions outside the United States and most of them will respond with *unit operation*. In the United States, companies have climbed on the unitization band wagon under the encouragement of the federal and state governments, and most discoverers of new fields move quickly to set up unitization programs.

Unitization agreements are drawn up by committees of representatives of all the lease owners, and detailed plans of operation are worked out which are acceptable to all parties. Then an operating committee is set up, or a unit operator is selected from the group, and thenceforth the field is handled entirely by the unit operator or the committee. Under the rules of these agreements, the operator makes all of the plans for field development, services the wells, keeps all of the records, pays the bills, and divides the income according to each owner's share of the production. Where the field includes federal land, the Government encourages these co-operative efforts by providing technical assistance and by releasing acreage contributed to the unit by each operator from his state quota.[3]

The great Salt Creek field in Wyoming was the leader and granddaddy of all unitization programs in the United States, the happy solution to 20 years of conflicting interests in oil claims, state leases, federal leases, withdrawn lands and private ownership. Since the Government was the royalty owner of most of the Salt Creek field, the U.S. Bureau of Mines really pioneered this unique plan through hectic battles until a form of unitization was finally achieved in the early 1920's. The Salt Creek field now has produced nearly 400 million barrels, mostly under the unit agreement. Reservoir pressures have been maintained by water and gas injection through careful, wise management over 40 years. It's no telling when the field might have *dried up* or how much unrecoverable oil would be left in the ground today if the early battles for production had continued unchecked. Salt Creek, once almost a bloody battleground, has long been a serene little oil town.

An early example of unitization without government intervention was the great Van field in Texas, where more than 111,000,000 barrels of oil had been produced in less than eight years, with a drop of only 165 pounds in the reservoir pressure. The gas-oil ratio was only 300.

---

[3] The Government limits the number of federal acres held by a company in each state.

Never heard of Van? That's because Van has produced oil instead of turbulence and lawsuits. Unit operation, like peace, makes few headlines.

The Essential Principles

To sum up the production problem, the job of producing the most oil at the least cost involves making the most efficient use of the gas and water pressure, and it also involves efficient mechanical operation. We have talked at length about conservation of pressure and have discussed some basic theories; now would you like to watch the mechanics of the job? If so, how about going back to the Johnson Roe No. 1? It has just finished drilling-in.

CHAPTER 9

Producing the Oil—
The Mechanics of It

Pressures

JOHNSON Roe No. 1 has turned out to be pretty gassy. Flowing wide open it will do 2,000 barrels of oil and 30,000,000 cubic feet of gas a day. The pressure at the wellhead is 2,200 pounds, which means that the reservoir pressure is perhaps 400 pounds greater. The pressure at the casing head or wellhead, when all valves are closed and no oil or gas is allowed to escape, is the *shut-in* or *closed-in* or *casing-head* or *well-head* pressure. It used to be called the "rock pressure," and the increase in pressure when the well is closed-in after being allowed to flow is still referred to as *rocking-up*. The true rock pressure, however, generally known as the *reservoir pressure*, or *formation pressure*, is the pressure at the face of the productive formation when the well is shut-in. It is equal to the wellhead pressure plus the weight, in pounds per square inch, of the column of oil and gas in the well.

The Well Connections

In a well with such a pressure and such a productive capacity, the Christmas tree will project as high as your head above the derrick floor. It will have one or more pipes leading out from it, and in each pipe will be a heavy valve or a choke or both. In addition to being tightly screwed or flanged onto the *production string*—the innermost and smallest string of casing—it may be anchored to the derrick foundations or the concrete lining of the cellar or both. The pipes, called "flow lines," lead from the Christmas tree to an *oil and gas separator*, sometimes (less frequently than in the past) called a "gas trap."

The rate of flow is controlled by means of the valves and the chokes. The valves are heavy-duty affairs of conventional types. A choke or

A CHRISTMAS TREE

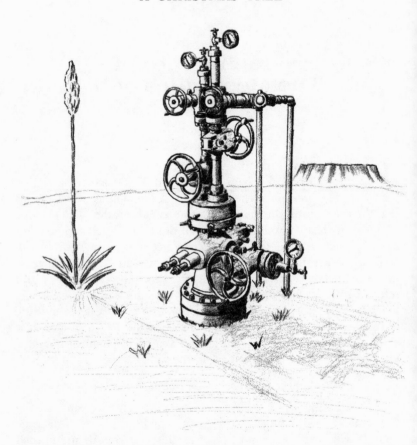

The flow of the well is controlled by a control head or "Christmas tree" screwed securely to the top of the production string of casing and, if the pressure is heavy, anchored to the concrete of the derrick foundation or of the cellar lining. The valves in the flow lines lead from the Christmas tree to the tanks. The two flow lines permit separate production of gas and oil from two different zones.

bean or *flow bean* is a heavy steel nipple with a small hole through it. It is inserted into the flow line, and the rate of flow—or the back pressure maintained in the well—is regulated by the size of the hole. In speaking of the production of a well, therefore, it is customary to mention the size of the choke orifice through which it is producing. We might say, for example, that Johnson Roe No. 1 is "doing 793 barrels through a one-inch choke."

Oil and Gas Separators

If the oil and gas were flowed directly from the well into a tank the gas would escape from the vent in the top of the tank—taking with it some of the lighter and more valuable parts of the oil. It might cause the oil to boil so violently as to spray some of it out of the tank.

The flow lines therefore take the oil and gas to an oil and gas separator, which is a vertical cylinder in which the gas is allowed to expand, and its velocity is checked. The reduced velocity and lowered pressure allow most of the gas to free itself from the oil. Instead of escaping into the air the gas is led away from the trap through a pipeline called the "gas release line" or, simply, "gas line."

Recovering the Casing-Head Gasoline

The gas from the separator is *wet*, that is, it carries considerable amounts of the light gasoline constituents, amounts varying from a fraction of a gallon to two or more gallons of gasoline per thousand feet of gas. This is a valuable product and should not be wasted. If the gas is compressed sufficiently and then cooled, the gasoline fractions will liquefy and can be recovered. If the gas is passed through an oil of the right character the oil will absorb the gasoline fractions, which can then be recovered by distillation, leaving the oil to be used again to recover more gasoline. Such gasoline recovered from the gas from an oil well is known as *casing-head gasoline*. It is one of the forms of *natural gasoline*, which we shall consider further in a moment.

The gas from the separator is taken to a *gasoline plant*, either to a *compression plant* where the gas is compressed and cooled until it drops its liquid fractions, or to an *absorption plant* where the liquid fractions are absorbed in oil, or to a *combination plant* where the liquid fractions are recovered by both compression and absorption.

Part of the *dry gas* from which the liquid fractions have been recovered may be used as fuel for field operation, including drilling, pumping, gasoline plant fuel, and heating. Or it may be pumped back into the wells in repressuring or gas-drive or gas-lift operations. The remainder, which too often is much the larger part, is sold if there is a market for it; otherwise it is wasted into the air. The burning flares that make some oil fields so spectacular at night represent dry gas for which there is no immediate local market; they are a sad sight when you know what they mean. The gas of one such flare would heat a good-sized town in a cold climate.

Here's another sad story. In some fields, not so many as formerly and usually the smaller and more isolated ones, no attempt is made to recover the casing-head gasoline; it is burned in the flares with the excess gas. The amount, presumably, would not pay for the installation of a gasoline plant, but it would run your car a lot of miles.

You would not do well, however, to use straight casing-head gasoline in your car. It is light, volatile, and has less fuel than the heavier gasoline fractions. You would find it the quickest starting gasoline you ever used, but your motor would lack power, your mileage per gallon would be low, and you would have plenty of trouble with "vapor lock." For these reasons responsible oil companies do not sell casing-head gasoline to the consumer unless they have treated and *stabilized* it for motor use. As a rule they use the casing-head gasoline to blend with heavier fractions obtained from the refining of oil and thus get a properly balanced gasoline that will give both quick starting and good mileage.

Wet Gas Wells

A gas well is a well that produces gas and no oil. Some wells produce dry gas from which no commercial amount of gasoline can be condensed, and some produce wet gas, rich in gasoline fractions. Wet gas from such wells is treated in the same way as gas from an oil well; it is taken to a gasoline plant and stripped of its gasoline content.

The gasoline recovered from the gas from a gas well is not called casing-head gasoline; that term is confined to gasoline recovered from the gas from an oil well. The general term that covers all gasoline recovered from gas, whether the gas comes from a gas well or from an oil well, is *natural gasoline*. All natural gasoline has the general characteristics of casing-head gasoline and is, or should be, used only for blending purposes.

Gas from some wells is so wet that liquids will condense from it on any sudden drop in temperature. At such a well, one or more expansion chambers—simply a joint or two of large pipe, or a series of sharp bends—may be put into the gas line, usually fairly close to the well, with valves for drawing off the gasoline that condenses. These devices are known as *drips* and the gasoline that condenses in them is called *drip gasoline*. It, too, comes under the general heading of natural gasoline.

Now let's get back to our oil.

The Field Tanks

On wells that produce only small amounts of gas, no separator may be used. In that case the oil from the well goes to *flow tanks* not far from the well. If a separator is used, the oil may go to the flow tanks from the separator, or may go to the lease tanks described below. For a small well, two flow tanks of a hundred barrels each suffice; for such a well as the Johnson Roe No. 1, the flow tanks would hold two hundred and fifty, or five hundred, or possibly a thousand barrels each, and there might be from two to a dozen tanks. When one tank is full the stream is diverted into another. The rate of production of the well is determined by measuring the oil that flows into the tanks in a given time.

When other wells have been drilled on the Roe lease there will probably be, in addition to the flow tanks at each well, a *battery* of lease tanks, into which the flow tanks are periodically emptied. The oil is taken from the lease tanks by the gathering lines of the pipeline system, of which we shall hear more later.

The flow tanks and lease tanks are, together, called *field* tankage. They, like the larger tanks used by the pipeline systems, are circular and are made of steel sheets either bolted, riveted, or welded together, and usually have gently sloping roofs. Walkways run either across the tops or on the sides near the tops, connecting the whole battery. In the top of each is sometimes a bolted manhole and a *thief hole* or *gauge hole* with a hinged lid, through which samples of the oil can be taken and the tank gauged.

BS & W

Most oil as it comes from the well contains a small amount of moisture. Some wells produce a considerable percentage of water with the oil; some even produce more water than oil. Such water produced with the oil may, particularly if the pressure is high, form an intimate physical mixture that is neither oil nor water but an *emulsion* of the two. In such a case, the water is said to "cut" the oil. The oil from some wells carries a small amount of fine sand or clay.

The water, the emulsion, and the mineral matter tend to settle to the bottoms of the field tanks, from which they are drained off from time to time through taps near the bottoms of the tanks. The emulsion and any mineral matter are known as BS, politely interpreted as *bottom settlings* or *basic sludge* or *basic sediment*. The basic sludge and the water appear on the scout tickets and the "run tickets" and in the trade journals as BS & W.

Not all of the BS & W collects in the bottoms of the field tanks; some of it goes to the pipeline, and the pipeline or purchasing company usually deducts from one to two per cent for BS & W before paying for the oil it has taken.

Commonly the field operator burns the pits where the BS & W is dumped, near the field tanks, creating a tremendous black cloud of smoke. This attracts attention, and wild reports of a disastrous fire in an oil field sometimes spread before someone casually comments on the cause of the spectacle.

Treating Cut Oil

Water produced with oil may settle out readily, leaving clean oil, providing the water has not *cut* or emulsified part of the oil. If a lease is producing a considerable amount of cut oil some method of treatment must be resorted to.

Some emulsions will break down if heated, and the water will then settle out, leaving clean oil on top. Some emulsions are treated in centrifuges which throw the oil out of the water in the same way that a cream separator throws cream out of milk. More stable emulsions are treated with chemicals which break down the emulsions leaving the water free to settle out. Some emulsions are broken down by shocking them with electric discharges.

Out of these various methods, the operator plagued by *cut oil* tries to pick the one that will give him the greatest recovery of clean oil at the lowest cost.

Tubing the Well

We must not spend too much time inspecting tankage, gas wells, gasoline plants, and treating plants, for Johnson Roe No. 1 must be *tubed.*

Most oil wells produce through a small pipe called tubing rather than through the production casing only.

The well could be produced through the production string of casing with which, you remember, it was finished, but in lifting the oil through so large a pipe, some of the gas would by-pass the oil and more gas would be used than necessary to flow the well. Accordingly a smaller string of steel pipe known as *tubing* is run in. The annular space between the tubing and the casing may be sealed at or near the bottom of the tubing with an ingenious steel and rubber device, a *packer*, but the annular space is usually left open except at the casing-head.

Tubing may be two, two and one-half, three, or four inches in diameter, depending on the size of the well.[1] The smaller sizes are most common. The tubing is screwed together and lowered into the hole, joint by joint, in the same manner as casing.

All Wells to the Pumps

Properly tubed and connected, our well has become a tame affair to watch. Day by day nothing happens around the rig except the occasional visit of a production man to check up, or to change the setting of a choke. Gaugers come and go about the tanks, but they do not concern us yet. Month by month the well puts into the tanks as much oil as its productive capacity and the proration rules permit. We'll assume that it is properly operated and its gas-oil ratio is kept as low as possible. You might think it would go on producing at a constant rate forever, but it won't.

Some day, next month or next year or ten years from now, it will fail to make its present allowable. Thereafter, unless the field is re-pressured by putting gas back into it, the well's productive capacity each month will be a little less than the month before. By that time, of course, its allowable will probably have been heavily reduced, so that when it is capable of producing two hundred barrels a day it may only be allowed to produce twenty. Eventually the field's potential production will become so small that it constitutes no danger to the market and there is no danger of gas waste. Then proration will be lifted or placed at a hundred per cent, which comes to the same thing.

Sometime during this downward course a radical change has come to the well. The pressure has become too small to lift the oil out of the well. The flowing stage is over. The well may still be good for a thousand barrels or more a day, or it may be down to fifty. Whatever its capacity, it is no longer a flowing well.

---

[1] The *size* of a well is the number of barrels per day it will produce, not its diameter or depth or any other dimension. The size of the hole, on the other hand, is its *diameter*.

Gas Lift and Air Lift

If the well's productive capacity is large enough to warrant it, and particularly if there are similar wells nearby so that the cost of the necessary installation is warranted, the pumping stage may be deferred for a time by the use of *gas lift* or *air lift*.

These are methods of supplying the deficiency of gas to lift the oil by pumping gas or air into the hole under pressure. Gas is used if there is a cheap supply available; otherwise air is used, though because of its corrosive action, the greater danger of explosions, and other reasons, it is decidedly inferior to gas for the purpose. The gas or air is compressed to the necessary pressure and injected between the casing and the tubing, thus forcing the oil up the tubing. In some fields gas or air lift has proved an effective method of producing oil more rapidly than mechanically pumping.

The gas or air *lift* should not be confused with *gas drive* or with *repressuring*, to be discussed later. In a gas drive, gas is pumped down a well or wells to drive the oil to other wells. In repressuring, gas is returned to the reservoir sand to dissolve in the oil and bring it to the wells. In the gas lift, little if any of the gas enters the sand or aids in bringing the oil to the well; the gas is merely used mechanically to lift the oil that is propelled into the well by other forces.

Gas or air lift is now being successfully used on small wells in old fields. In some fields it is used for many years. The percentage of fields in which it is used is comparatively small, however, and in few fields is it used for more than three or four years.

On the Pump

When gas or air lift is not used, or when it has been used and abandoned, the field goes on the pump.

This may not happen to the whole field at once, you understand. The smaller wells, the wells at the margin of the field, and the wells in tight spots in the sand are likely to quit flowing first. Months or even years may elapse between the time the first and the last wells in a field go on the pump. Sooner or later, however, practically all of them become pumpers.

The simplest form of pump is a *working barrel* with appropriate valves. The working barrel is screwed onto the bottom of the tubing.

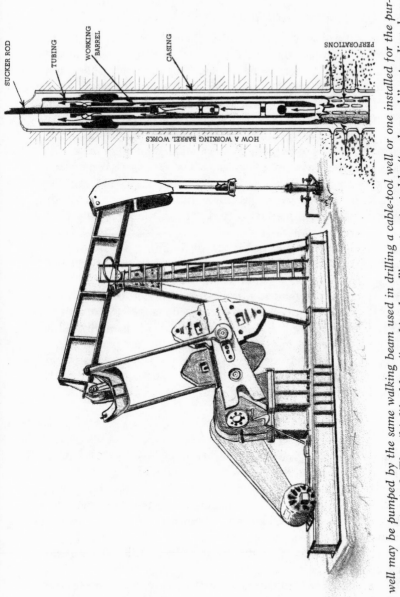

SUCKER ROD

TUBING

WORKING
BARREL

CASING

PERFORATIONS

HOW A WORKING BARREL WORKS

The well may be pumped by the same walking beam used in drilling a cable-tool well or one installed for the purpose, or by a pumping jack. The oil is lifted by a "working-barrel" pump actuated by "sucker rods" extending down
from the surface, although electric and hydraulic submerged-type "bottom-hole" pumps are coming into use.

The valve mechanism inside the working barrel is actuated by a string of steel rods, called "sucker rods," screwed together length by length, after the fashion of casing or tubing. Up-and-down motion of the string of sucker rods opens and closes the valves and lifts the oil to the top of the hole.

The up-and-down sucker rod motion may be imparted by hooking the string to a walking beam (possibly our old friend of the cable-tool rig, but probably a special one installed for the purpose); then the well is said to be "on the beam." Usually, however, the rods are actuated by a device four to fifteen feet high, known as a *pumping jack.*

The original drilling derrick may be used for pumping, or it may have been replaced by a smaller derrick known as a *pumping derrick* or *pumping rig.* Or, no derrick at all may be used and the only sign of the well is the casing-head protruding from the ground and sur-mounted by a relatively inconspicuous jack. Hundreds of pumping jacks can be seen from train windows or highways in many parts of the country; their sleepy nodding is typical of settled, producing oil fields.

Not all pumps are of the kind just described. Hydraulic pumps and bottom-hole electric pumps are also used, especially for deep wells. They differ from such pumps in ordinary industrial use chiefly in their adaptation to operation in the confines of wells a few inches in diameter. They are not common, however, and by no means monopo-lize deep-well pumping. Most wells, including many of them more than fifteen thousand feet deep, are pumped with sucker rods.

Power *and* Powers

Power for pumping may be supplied by an electric motor, a gas or gasoline or diesel engine at each well, or it may come from a *power,* pumping a number of wells. A power usually consists of a geared unit or a large horizontal wheel which turns a smaller horizontal eccentric wheel. Steel *pull rods* are attached to the circumference of the eccen-tric wheel; they lead out across low supports to the jacks on the in-dividual wells.

Twenty wells, or even more, may be pumped from a single power. The method can be used wherever the wells are small, not too deep, not too far apart, and on not-too-rough terrain. Under these circum-stances it may be the cheapest method of pumping, although individ-ual pumping units have almost driven powers into extinction.

Bailing and Swabbing

Now let's go back a bit. Johnson Roe No. 1 was a flowing well, and we have watched it until it is now a pumper. What would have happened if, like many another well, it had not flowed at all?

The production might first have been tested by bailing with the bailer, in exactly the same manner as water and cuttings are bailed out of a drilling cable-tool well. If the bailing had shown the well's production to be small enough for adequate handling by a pumping unit, the well would have been put on the pump forthwith. Most of the wells drilled in the United States go on the pump without ever having flowed.

If the well had shown a somewhat larger production, or if there had been some delay in installing pumping equipment, the well might have been produced by *swabbing*. A swab is a steel contraption a few feet long with rubber cups that fit tightly into the casing or tubing; it is lowered to the bottom of the well by the sand line. Valves in the swab permit oil to flow up through it while it is being lowered, then close when it is lifted so that it lifts out of the well practically all the oil that is in the casing. Thus swabbing may be an effective means of producing more oil than ordinary pumping, although it is almost never employed as a permanent production method.

Cleaning and Repairing

An oil well does not produce oil without constant work and attention. Mechanical equipment wears out or breaks; wells sand up, mud up, and paraffin up. Repairing equipment and cleaning out wells keep production crews on the go.

On the mechanical side, leaks in tanks, flow lines, and lease lines have to be watched for and repaired and the power and pumping equipment must be kept in good running order, but most of the work centers on the well itself. The working valves wear out and have to be replaced, which necessitates shutting down the well while the sucker rods are pulled and unscrewed, the repairs made, and the pump and sucker rods replaced in the hole. The working barrel may need repairs or the tubing begin to leak, in which case the well is shut down while both rods and tubing are pulled and rerun. Thus the round of mechanical work goes on from well to well as needed.

Cleaning out is another constant job, frequently required in some fields, less frequently in others, with the frequency varying from well to well in each field.

In some fields sand sloughs off the wall into the well, or the oil brings sand into the well. In time, enough packed sand accumulates in the bottom of the hole to impede the production of the oil. Then the well must be shut down, the rods and tubing pulled, and a "cleanout string" of tools used to clean out the sand. The cleanout string may be either light cable or rotary tools or a set of special tools designed for the purpose.

In other fields mud plasters up the sand and impedes or stops production. It may be a residue of the mud used in drilling, it may be mud from an uncased part of the well, or it may be brought by the oil from shaly parts of the reservoir bed. Whatever the source of the mud, the well must be cleaned out. Rods and tubing must be pulled first. Then it is sometimes possible, though generally not, to wash down the casing. If not, the mud is removed with a cleanout string, possibly equipped with an under-reamer.

Another great bugbear is paraffin. Fields with asphalt-base oil are not troubled by it, but most others are. When a paraffin-base oil is full of gas the paraffin stays in solution, but in a pumping well, gas escapes and cools the oil. More cooling takes place as the oil nears the surface, and some of the paraffin is likely to congeal and deposit. It may seal off the producing sand as effectively as a layer of paraffin seals a jar of jelly; it may fill the tubing until the only hole left is that occupied by the sucker rods.

Many methods have been devised to keep paraffin from depositing or to melt it or to remove it by solution. But getting effective heat at the bottom of a cased and tubed well is not easy, and paraffin is not readily soluble in solvents that are easy to obtain and use. When such methods fail the only effective way to remove the paraffin is to pull the rods and tubing, steam the paraffin from them, and use a cleanout string to clean the face of the sand. In some fields, paraffin cutters—run inside the tubing and expanding in the fashion of an under-reamer—are used successfully.

You have already noticed, no doubt, that all of these repair and cleanout jobs require shutting down[2] the well until the job is com-

---

[2] *Shutting in* and *shutting down* are terms widely used in the oil business. Any operation that is suspended, whether the drilling of a well, the producing of the

pleted. In consequence they involve a double burden on the operator —the amount he expends on the repair or cleanout job and the loss of income while the well is shut down. In fields where sand or mud or paraffin is really bad, wells may be off production from a quarter to half of the time, and because clogging begins again as soon as production is resumed, they may be on maximum production only a small part of the time. These are the things that bring gray hairs to the producer trying to keep down his costs.

Stripper Wells

Don't think because a well has stopped flowing that its production has become stationary. A pumping well declines the same as flowing wells, and usually at about the same rate as it declined during the flowing period after the first flush production was past. It doesn't fall off so fast in barrels per day—it couldn't—but each year it produces about the same percentage less than the year before. In time (and the time may be months or years) the production gets down to a few barrels a day, and eventually to a few barrels a week. Such wells are known as *strippers*. The production from each seems negligible, but the aggregate stripper production constitutes a goodly percentage of the total production in the United States.

The importance of the production from stripper wells is one reason why periods of demoralized prices for crude oil are unfortunate; they bring nearer the day we may all have to pay more for gasoline than we do now. The cost of producing a barrel of oil from a stripper is of course greater than the cost of producing a barrel from a flowing well. When the price of crude gets too low, many stripper wells must cease to operate. Equipment deteriorates, casing rusts, water may break in, and if the low-price period lasts long enough, many strippers may be plugged and abandoned. When the price goes up again, the wells are gone, and no new ones will be drilled because the production that would be obtained would not pay the cost of drilling. Some of the oil

---

oil from a well, or the conducting of an office, is a *shut down*. This must be distinguished from the term *shut in*, which is applied to flowing wells only. If the control valves on a flowing well are closed so that no oil is produced, the well is shut in. The amount that the well might produce is called "shut-in production," and this expression is sometimes used for the amount a well would produce above the amount that it is being allowed to produce. In this latter sense, shut-in production may be the difference between the well's potential and its allowable.

may some day be recovered, when the price of oil is again high, by the institution of secondary recovery methods—but most of the oil left in abandoned stripper areas is lost for good, so far as concerns its availability for human use. When you remember that the average production per well per day in the United States is less than twelve barrels, despite the thousands of wells capable of producing from a thousand to ten thousand barrels a day, you get some concept of the number of strippers there must be, the amount of oil there must be in the stripper areas, and the disadvantage to the public of having the stripper wells abandoned.

The Life of a Field

Assuming that price demoralization does not cause its abandonment, how long may a field last? The answer is: a long, long time. Some fields go to water in a few years, but most fields keep on producing year after year, although after the proration stage is past the production diminishes each year. Many fields have passed their fiftieth birthdays. Very few fields have been abandoned entirely, although many of them have declined to an almost negligible production per well.

This is even more true for recent fields in which gas conservation and good production engineering have been practiced. Such fields have not been allowed to produce their oil so rapidly, which means that they will produce for a longer time. Unless our theories are all awry, these fields will also produce more oil before they are abandoned, which is another way of saying that they will produce a bigger percentage of the oil that is in the reservoir bed and will leave less there when abandonment comes.

Pressure Maintenance and Gas Cycling

One of the best things being done in the oil business today is putting back into the producing sand a part or all of the gas taken out of it. The idea originated within 20 years of the Drake well discovery and was born out of the desire to extend the life of declining wells. Earlier attempts to increase production by sucking the oil from the reservoir were quickly supplemented by efforts to push the oil to the wells. In 1884 the Richards patent proposed to force air, gas, or other fluid into the oil-bearing reservoir and by maintaining pressure, to force the oil into adjoining wells. The first direct application of the idea appears to have been put to work by a fellow named James D. Dinsmoor in

1888. He was successful in several tests on the Third Venango sand in Venango County, Pennsylvania, where he combined both vacuum pumping and repressuring to increase production from dying wells. More recently many fields in the United States and abroad have been repressured from the start. Because such early repressuring tends to maintain the original pressure, instead of depleting it and then rebuilding it, the method is called *pressure maintenance*.

The gas produced with the oil is stripped of its gasoline and compressed to a pressure enough higher than the reservoir pressure in the field to force it down *input* wells into the producing sand. It makes its way into the sand, where part of it is probably redissolved in the oil. It again does its job of bringing oil to one of the producing wells, when it is again stripped, compressed, and sent down an input well to do the same job once more. In some fields, additional gas, not produced with the oil, is also forced down the input wells.

Fields have already been operated for many years with only slight decreases in their reservoir pressures. There is little doubt that a field properly repressured from the start will produce by flowing, without resorting to pumping, a high percentage of the oil originally in the sand. This means not only a greater ultimate percentage of recovery of the oil contained in the sand; it also means that more oil is produced by low-cost flowing methods instead of by high-cost pumping.

During the 1930's, when pressure maintenance had become a standard practice, operators found that some fields produced clear, almost colorless oil, a product of the condensed gas. These fields are known as *gas condensate* fields. Under the very high temperatures and pressures within the reservoir the oil is present entirely in gaseous form. When this gas is brought near the surface, part of it changes to a liquid and is produced as condensate. The remainder is very wet gas. The wet gas from a gas condensate field may be processed through an extraction plant, where the liquids are removed. The dried gas is then reinjected into the reservoir under high pressure to chase out more wet gas, and thus the gas keeps returning to the reservoir. Quite naturally the oil man calls this *gas cycling*. The Tidewater Oil Company and the Seaboard Oil Company put the first cycling plant into operation in 1936 in the Coyuga field of East Texas and thus developed another application of pressure maintenance. Pressure maintenance is feasible primarily in fields operated as units or with co-operation so close among all the producers as to amount to unitization. Otherwise, no one could afford to compress gas to build up reservoir pressure for the benefit of someone else's wells.

Secondary Recovery Methods

As a means toward recovering more of the oil that is in the producing sand, and toward recovering it a little sooner, there is an increasing adoption of methods of assisting nature to move the oil out of the sand to the well. These are known as *secondary recovery methods*. They include repressuring with gas, driving the oil out of the reservoir with gas, air, or water, sinking shafts and driving drifts through the sand to provide greater *bleeding surface* for the oil to escape from the sand, and even mining the sand to recover the oil from it.

Repressuring

We have already seen what pressure maintenance may do to prolong the flowing life of a field and increase its ultimate recovery. Even where pressure has been depleted and the wells are approaching the pumping stage, a field may be rejuvenated to some extent by repressuring it with gas. The procedure is practically identical with that used in pressure maintenance; in fact, as we have said, pressure maintenance grew up from repressuring, by starting earlier in the game.

As with pressure maintenance, the method is attractive primarily in fields operated as units.

Gas Drive

In some fields where multiplicity of lease ownership has precluded repressuring, but where surplus gas is available, a gas drive is used with beneficial effect. The gas is compressed and forced down input wells to drive oil through the reservoir bed to producing wells. How much of the gas dissolves in the oil and increases its mobility is a question, though some solution probably takes place. The method is used only where a single operator or co-operating group of operators controls a considerable area. A small tract would not warrant the installation and compression expense involved, and a drive on a small tract would probably drive oil from the driver's land to his neighbor's wells.

Air Drive

In some old fields where gas is not available for a gas drive, air has been used. The air is forced down input wells and drives some of the oil to the producing wells. The effect is purely mechanical; the air does not readily dissolve in the oil and thus cannot lessen the oil's viscosity or increase the ease with which it is moved; on the contrary it tends to oxidize the oil, render it more viscous, and increase its adhesion to the sand grains. For this reason it is used only where gas drive is not available.

Water Drive or Water Flood

The Bradford field of Pennsylvania was discovered in 1871; for many years it was one of the outstanding fields of the United States. By the 1920's all its great area that had not been abandoned was dotted with stripper wells good for a barrel or two or ten a week.

Not far below the surface is a bed of gravel, full of water and always carefully cased off in drilling. Many years ago an occasional operator discovered, probably by accident, that soon after he let water from the gravel bed get into a well, the oil production of his nearby wells began to increase. The water, under the *head* of the vertical distance between the gravel bed and the oil sand, was driving the oil ahead of it through the oil sand to the pumping wells.

Nothing was said about it, for letting water into the oil sand was contrary to law and custom, but in all probability the practice was followed purposely by a few operators as far back as the 'nineties. In the 'twenties, guided by competent engineering, it was sanctioned by law and came out into the open. Today, the Bradford field is under water drive; it has produced more oil by this method than it did before. In 1937, under the water drive, it produced 16.5 million barrels for its second peak production year since the heyday year of 1881 when it gave up 23 million barrels. The old methods of natural flowing and pumping had recovered perhaps 25 per cent of the original oil content; the water drive will probably recover over 30 per cent more. The remainder will still be in the sand, adhering to the sand grains, or trapped by water that has by-passed it.

The water drive method is being used successfully in many fields throughout the country.

Miscible-Phase Displacement or Solvent Flood

In the 1950's and early '60's, spectacular secondary techniques were developed. Earlier, in the '30's and '40's, engineers were mostly concerned with improving water flooding efficiency, but water flood limitations, the challenge of higher recovery efficiencies, and the increasing scarcity of new major fields have brought about ingenious new developments. Several of them may be grouped under one general heading, *miscible-phase* displacement. Under this family name the several processes differ in detail, but they get the same job done underground. Essentially, the engineers use a solvent and a driving agent to carry the oil out of the reservoir. The solvent, driving through the pores, dissolves some of the crude oil and sweeps forward, pushing more crude ahead of it to the producing wells. We might say that the engineer dry cleans the reservoir.

The first thought in miscible-phase displacement was to fill the reservoir with a solvent such as butane or propane and let it act as both solvent and driving agent. That turned out to be too expensive.

Next was to inject dry gas under very high pressure so that it practically became liquid and dissolved in the oil readily. The injected gas drives the lightened oil just like any driving agent. The Atlantic Refining Company has been applying this method with great success since 1949 in Crane County, Texas.

Modifications of "H.P. gas injection" for lower-pressure reservoirs have been worked out by the Humble Oil and Refining Company. A very wet gas at lower pressure produces about the same effect as high-pressure, dry gas injection. Humble has found that propane added to already wet gas increases recoveries in the company's Selligson field in South Texas to 475 per cent of what was expected by primary methods.

If wet gas can do such a fine job, the next logical experiment is to inject a *slug* of liquefied petroleum gas first, then to drive it along with dry gas. This has been dubbed the LPG-slug method and it has worked successfully in many fields in Oklahoma, Texas, California, and New Mexico.

All of the miscible-phase techniques and their combinations are looked upon as new tools in the engineer's bag of tricks to get more of the oil from the reservoirs of the world. They constitute a fast-growing science.

Fire Flood or In-Situ Combustion

One of the oldest ideas in secondary recovery probably saw the application of an underground fire as far back as the turn of the century. Both American and foreign oil men tried out the principle of inducing a heat wave in the reservoir to free the oil, make it more mobile, and thereby to drive more of it to the producing wells. Early attempts failed because the industry lacked the necessary tools and knowledge of just what was going on underground. New devices and decades of research show that an underground fire can provide the heat and that the problem is far more complex than merely heating the oil to make it move.

The Sinclair Oil Company operates a successful full-scale operation in the Humbolt-Chanute field in Allen County, Kansas; the fire sets up an actual underground refining process. The fire, started in the *pay zone*, is fed air or a mixture of gas and oxygen in delicately controlled proportions. The fire advances through the reservoir, consuming about 10 per cent to 15 per cent of the crude as fuel. As heat from the fire reaches out through the reservoir toward the producing wells it first drives off the light hydrocarbons and other volatile gases along with water. With the continued temperature rise the asphalts and tars break into more light hydrocarbons and coke (carbon), and the coke becomes the fuel for sustaining the fire. The fire advances, leaving the cleansed sand in its wake; the distilled or refined products have been driven ahead by expanding gas and steam, and these in turn have driven more crude oil ahead to the producing wells.

Fire drive is one of the most promising secondary recovery techniques yet devised. Where it is applicable it may turn out to be economical and extremely fast.

Underground Drainage

Obviously one of the restrictions on recovering oil from wells is the small surface area of the hole through which the oil must squeeze from the reservoir bed into the well. From ten to eighty acres of reservoir yield their oil into a six-inch or eight-inch hole. More holes, or larger holes, can be drilled, but on most tracts the additional *sand face* is not great enough to pay for the drilling.

At Pechelbronn in Alsace, once in Germany but now in France, the problem has been solved by mining shafts to the oil sand and then mining tunnels through it. The oil seeps into the tunnels and is pumped to the surface through the shafts. The field is an old one; the mining of asphaltic sands began in 1742 and the drilling of wells in 1881. About 55 per cent of the field's production still comes from wells; the remainder comes from shafting and tunneling.

Somewhat different methods of increasing drainage and recovery by underground methods have been devised in the United States, but they have not yet had an adequate test on a commercial scale. The time will doubtless come, with increasing demand for oil, when they will be tried out, with probable success.

Mining Oil Sand

All of the methods so far described, including that used at Pechelbronn, leave part of the oil in the sand. The only way to recover all of it is to mine the sand itself and then remove the oil from it. This has been done at Pechelbronn with the sand recovered from the tunnels, and it has been done on a small scale in one or two other places in Europe.

At Mildred Lake, about 300 miles north of Edmonton, Canada, a pilot plant has been tested which mines and processes the famous oil sands of the Athabaska area. Often referred to as the *tar sands*, the Athabaska deposits are estimated to hold 300 billion barrels of recoverable oil. Surface exposures where the streams have cut their valleys in a 30,000-square-mile area show that the tar sands embody a locked-up supply equal to the rest of the world's known reserves. This is the equivalent of an oil field the size of South Carolina. A four-company team, composed of Imperial Oil Ltd., Cities Service Athabaska, Inc., Richfield Oil Corp., and Royalite Oil Company, has pressed the attack on the mining and extraction problem. It now seems probable that the oil sands can compete for North American markets. These companies' successful 3,000-barrel-a-day pilot plant will be replaced with a 100,000-barrel daily operation, connected by a 310-mile pipeline to Edmonton. The technical battles of nearly fifty years have apparently been won by employing a German mining wheel and an extraction plant, both set on a mobile carriage. Buckets mounted on a large wheel gouge into the open pit face and dump their loads on a conveyor belt which in turn carries the sand to the attached extraction unit. Here the bitumen is diluted with a solvent extracted from previously treated sands, then water is added to force the sand grains to settle. All of the petroleum fluids, now free of sand, separate from the water like cream from milk, and are poured off. The water returns to the head of the line for re-use and the bituminous material flows to a refinery where the fluids, gases, and solids are separated. Oh yes, the clean, round, oil-free sand grains are conveyed back to the outcrop, where they may someday be used as the ideal raw material for a glass factory.

Still other areas of bituminous sands seem to offer great promise for future oil production. Southwest of Vernal, Utah, the curious deposits of Asphalt Ridge are being mined from open pits. Although the county road department finds that this material, mixed with river sand, is just right for highway surfacing, the deposit represents a reserve of almost two billion barrels of oil. (The oil content averages a barrel per cubic yard of sand.) Standard Oil Company of Ohio has been studying procedures to exploit these asphaltic sands, both by mining and by underground extraction.

In Czechoslovakia, oil sands 260 feet below the surface are available to underground mining, but mining these will probably prove to be a very expensive venture. Tremendous scale mass production mining, even at the surface, will probably always find stiff competition from conventionally produced oil. Any additional step required in winning the solid raw material, which is not required in the production of fluid crude oil, increases the cost of production.

One of the bright spots about oil sand research and the development of oil sand reserves lies in the ready availability of the oil sands for national emergencies.

Mining Gilsonite

The first commercial gasoline production from solid raw material began in June, 1957, near Grand Junction, Colorado. Black, light weight, coaly-looking gilsonite is mined from veins in Uintah County, Utah. There it is ground up and transported as fine chips in a water pipeline, 72 miles long, to the refinery at Fruita, Colorado. This plant, built and operated by the American Gilsonite Company, is a combined mineral-beneficiating mill and petroleum refinery; it is capable of producing 1,600 barrels of gasoline, 1,300 barrels of diesel fuel, and 350 tons of coke a day.

Gilsonite, also known as *uintahite*, is one of the species of asphaltites which have become hard and brittle through a great span of geologic time. Laboratory tests have indicated that it will produce 19 per cent gasoline, 13½ per cent kerosene, 8 per cent gas oil, and 45 per cent motor oil, in addition to high-grade coke, which is the left-over solid matter.

Because the gilsonite fractures easily and is only four per cent heavier than water, the company has been able to devise a revolutionary mining technique to extract the "ore." It has combined the ancient process of hydraulic placer-mining with oil-well drilling to make an explosion-and-fire-proof operation out of a formerly dangerous and laborious digging and blasting task. First, a deep shaft is dug into the vertical vein; then from its bottom long drifts are driven slightly upward along the vein. Next, an oil-well drilling rig is set up on the vein and a six-inch hole is drilled down the vein into the mine drift. A pair of high-pressure nozzles are then substituted for the drill bit. The pipe and nozzle are rotated and raised slowly while discharging a terrific blast of water at 2,500 pounds per square inch. (An average fire hose and nozzle operate at 150–200 pounds pressure.) The high water pressure directed against the walls of the hole tears and blasts the fractured gilsonite free so that it falls to the bottom of the drift. Here it is picked up by a flowing stream of water which carries it down the inclined drift to the shaft. Heavy-duty pumps lift it to the surface where it is cleaned and processed before being run into the pipeline to the Fruita refinery. Since the method eliminates the need for men underground to *muck-out* the ore, the company may hang a closed circuit television camera so that the operator can monitor the mining from the surface—a classic wedding of old and new sciences to conquer the plaguing problems of a once difficult and hazardous operation.

Mining Oil Shales

Other rocks also contain the parent materials of oil called *kerogens* which cannot be produced as fluids from a drilled hole nor from drainage into underground mine workings. Heat must be applied to convert them into oil. We have read about these curious substances in Chapter 3, "Picking the Place to Drill—The Theory."

Oil reserves locked up in the Green River shales of Western Colorado, Wyoming, and Utah have been estimated at over 1,500 thousand-million barrels. This is much more than the world's total production to date. The problem is to extract the oil at low enough cost that it can be marketed at prices competitive with conventionally produced oil.

The U. S. Bureau of Mines, The Union Oil Company of California, and other major oil companies have researched for years on high-rate mining methods and refining techniques in attempts to find the key to economic shale-oil production. Improved methods make it seem likely that gasoline from shale oil will soon be competitive in price with the conventional product.

Shales of 10 to 34 gallons of oil per ton are mined by large-scale underground methods. The raw material is extracted by digging rooms within the shale bed, which is up to one hundred feet in thickness. Each successive room is separated from its neighbor by a *pillar* of shale which is left to support the roof. The mining process is therefore known as the *room and pillar* method of mining. Mined shale is then crushed, screened, and passed through a specially-designed retort which distills-off the fluid petroleum products.

Underground atomic blasts and other applications of nuclear energy are also being considered as tools to force the shale to release its oil or to simplify mining operations.

Many of our natural resource economists consider the oil shale our "ace in the hole" for petroleum for national emergencies. Oil shale production may someday be important to the economies of Brazil, Manchukuo, Japan, Great Britain, France, and Thailand, too.

The Brazilian deposits hold in reserve the largest volume of oil shales in the world outside the United States. Two principal areas have been known for almost 200 years and have been worked sporadically and on a small scale for the last 50 years. In the very heart of the industrial area of Brazil, the Paraiba Valley oil shale deposits are thought to contain over 2,000 million barrels of oil. Unfortunately, the Paraiba Valley shales present one of the most difficult refining problems of any in the world. The oil yield is much lower than that of the Colorado shales and they contain 33 per cent to 35 per cent water besides, which complicates the distillation process.

The other major Brazilian deposit, the Irati oil shale, is of fantastic proportions, extending for over 1,000 miles along the outcrop. It is so large and so remote from transportation and from the industrial centers that development has been slow. New processes now going into operation indicate that Irati shales can be handled by an economical refining process, more so than the Paraiba Valley deposits. They have been estimated to hold very much larger reserves of oil—anywhere from 2,000 or 3,000 million barrels to 200,000 to 300,000 million barrels. These deposits are unique for their ability to give up dry gases, liquefied petroleum gas, sulphur, and a lower pour-point oil than their counterpart in Colorado. An oil shale, petrochemical, and shale-gas business seems assured for the nationalized Brazilian industry.

The Cost of Producing Oil

You have learned by this time that producing oil involves something more than turning a valve and letting oil flow into a tank. Have you any idea what it costs, on the average, to produce a barrel of oil? Let's consider the items involved.

First there is the cost of finding the oil, which includes the geologic work, geophysical surveys, the leasing of the lands and the payment of rentals, legal fees, and the cost of drilling. In the drilling cost must be included the costs of dry holes; wells have to be paid for whether they produce or not. Then comes the cost of equipping the well and the lease, including roads, field lines, tanks, and living quarters, and later the cost of pumping equipment. Then comes the cost of labor and materials used in producing, repairing, cleaning out, and all the multitude of other jobs necessary to keep the field operating. Then there is the value of the royalty oil which the producer pays the royalty owners. Over and back of it all is the depreciation of the equipment, and the depletion of the value of the property with every barrel of oil taken out. And we can't forget the taxes imposed on oil, gas, and other minerals when they are removed from the ground; most states levy some kind of a severance tax on the production of natural resources. Finally, there is interest on the investment at, say, 6 per cent. All these things, according to an estimate by the United States Petroleum Administrative Board for the years 1931–1934, added up to 79.8 cents a barrel. By 1954 the figure was $2.28, and in 1960 costs had reached the all-time high of $3.19 per barrel.

Different wells and different fields vary widely from the average, of course. A shallow flush well, putting thousands of barrels a day into the tanks, produces oil at a few cents a barrel; oil from a small, deep pumper may cost two or three dollars a barrel. The most worrisome fact in an oil producer's life is that as production goes down the cost per barrel goes up, for it costs nearly as much to pump a well when it is producing ten barrels a day as when it is producing fifty, and there are only one-fifth as many barrels to which to charge it.

The average cost varies from year to year and from district to district, depending on the proportion of the production that comes from flush fields.

The average figures include no storage or handling charges, and nothing for transporting the oil from the lease to the refinery. The above averages are only the costs of finding and getting the oil out of the ground into the lease tanks. Getting it from there to the refinery is a transportation job. If you are still interested we'll watch that for a while.

CHAPTER 10

Transporting the Oil

Oil in the field tanks is like a fat steer on the range; it needs to be taken thence and made into something useful. The task of taking it usually falls to a pipeline or to sea tankers, though tank barges, tank cars, and tank trucks bear an important part as we shall see. These agencies transport the oil from the field to the refinery, and the same types of transportation are used to carry the refined products to their points of use. All these transportation facilities will interest us; let's concentrate on pipelines and tankers, which do the biggest part of the job.

Pipeline Systems

A great pipeline system has much in common with a great railway system; it picks up freight from a multiplicity of points on its branch lines, gathers it to its trunk lines, and transports it long distances to a few great centers. Like a railroad, it transports goods at published tariff rates. Interstate pipelines, like interstate railroads, are common carriers, as are many intra-state pipelines also, and all interstate lines are subject to the Interstate Commerce Commission.

Unlike a railroad, the pipeline system carries no freight back from the centers to the gathering points. Practically all of the pipeline traffic moves in one direction; the counter-movement of finished products is done by separate pipelines and by tank cars and tank trucks. In addition to domestic lines there are many international pipelines in operation today. For example, the Trans-Arabian pipeline originates in Saudi Arabia, traverses the territories of Jordan and Syria, and finally terminates in Lebanon. The political problems associated with international trunk pipelines are more complicated than those of domestic

178

lines within the United States because there is no Interstate Commerce Commission to regulate their operations and no common taxing authority to determine the division of tax revenues between the governments involved.

Some trunk pipelines are almost as long as any single railroad trunk line. One system can pick up a barrel of crude in Wyoming and deliver it to St. Louis, Kansas City, or Chicago. A barrel of oil from Oklahoma, Kansas, or North Texas may wind up in Chicago or Toledo or Bayonne.

Not all pipeline systems are extensive, however. Out of several hundred operating pipelines in the United States, only 82 are interstate and only about three-fourths of the interstate companies operate far-flung systems. Most pipeline companies operate a few dozen or a few hundred miles of line, connecting wells in one or two or a half-dozen fields with a single refinery or perhaps two or three or four refineries.

The essentials of construction and operation of a pipeline system are the same whether the system is large or small, and we shall examine them both. Before we start our study, however, let's give a moment's thought to the ownership of the oil transported.

Crude Oil Purchasers

A pipeline system is usually owned and operated by a pipeline company. Even in the integrated organizations that produce, transport, refine, and market, the pipeline department is usually a separate corporation, the stock of which is held by the integrated parent company.

Some pipeline companies are purchasers as well as transporters of crude oil, but as a rule a pipeline company confines itself to transportation. The oil is purchased either by the refiner for whose use it is destined, or by a crude-oil purchasing company which in turn sells it to the refiner. In the case of the integrated organizations, the purchasing of the crude is done by the refining department or by a separate crude-oil purchasing department or corporation created for the purpose, and the pipeline department or corporation attends solely to the transportation of it.

The purchasers of crude, whether pipeline companies, or refiners, or crude-oil purchasing companies, keep close track of the needs of the refineries they serve and the wells from which they gather. They see that the refineries are supplied with oil of the character they want, from the wells from which such oil can be most cheaply obtained and delivered. As a matter of policy they also give consideration to an equitable distribution of their *takes* among the various producing districts that they serve, watching to see that each district has an outlet for a fair share of its production. This is not philanthropy; it is merely good business. A company might save money this year by taking all of its oil from an advantageously-located flush field, letting the operators in other fields make out as best they can. But next year it might need oil from the neglected fields and find that the operators had tied-up to some other purchaser. When a pipeline serves to move oil produced only by the parent company or another affiliated company, reservoir engineering and prudent oil field operation will also dictate dispersal of offtake among several fields rather than concentrating withdrawal from one field. A crude oil pipeline is expected to last 20 years or more, and unless it can operate close to its capacity throughout this whole period it will not remain a paying proposition. Therefore, any purchaser not far-sighted enough to adhere to a policy of multiple crude sources for a pipeline would soon find himself in trouble with his crude suppliers, the proration authorities, and his own reservoir engineers.

The Gathering System

The initial pipeline task is to get the oil from the lease tanks, scattered about the field on the various leases, to a central pump station at the beginning of the main line. This is done through *gathering lines* of two inches to eight inches diameter, usually buried from one to three feet deep. In warm climates and in fields where the oil does not congeal when cold, some of them may be above ground. The oil may flow through them by wellhead pressure or gravity or be forced through by *field pumps* or *gathering pumps*. In a good-sized field there are many miles of gathering lines, and the investment in them and their pumps and auxiliary tanks may run into hundreds of thousands of dollars. The total length of gathering lines in the United States is about 85,000 miles and they gather 171,000 barrels of oil every hour.

Gathering and Gauging

You may have wondered when the producer gets paid for his oil. Unless he is also the transporter and refiner of the oil, his payoff comes when the oil is run from his field tanks into the gathering system. At

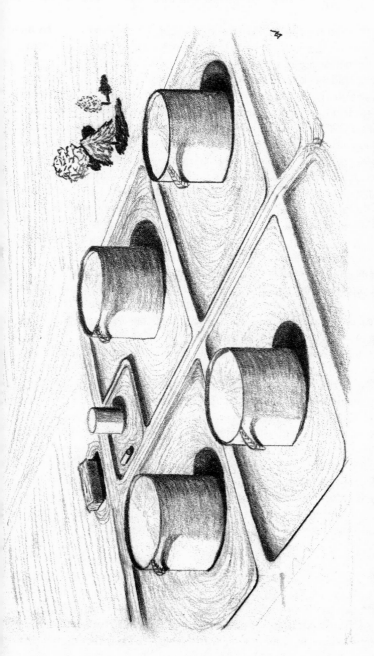

The oil from the wells goes to field tanks, from which, after being sampled and measured, it is taken by the pipeline gathering system.

this point the purchaser buys the oil. At this point, therefore, the oil is measured or *gauged*, and this is done even though producer, refiner, and transporter are one and the same.

Every field tank is *strapped*, or *calibrated*, which means that it is measured so that its capacity per vertical inch is known accurately. Then these measurements are put on a chart called a tank table. When a field tank is to be run, all oil lines leading into it are closed and the pipeline company's gauger measures the height of the oil in the tank by lowering into it a graduated stick or rod known as a gauge. In a modern tank farm some of the storage tanks may be very large indeed, storing between 250,000 and 300,000 barrels of oil. In these larger tanks the gauger uses a weighted steel tape for measuring the height of the oil level. Then he opens the gathering line leading from the tank. When the tank is nearly empty the gathering line is closed and the gauger again measures the level of the oil in it. By subtraction he then has the number of inches the oil level in the tank has fallen. Meanwhile, another tank or tanks may have been run. At the end of the day the gauger makes out a *run ticket* showing the amount of oil run from each tank on the lease, and copies of the run tickets are furnished by the pipeline company to the producer, who may check them against tickets made by his own gauger. Monthly payments for the oil are made from the run tickets.

The gauger may also *thief* a sample of the oil. His *thief* is a complicated, slender dipper attached to a long tape. He lowers it into the tank, through the manhole or a *thief-hole* in the roof, and with it brings up samples from various levels of the tank. In the most modern practice, he has testing apparatus in his car with which he determines the specific gravity, BS & W content, and other characteristics of the oil. Where this practice has not yet been adopted the samples are sent to a laboratory for a check on the specific gravity of the oil and its BS & W content. The facts thus learned about the oil are noted on the run ticket and are used in computing payment for the oil.

LACT

The Latin experts and the chemists should not associate LACT with milk or its derivative products. LACT stands for *Lease Automatic Custody Transfer*, a jaw breaker name for automatic gauging.

After World War II, when labor was short and the cost of crude production became high, much effort was devoted to streamlining production, gauging, and crude oil transfer.

Mechanically-minded landowners were more willing to trust their fortunes to automated equipment than to human observation. The first Lease Automatic Custody Transfers were by means of calibrated tanks into which oil was flowed in a controlled manner up to a certain level. When the measured level was reached, flow was automatically stopped and incoming crude was diverted to another identical tank with the same automatic equipment. As a check on the system, recording flow meters were also installed. Nowadays, oil operators and royalty owners have gained such trust in the meters that most LACT-equipped fields rely primarily on meters. The meters are *proved* periodically by flowing oil through them into calibrated tanks. The automatic features of LACT go beyond measurement of oil and include the running of the gas-oil separators, removal of BS & W, and sampling of the oil for gravity determination. Perhaps in future years, manual gauging of tanks will be done only in very large tank farms and refineries and during the first few days after the installation of lease tanks at the well.

Tankage

The gathering system delivers the oil into big steel tanks at the main pump station, where it is stored until it goes into the main line. Each tank may hold 250,000 or 300,000 barrels, and around each is an earthen dike to hold the oil that would escape if the tank should spring a major leak or should catch fire and burst.

A group of storage tanks is known as a *tank farm*. There are nearly always one or more tank farms at the refinery, and there may be others at the main pump station or at intermediate pump stations along the line, though the tendency is away from much tankage at intermediate pipeline stations. Tank farms may also be located at other points to store oil against future needs or to care for surplus production or as a speculation in the hope of higher prices.

One of the most important uses of oil storage is to meet seasonal fluctuations in demand. In North America, oil consumption during the winter months is greater than the summer consumption. Moreover, demand for gasoline is highest in the summer months, while distillates and fuel oils are mostly consumed in the winter. Hence the need for tanks for storing oils during the *off peak* periods. Needless to say, defense considerations also dictate the keeping of substantial oil stocks.

Storage tanks may be owned by a refiner or a crude-oil purchasing company affiliated with the pipeline company or a refiner or producer.

In some cases they are owned by the pipeline company itself, but as a rule they are owned by a refining or producing company or a purchasing company affiliated with it. Tank farms holding more than a million barrels of oil are not uncommon.

The trend in tank farms has been toward larger tanks. Twenty years ago a 100,000-barrel tank was regarded as a giant container, while nowadays oil men consider such tanks only medium in size. Tanks holding 250,000 to 300,000 barrels are generally used for crude storage. In some localities, notably in California, the desire for large economical storage units has led to the construction of huge underground concrete-lined reservoirs, some of which hold a million barrels or more.

Most large storage tanks have some sort of floating layer over the liquid to decrease evaporation losses. Until recently it was customary to equip crude oil and gasoline tanks with floating roofs made of steel with a special ring of sealing material on the edge. Of more recent use are plastic materials such as tiny *micro-balls* or plastic foam sheets. The lack of an air space above the liquid surface lessens the evaporation of the light fractions of crude oil and light refined liquids in the gasoline range.

Lighter liquids, including propane, butane, and ethylene, must be stored under pressure to keep them liquid. Propane and butane—together lumped into "LPG" (the initials stand for Liquefied Petroleum Gas)—were of little interest or value twenty years ago. They are now the foundations of the LPG industry, purveyor to the public of convenient liquid fuel that burns as a gas. Above-ground LPG tankage costs twenty times as much as tankage for the less volatile products, so most LPG storage (over 80 million barrels of it) is now underground in caverns dissolved in salt domes or in natural rocksalt beds, and in other caverns mined in shales.

Underground storage is not limited to storage for LPG. Near Wind Gap in Pennsylvania an abandoned quarry serves as a fuel oil tank, and, in Sweden, fuel oil and other products are stored in abandoned mines.

Oil in Storage

Tank farms are reminders of the fact that not all oil produced is refined immediately. A pipeline must be full of oil in order to operate. The oil travels about five miles an hour through a trunk line, so it is in the line for some time. The oil required to fill all the lines in the

United States is about 70 million barrels. In addition to oil in the lines, working stocks must be kept on hand at terminals, trans-shipment points, and refineries. When an owner is unable to find a market at a price he is willing to accept, the surplus oil is sent into tank farms. All of these items taken together constitute *oil in storage*.

The Bureau of Mines has estimated the refinable crude oil in storage in the United States in 1960 at 240 million barrels, of which part was on leases, some was in tank farms, 25 per cent was enroute in pipelines, and the remainder was at refineries.

Oil in storage, however, is not a pipeline matter, with the exception of having enough oil for working stocks and filling the lines. The pipeline's job is to gather and deliver oil, not to store it.

The Pump Stations

The trunk line pump station to which the gathering system delivers the oil is equipped with powerful pumps for propelling the oil through the line. If the line is long, similar pump stations will be located at intervals along it, usually about 100 miles apart in level country and closer together where the line crosses mountains.

The pumps are generally of the centrifugal type, but some reciprocating pumps are in use in places for special viscous oils. Pumps are driven by electric motors or by diesel or steam engines or gas turbines. The main advantage of electric motors and gas turbines is their suitability for remote control operations.

One of the largest pump stations in the world is equipped with a 5,000-horsepower gas turbine. It drives a centrifugal pump at 5,000–6,000 revolutions per minute and puts 450,000 barrels per day through the Trans-Arabian pipeline from the Persian Gulf fields to ports on the Mediterranean Sea.

Most of the pump stations in the United States use pumps capable of pressures of 2,000 pounds or more, but the actual working pressures are seldom more than 1,200. A pump station may cost from $25,000 to almost $1,000,000.

A few short lines are so fortunate as to have their upstream ends considerably higher than their outlets so that they flow by gravity, without pumping. This is the pipeline man's dream—one that seldom comes true. Nevertheless, one of the largest fields in Iran, Agah-Jari, empties its whole production into a sea terminal on the Persian Gulf by gravity flow through several pipelines.

The Trunk Line

The trunk line is to a pipeline system what the main line is to a railroad, with the gathering system comparable to the spurs and sidings.

Have you ever watched a water main being laid beneath a city street or alley? Imagine such a line running cross-country for ten or a hundred or a thousand miles and you will have a mental picture of a pipeline. The pipe is steel and may be anywhere from four to forty inches in diameter. It comes in lengths or joints averaging either twenty or forty feet long, twenty-foot joints being used in short connecting lines and in extremely rough country, and forty-foot joints on practically all other main lines. The laying of such a line is worth describing.

Surveying and Clearing

The line is first located and surveyed, a route being picked that so far as possible avoids high divides, rough ground, swamps, and solid rock, though many a line must cross mountains or badlands or marsh or jungle. Aerial photography may help in the choice of the route, but it may not eliminate the necessity for careful instrumental surveys. When the route has been chosen and surveyed, a *right-of-way gang* clears off trees, brush, and other obstructions from the path along which the line is to be built, doing the job so thoroughly that trucks may be driven to almost any point along the way. The one thing that the right-of-way gang is careful not to remove is the survey stakes.

Stringing

When the right-of-way has been cleared, heavy trucks loaded with pipe are driven alongside the survey stakes, and joints of pipe are laid-off as they go. Watching this *stringing,* you may think the joints are thrown off carelessly and haphazardly, but when the joints are put together in a finished line you will notice that almost exactly the right number of joints have been strung and that each joint is within a few feet of where it will be needed. Sometimes two or three joints of pipe are welded together at mobile stations before they have been loaded on special trailers capable of carrying such long loads. They are then *strung* in the usual way.

Laying

Laying the line, which means putting the joints together into a continuous line of pipe, may be done by welding or by screwing them together with a coupling or collar in the same manner as the casing in wells. The newest generation contends that a welded line is better than any screwed line, whether laid by hand or by machine. In a screwed line, the collars are the weakest points, whereas in a welded line the pipe has no collars or threads and the welds are as strong as the joints of pipe between them. Welding, moreover, is both faster and cheaper than screwing. The best proof of its superiority under most conditions is the fact that nearly all new lines are welded.

WELDING A PIPELINE

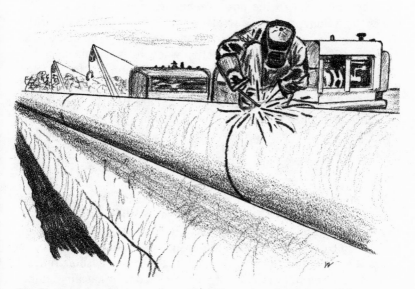

Pipelines used to be screwed together like casing, but nearly all lines nowadays are welded.

Welding

Either oxy-acetylene or electric welding may be used, but electric welding is by far the more common method. The electricity is furnished by gasoline or diesel-driven generators mounted on trucks or trailers or caterpillar tractors that travel along beside the line. Recently, automatic welding machines have increased in use. The conservatives employ them only in the mobile workshops where two or three joints of pipe are welded together before being strung, but many pipelines (primarily 10 inches and smaller) are welded automatically in the field. The welding unit is suspended from a side-boom tractor and is attached to the ends of the two joints to be welded together. Then the unit performs the welding automatically by fusing several layers of molten metal in the gap between the joints.

Where the line is long and the terrain is not too difficult, the job is done by the "firing-line" method, in which the work may be strung out over several miles. About two hundred feet of pipe is placed on low skids on which are rollers known as *dollies*. The joints are lined up and fastened with just enough weld to hold them together by tack welders, who then move on to the next two-hundred-foot section. Presently comes the welding crew, one welder for each junction or weld of the section. Men with tongs slowly rotate the pipe on the dollies, and as it turns, each welder welds the junction where he is stationed.

After the welding crew has moved on, the two-hundred-foot section is picked up by a group of tractors and is lowered into the ditch. A *bell hole* is dug where the section is to be joined to the line already laid, deep enough to permit access to the entire circumference of the pipe, and a third welding crew then welds the section to the line already laid.

Coating and Wrapping

Most lines are coated with a protective paint—usually of bituminous base—and then coated with glass wool and plastic or asbestos tape. Another coat of asphaltic enamel may be poured over all the wrappings, then covered with felt. Without such precautions, a line laid in corrosive soils would be eaten through in a surprisingly short time.

All the coating and wrapping of pipelines is done by machines riding the pipe or suspended from side-boom tractors; they move along the line after the welding is completed. A whole caravan of tractors is used where the line is welded automatically. The first tractor holds the joint in position, the second one carries the automatic welding machine, the third one the wrapping machine, and the fourth one holds the line off the ground. At times, half a dozen tractors or more are operating along the *firing line*.

Digging the Ditch

In normal soils a ditching machine, employing a large wheel with cutting edges, digs a ditch from 2 to 6 feet deep and about twice as wide as the diameter of the pipe. In country that is too rocky or rough, a crane with a dragline may dig the ditch, but in extremely rocky ground, blasting the ditch with the help of compressors and explosives may be necessary. The dirt from the ditch is piled in a neat ridge across from the side where the pipe is strung. The survey crew has, of course, done its best to stake the line where it will avoid the tough digging and blasting.

Lowering In

In the cool of some later morning, probably the next one, the pipe is *lowered in*. The lowering-in gang is up before the sun. On small jobs the pipe may be pushed and pried over onto wooden skids lying across the ditch; then levers, windlasses, or chain hoists, hanging from tripods or from wooden horses, raise a stretch of the pipe. The skids are removed and the pipe is lowered carefully into the ditch. On jobs of any size the pipe is picked up and lowered into the ditch by tractor cranes, sometimes immediately after the coating and wrapping.

Even in the cool morning the line probably has so much slack that it has to be crowded and jammed. A bystander, if one is there at such an unearthly hour, is likely to wonder how the ditch can be made to hold so much pipe and why so much slack is laid. When the job is done, however, he sees that it is all in, and if he could see the line when cold oil is in transit, he would probably see that all the slack is out and the line is smooth and straight. The foreman of the laying gang, from long experience and without the aid of expansion tables, has allowed correctly for the change in temperature and the resulting contraction.

Back-Filling

Next come the back-fillers, whose business it is to refill the ditch with the dirt previously taken from it. On small jobs it may be a hand gang, doing the job with shovels, or a man with a tractor and a flat wooden scraper. On larger jobs the gang will be equipped with a *bulldozer*, a large scraper blade on the front of a caterpillar tractor, or a drag-line scraper.

Whatever form of back-filling is used, the crew follows two rules: First, all rocks must be discarded, for one of the functions of the filling is to prevent water from seeping into the ditch. Loose rocks do not exclude water, and the fill must not consist of sharp objects which may damage the protective paint and wrapping. Secondly, the fill must be built up higher than the ground level so that when it settles there will be no depression in which water will collect and seep down onto the pipe.

PIPELINE READY TO LOWER IN

This line has been welded and is ready to lower into the ditch.

Water Crossings

Crossing swampy grounds, rivers, or even stretches of lakes and seas is small hindrance to the modern pipeliner; there are many submarine lines in operation today. The pipe joints are carried on a special craft equipped with machines for welding and coating the pipe with cement to weight it and to protect against corrosion. The pipe is laid to rest on the bottom of the sea or lake gradually as welding and cement-coating operations progress. If the crossing is short, the pipe may be welded on shore, then rolled to the water by tractors with the ends of the pipe closed to permit floating it to its final location. Then the two plugs at the ends are removed and the pipe is submerged to the bottom of the river or sea, as the case may be. It is up to the pipeline engineer to determine whether it is preferable to cross a gully or river by means of a special bridge holding the pipe in suspension or to bury the pipe and submerge it under water.

Testing and Clearing the Line

When the line is finished it is given a pressure test. Water or compressed air is pumped into it at the pump station, the other end is closed, and the pumps build up a pressure considerably higher than the line is likely to have to withstand again. The pressure is maintained for a day or two, while men patrol the line to look for leaks, which show up as damp spots or trickles or washouts on the *back-fill* if water is being used, or as sizzles through the back-fill if air is being used. When a leak is found, a repair gang digs down to the pipe and fixes it. Usually it is small and a few additional inches of weld will cure it, but now and then a joint bursts. When this happens the pumps must be shut down while a new joint is put in.

Before the water or air is put into the line for testing, a *go-devil* or *pig* is inserted, a contraption of rubber and steel rollers and steel springs that fills the pipe but is flexible enough to round any curve on the line or to pass any minor irregularity on the inside of the pipe. The water or air pushes the pig and the pig pushes along any loose objects that may be in the pipe. The pig makes a slight scraping noise and a man with good hearing walks the line and keeps it company. But sometimes pigs do get lost in pipelines, and it is becoming customary to insert inside them capsules loaded with radioactive isotopes. The line walker then carries a Geiger counter; if the pig stops, he knows that a major obstruction has been encountered and he calls the repair gang to dig up the line and remove it. When the pig reaches the far end of the line a special trap is opened to catch it. Out comes the pig, usually preceded, despite much care to keep the line clean, by an assortment of sticks, brush, and rabbits.

After the line is in operation, it is good practice to put through a pig every now and then; pigging clears out paraffin or sludge that may have collected.

Pipelines may go over or under streams and gullies. The pipeline engineer decides whether to use a suspension bridge, similar to the one shown here, or bury the pipe.

Moving the Oil

When the line has been tested it is filled with oil. The pumps in the pumping stations draw the oil from tanks by suction and gravity and push it forward into the line. To avoid mishaps and assure co-ordination of operations, the pump stations communicate by radio, and usually they are also linked by teletype. Only quick communication can assure immediate stoppage of pumping when a leak or other trouble is discovered. Most pipeline leaks are detected by undue pressure drops at the pump stations.

Pump station control has become so precise that many stations are operated without human supervision by remote control from other pumping stations; gauges and meters are read and pumps and engines are switched or stopped by electronic remote control.

Heaters

Although not all oils are equally viscous, they are more viscous (less fluid) when cold than when hot. The more viscous an oil is, the harder it is to move it through a pipeline. In cool climates, therefore, some pipelines are equipped with heaters by which the oil is heated without interrupting its passage through the line. In most heaters the oil passes through a horizontal chamber two or three times the diameter of the pipeline. There it is in contact with numerous small, longitudinal pipes containing steam or hot flue gases. In the California type, the oil passes through tubes in a heated chamber. Heaters are placed at whatever intervals may be necessary to keep the oil warm enough for efficient operation of the pipeline, a matter that is obviously dependent on the climate, the character of the oil, and the operating pressure of the line.

Loading Racks

If the line should lead to a railroad instead of to a tank farm or refinery, as a few short lines do, there may be a *loading rack* for convenience in loading the oil into tank cars. Next time you drive through the oil country you may notice along a siding a platform eight to ten feet above the ground with a series of two-inch or three-inch pipes projecting above it. The pipes rise vertically from below the ground to several feet above the platform, turn at right angles for several feet, and terminate in shorter, downward sections. When a string of tank cars is *spotted* on the siding the pipes are swung around and lowered until the ends of the short sections are inside the domes of the tank cars, valves are opened, and the cars are filled. If cars are to be emptied instead of filled, the oil is sucked from the tank cars through the loading rack lines; the necessary suction is supplied by suction pumps.

Very few of the loading racks you see are for crude oil. Most of them are at refineries and bulk stations and are used for loading and unloading gasoline, kerosene, and other refined products.

Operation and Maintenance

Once the pipeline has been built it must be maintained and operated. Operation is somewhat similar to the operation of any other transportation enterprise: the oil is gathered and gauged and fed into the trunk line, the pumps are kept in operation, pressures are watched constantly in the pump house to detect the development of obstructions and leaks in the line, the line is walked or ridden or *flown* at frequent intervals to detect leaks, pigs are put through the line now and then to remove accumulations of dirt or paraffin, heaters are started up or shut down with seasonal changes in temperature, and nothing is overlooked that will help keep the oil moving smoothly, continuously, and efficiently.

Maintenance presents an occasional special problem over and above the routine matters involved in the maintenance of any operating equipment. The chief of such problems is line corrosion; despite all the coating and wrapping that can be done, line pipes deteriorate. The most effective preventative so far found is known as *cathodic protection*.

The theory is that the corrosive action of the chemicals in the soil on the pipes sets up slight electric currents. The electrical action and the chemical action are interdependent. By neutralizing the electrical action, the chemical corrosive action is reduced. If, therefore, electric currents are applied to the line in such amounts as to neutralize or reverse the stray currents caused by corrosion and oil friction, corrosion should not take place. Very often, to induce the current an *anode*, usually made of magnesium, is wired to the pipeline and buried in the ground. The anode, instead of the steel pipe, will take the brunt of the corrosion and will be gradually eaten away by the electric currents.

Cathodic protection is old hat now; under most conditions it works most satisfactorily. In fact there are few pipelines nowadays which are not equipped with cathodic protection besides their bituminous coatings and careful wrapping.

Product Pipelines

There are 45 thousand miles of refined-product pipelines in the United States, handling 20 per cent of the product traffic. Most of the product pipelines are from refineries to marketing centers, but some of them originate in tanker terminals. The construction of a product pipeline is similar to that of a crude pipeline but the operation is much more involved. Normally, a product pipeline transports several products of different specifications and characteristics. As a matter of fact, most product lines transport aviation gasoline, several grades of motor gasoline, jet fuel, kerosene and gas-oil, and other distillates in succession. Great care must be taken to avoid contamination of one refined product with another. This is done by maintaining *turbulent flow* in the pipeline all the time. If the flow of oil in a pipeline is slow, or *linear*, there will be a substantial contamination of products moving in succession inside it. Paradoxically, higher rates of flow generate turbulence, and mixing becomes insignificant in turbulent flow. Sometimes pigs are used to help maintain the separation between the different product batches in a pipeline.

The art of successful operation of a product pipeline depends on clever dispatching as well as on the maintenance of turbulent flow, and the pipeline dispatcher's responsibilities and ulcer tendencies are as great as those of his counterpart on a railroad.

What Pipelines Do

Now our line is completed, tested, and in operation; let's see what the pipelines in the United States do. There are about 160,000 miles of them taking oil from 600,000 wells in 32 states and delivering it to 298 active refineries in 40 states. In 1960, they transported 5,600,000 barrels a day, enough to fill a train of tank cars extending from Washington, D.C., to Philadelphia, or enough in one year to fill a river 100 feet wide, 2 feet deep, and as long as the Mississippi. They carried over 75 per cent of all crude oil transported to refineries. The oil carried during the year was worth more than four billion dollars at the refinery price. The amount invested in carrier property by more than 80 pipeline companies engaged in interstate commerce was nearly 3 billion dollars. Transporting oil by pipeline is no puny business.

And if you think pipelines are tame cats sticking close to the North American hearthstone, take a trip to such places as Colombia or Venezuela or down the west coast to Ecuador and Peru and around the

Horn to Argentina, or to ancient lands like Persia and Mesopotamia, now Iran and Iraq, or into Russia, or east to Burma and Borneo and Sumatra. You will find oil moving through pipelines from the Andean foothills to the Caribbean Sea, from the Caspian Sea to the Black Sea in the land of the Soviets, from the Tigris, below the ruins of Nineveh, to the Mediterranean in ancient Phoenicia, and into Rangoon along the road to Mandalay. White men and black men, red men and yellow, Arabs and Caribs, peons and mujiks, head-hunters and Patagonians, horses and camels and mules and elephants and llamas—all have borne a part in laying the world's pipeline systems.

Tank Cars

Extensive as they are and as much oil as they move, pipelines by no means handle all of the oil produced. Great volumes of crude, and nearly all refined products, are moved by tank car, tank truck, or tanker.

You have seen plenty of tank cars being hauled in freight trains, oftentimes solid trains of them, or spotted on sidings in every railroad town and village. They are the huge black or aluminum-colored steel cylinders, with a little dome protruding from the top, that you may have wondered about. Most of them are filled with gasoline, kerosene, or some other refined product, but many of them are taking crude from the well to the refinery.

Some tank cars are provided with steam coils so that heavy oil may be heated to make it flow more easily when the car is to be emptied. Some are insulated to prevent heating and evaporation. Some of those used for refined products are divided into compartments so that more than one product may be transported in them. All of them have the dome on top, resembling the steam dome of a locomotive, in which gas may collect. They are loaded through this dome, the cover of which is removable. All, of course, have outlet pipes and valves for unloading.

Tank Trucks

We won't spend much time discussing trucks; you have probably wasted plenty of time trying to pass them on the highway. Their service in transporting crude oil is mainly in serving fields and wells not yet reached by pipeline, but they play a major role in the distribution of refined products.

Did you ever wonder why nearly every tank truck has a chain dan-

Crude oil and refined products are loaded into or unloaded from tank cars at loading racks. You can see one or more loading racks in almost any town or village.

gling, with the end dragging on the pavement? Its purpose is to discharge the static electricity that might otherwise collect on the tank and cause an explosion.

Barges

Barges have been hauling oil since 5 years after the Drake well came in. In 1864, 3,000 barrels daily went down the Allegheny to Pittsburgh. Barges were wooden and so were most of the paddle-wheeled steamships that pushed them.

Inland and coastal water movement of crude oil and its products amounted to 128 million short tons in 1957. More than 15,000 vessels were engaged in the trade. Towboats move barges loaded with 175,000 barrels of oil to markets and refineries on the Mississippi-Ohio river system, on the Great Lakes, and in ports on the Atlantic. A modern towboat is all steel and efficiency. It may have 5,000 horsepower, and it pulls rather than pushes its load of barges.

Tankers

The pipeline's great competitor in the transportation of oil is the oil tanker. The cost of transporting a ton of oil per mile is much less by tanker than by pipeline. It is not surprising, therefore, that if the sea distance between two points is not much greater than the land distance, tanker transportation is preferred to a pipeline. True, there are geographical circumstances which allow pipelines to beat sea transportation in places. The land distance between the oil fields of Saudi Arabia and the Mediterranean Shore in Lebanon, for example, is about 1,100 miles, while connecting these two points by tanker involves a 7,000-mile round trip; the building of the Trans-Arabian pipeline was therefore considered economical. Nevertheless, in most instances tanker transportation is better all around and it is used almost everywhere it can be. Mid-Eastern oil is transported even to European markets by tankers.

More than 44 per cent of all water-borne cargo in the world today is made up of petroleum and petroleum products. Venezuelan crude goes by water to New York and Montreal, London and Liverpool, and most of Western Europe and South America. American oil goes by water from Houston to New York and from Los Angeles to San Francisco. Petroleum and its products account for over 37 per cent of all freight traffic handled in the seaports of the United States.

Tanker Construction

The average tanker size exceeds by far that of the average dry-cargo ship; the biggest ships afloat are tankers. Gone are the days when transatlantic passenger ships like the *Queen Mary* or *Queen Elizabeth* held the heavyweight championship of the sea. No other ship afloat exceeds the dead-weight tonnage of the giant tankers.

Like many other objects in the oil industry, tanker size has grown immensely during the last two decades. Just after World War II, ships over 20,000 tons were considered *super tankers*, but soon the *giant* label was applied only to those of 30,000 tons or more dead weight. After the Suez Crisis in 1957 giant tankers were considered those in the 45,000–65,000-ton class; today there are tankers bigger than 130,-000 tons in service, and even bigger nuclear-powered tankers are being planned on the drafting boards.

The main advantage of bigger tankers lies in the fact that the size of the crew does not grow proportionately with tonnage. By the way, "dead-weight tons" does not mean the weight of the ship empty; it is the amount of oil the tanker can carry. It is the *dead weight* that earns the ship its livelihood.

But let's return to tanker construction. Most modern shipyards can build seaworthy tankers if they have large enough berths. Usually the pumps and the turbines are of standard designs and come from the same manufacturers, regardless of the country of construction. For example, one can find Babcock & Wilcox or Foster Wheeler boilers in tankers built in the United States, Germany, and Japan. Similarly, De Lavalle turbines and Burgmeister & Wayne diesels are in Scandinavian and French ships. Typical tanker construction places the engines aft and the bridge forward. Most of the deck space is covered with hatches which are the superstructure openings for the oil tanks below. All tankers have separate compartments for storing the oil—usually between 15 and 30 of them. Since all the deck space is free, there is plenty of room for spacious crew accommodations. There are very few tankers where there are more than two seamen to a cabin, and generally officers as well as able-bodied seamen have separate rooms for themselves.

All modern tankers are fully air-conditioned and most of them are equipped with swimming pools. This luxurious accommodation for the crew is designed to help offset the tedium of the lengthy trips and the short leaves ashore; the average tanker spends about 320 days per year at sea.

Tanker Operations

Tankers are unique among cargo ships in other ways, particularly since their operation includes extremely short stays in ports. Literally speaking, many tankers do not touch ports at all but load offshore through submarine lines at distances from land that allow for a sufficient depth of water to accommodate the big ships. At the ends of the steel submarine pipelines, flexible rubber hoses are connected to buoys. The buoys, equipped with hooks at the tops, and anchored to the sea floor by heavy chains and anchors, provide the berths for the tankers. The tanker is lashed in position with heavy ropes and steel cables to the hooks of the buoys at sea.

Of course, in many instances, tankers are brought inside harbors and docked at piers like other ships. Then the flexible connections are moved to the ship's pipe-manifold by means of special hoists.

Whether at sea or at the pier, the ship is loaded by shore pumps; to unload, pumps on the tanker discharge the oil. The pumping operations for either loading or discharge usually do not exceed two days, and at times may last even less than 24 hours. During loading and discharge operations, the crew is usually kept busy in berthing the ship and in operating the discharge and loading facilities on board.

We can see that shore leaves are brief for members of the crew, and on many occasions no shore leaves are granted at all. No wonder that tanker crews are not as happy a lot as most other sailors are; their lives are mostly work with little play. Only characters accustomed to the isolated life at sea withstand the rigors of working on a tanker.

So Much for Transportation

So much for transportation. In studying the facilities for transporting crude from where it is produced to where it is processed, we have leap-frogged ahead somewhat and let ourselves discuss the transportation of refined products. Let's go back to the end of our pipeline and see how our crude oil is transformed into the products we have been taking so gaily from Bayonne to Bombay. Take a deep breath, for when you enter the refinery, you enter the domain of the chemical engineer. He's a thoroughgoing scientist, and we may have trouble persuading him to talk understandable English.

Crude and refined products are carried across the Seven Seas and on many inland waters by tankers. Tankers constitute a third of the American merchant marine tonnage.

CHAPTER 11

Refining the Oil—The Chemistry of It

Refineries in General

THE refiner is a manufacturer. His raw material is crude oil. His finished products are gasoline, kerosene, fuel oil, lubricants, asphalt, petroleum jelly, ethers, alcohols, and more than a thousand others. No refiner makes all of these products; some make only a few, others make many.

Because of the wide differences in crude oils, no two refineries are alike and no two refineries employ exactly the same combination of processing methods. Some refineries are very complex; others are simple. The site of a refinery and the number of processes it uses depend on many things—the amounts and types of crude oil available to it, the kinds and variety of products needed to supply its market, and the prices its competitors charge for the same products.

Since refining is a competitive business each refinery must be carefully designed for the crudes to be processed. A large plant may process different oils in different sections, each section producing a particular product or several similar ones.

In designing a plant, one of the first things the engineer must consider is the character of his crude. To get an idea of his problem let's review briefly the chemistry of petroleum.

Organic Substances

All substances that have or have had life, whether animal or vegetable, are known as organic substances. So are certain mineral substances such as coal and oil that are derived from previous living substances. All organic substances are made up chiefly of carbon and hydrogen. Hydrogen is the very light and highly flammable gas that was once used to inflate balloons and dirigibles. Carbon is a solid at ordinary temperatures, and is known to some of us in the form of charcoal, to most of us in the form of the graphite in our lead pencils, and to a few folks in the form of diamonds.

Organic substances composed almost wholly of hydrogen and carbon, with or without an atom or two or three of oxygen, nitrogen, or sulphur, are called hydrocarbons.

Elements, Compounds, and Mixtures

Both carbon and hydrogen are elements; that is, they are simple homogeneous substances that cannot, by ordinary methods, be broken down or decomposed into simpler substances.[1] Every element is made up of atoms, each of which is identical in character with every other atom of that element, but is different from an atom of any other element.

Atoms combine to form molecules, and the character of a molecule depends on what atoms combine to form it. If a molecule is composed solely of the atoms of a single element, the molecule has the same characteristics as the atoms, and any amount of the element has the same characteristics as any molecule. The character of gold is the same, for example, whether you consider an atom, a molecule, a pound, or a ton of it.

But what if atoms of one element combine with those of another element? The result will be a new substance, the molecules of which will all be essentially alike, but will be unlike those of either of the constituent elements. The resulting substance is known as a chemical compound, and its chemical composition is fixed and uniform, which is only another way of saying that its molecules are always the same and that each molecule contains the same number of atoms of each constituent element.

One thing should be noted: The character of a chemical compound may be radically different from that of its constituent elements. Many examples come to mind. Sodium is a highly combustible metal,

---

[1] By breaking down atoms into their components certain elements may be transmuted into different elements, but this goes beyond the needs of our present discussion.

chlorine is a poisonous gas, but an atom of sodium and an atom of chlorine form the molecule of the salt that seasons our food. Hydrogen is a gas that burns so readily that passenger balloons no longer dare use it, and oxygen is a gas that accelerates and supports the burning of other substances. When hydrogen burns, two atoms of hydrogen combine with one of oxygen and the resulting molecule is water; two gases that are capable of making one of the fiercest fires known unite into a liquid that puts out fires. Oxygen is the beneficial ingredient of the air you breathe; carbon is a harmless combustible solid; but burn carbon with a limited amount of oxygen and you get carbon monoxide, deadly exhaust gas that kills people in their garages. So it goes. We live or die because the atoms of dangerous or beneficent elements combine to form molecules of beneficent or dangerous compounds. Much of man's progress is due to his ability to break chemical compounds into their elements, or to take the elements themselves as they occur in nature, and then to combine the elements into useful compounds.

Suppose now we mix two chemical compounds. One of two things will happen. The two compounds may react on each other, breaking down their molecules and forming new and different molecules, in which case we have a new and different chemical compound. Or the two compounds may not react on one another, in which case each compound remains unchanged and the result is merely a mixture, the character of which will depend on the relative amounts of the two compounds in it.

Every substance in the world belongs to one of the three classes we have named; it is either an element, a compound, or a mixture. There are no others.

Another point should be noted, for most of the chemistry of oil is wrapped up in it. The atoms of the same two elements do not always combine in the same number to give the same result. If they combine in one proportion the result is a certain compound, but if they combine in a different proportion an entirely different compound will result. The addition of more oxygen to deadly carbon monoxide forms carbon dioxide, a compound necessary for plant growth and hence essential to food production.

Chemical compounds composed of atoms of the same elements, in different proportions in each compound, may exhibit certain common chemical characteristics, and their physical characteristics may vary progressively as the number of atoms of one element increases in proportion to that of the other. Such a group of compounds is called a chemical series, and one of the best examples is the paraffin series, which constitutes a large part of many oils—and of which much more, anon.

Symbols and Formulae

To simplify his work, the chemist uses symbols for the elements and combines the symbols into formulae to show the constitution of chemical compounds. Every chemical element has a symbol, in most cases the initial of its name, in some cases the initial of its Latin name, and in some cases the initial followed by a small letter. Hydrogen's symbol is H, oxygen's O, carbon's C, nitrogen's N, sulphur's S, lead's Pb, and so on.

The formula for a compound is a statement in chemical symbols of the atoms that compose its molecules. Thus water, with two atoms of hydrogen and one atom of oxygen, is H_2O, and carbon dioxide, with one atom of carbon and two of oxygen, is CO_2. These are often written graphically as H-O-H and O=C=O.

What Is Oil?

Now back to our oil.

Oil is not an element; far from it. It is not a definite chemical compound; if it were, it would be of uniform character—which it decidedly is not. It must, therefore, be a mixture of chemical compounds. Moreover, each oil is a different mixture. Some oils contain only a few chemical compounds, other oils may contain millions of them. Those that contain the same compounds may contain them in different proportions. The number of possible combinations of compounds is almost limitless—and the multitudinous oils of the world use a large number of the possibilities.

Each of these mixtures is homogeneous and relatively stable at ordinary temperatures. If oil is stored in a sealed tank, the heavier compounds do *not* all settle to the bottom leaving the lighter compounds on top; as a rule the oil will remain the same throughout the tank. The reason is, the heavier compounds are dissolved in the lighter ones. The mixture is not a simple mechanical mixture; it is a solution.

Perhaps the quickest knowledge of these various mixtures can be gained by studying some of the compounds that go into them.

The Paraffin Series

Chief among the petroleum compounds are those known as the paraffin series. At the light end is the gas known as *methane* or *marsh gas*, the molecule of which is made up of one atom of carbon and four of hydrogen (CH_4). Toward the heavy end of the series is a solid called "deca-contane"; its molecule contains 100 atoms of carbon and two hundred and two atoms of hydrogen ($C_{100}H_{202}$). Between the two are a host of compounds—gaseous, liquid, and solid.

$$H-\overset{\displaystyle H}{\underset{\displaystyle H}{C}}-H$$

A chemist's diagram of methane

Methane and the gas next heavier, ethane (C_2H_6), are *dry* gases; they are never liquids at ordinary temperatures and pressures. Next heavier are propane (C_3H_8) and butane (C_4H_{10}), which are liquefied under moderate pressure, placed in containers, and sold the country over to burn in stoves and domestic lighting plants, and in small power units, buses, tractors, and the like.

$$H-\overset{\displaystyle H}{\underset{\displaystyle H}{C}}-\overset{\displaystyle H}{\underset{\displaystyle H}{C}}-H$$

ethane

Next in order of increasing heaviness come the gasoline hydrocarbons, pentane (C_5H_{12}), hexane (C_6H_{14}), heptane (C_7H_{16}), octane (C_8H_{18}), nonane (C_9H_{20}), decane ($C_{10}H_{22}$), and undecane ($C_{11}H_{24}$). Some butane is also included in the gasoline group, held in its liquid phase by the heavier members of the group. The gasoline hydrocarbons are liquids at ordinary temperatures, but

$$H-\overset{\displaystyle H}{\underset{\displaystyle H}{C}}-\overset{\displaystyle H}{\underset{\displaystyle H}{C}}-\overset{\displaystyle H}{\underset{\displaystyle H}{C}}-\overset{\displaystyle H}{\underset{\displaystyle H}{C}}-H$$

n-butane

they boil, which means become gases, at comparatively slight increases in temperature. Butane and pentane boil first and the succeedingly heavier compounds boil at successively higher temperatures. All of the gasoline hydrocarbons are highly flammable and when mixed with the proper amounts of air and put under pressure are highly explosive. That's what makes your car go.

iso-butane

There are even heavier compounds than those in gasoline. At the light end are the compounds included in kerosene or *light distillate*, and in the successively heavier portions are diesel fuel, furnace oils, range oils, lubricating oils, and *heavy fuel oils*. All of these compounds burn, but from the lighter ones to the heavier ones, they are less and less flammable. Nearly all of the heavier portions contain paraffin waxes that can be separated out and solidified.

Finally comes the solid end of the series. Some of these compounds are fairly soft and melt at moderate temperatures; others are hard and will melt only under high heat. About the heaviest and hardest so far recognized is decacontane with its hundred atoms of carbon and two hundred two of hydrogen.

Do you know how to recognize a paraffin molecule when you meet it? They all follow an infallible rule: The number of hydrogen atoms is two more than twice the number of carbon atoms. Look back at the formulae already given for methane, hexane, decacontane, and the rest; you will see in each case that if you multiply the number of carbon atoms by two and then add two, you have the number of hydrogen atoms. This is the distinguishing atomic relationship of the paraffin series. The

$$H-\underset{\underset{H}{|}}{\overset{\overset{H}{|}}{C}}-\underset{\underset{H}{|}}{\overset{\overset{H}{|}}{C}}-\underset{\underset{H}{|}}{\overset{\overset{H}{|}}{C}}-\underset{\underset{H}{|}}{\overset{\overset{H}{|}}{C}}-\underset{\underset{H}{|}}{\overset{\overset{H}{|}}{C}}-\underset{\underset{H}{|}}{\overset{\overset{H}{|}}{C}}-H$$

n-hexane

chemist expresses it by saying that the formula of the paraffin molecule is C_nH_{2n+2}. Not all of the members of the paraffin series have been found in crude oils, although many of them have been, and many others have been produced by the treatment of crude oils.

With minor exceptions, all petroleum gases so far found in nature and identified belong to the paraffin series.

Notice one thing: The solid members of the series are soluble in the lighter members, so that an oil may contain a considerable proportion of the heavier compounds and still be liquid.

The Various Asphaltic Series

Another series widely found in oils is the naphthene series. In this series the molecules are more complex than those of the paraffin series, and they include many derivatives containing one or more atoms of oxygen, sulphur, or nitrogen. The general formula for the series is C_nH_{2n}, meaning that there are twice as many hydrogen as carbon atoms in each molecule, but not all the compounds of the series follow the formula, and even those that do are by no means so simple as the formula would indicate. On the contrary, the atoms of this series are hooked together in rings in all sorts of odd fashions. The series is so complex that chemists are far from having worked it all out.

methyl cyclo-butane

The lightest naphthene so far identified in nature is methyl cyclo-butane, a gas reported in oil from the Balakhani field, Russia, but about a dozen lighter gases are produced synthetically in the refining of naphthenic oils. Most of the naphthenes found in crude oil are liquids or solids, and they increase in abundance from the gasoline range to heavy solid asphalts.

A closely related and even more complex (but less common) series is the naphthalene series. Its molecules make those of the naphthene group look simple. All of its members so far found in oils are solids. Another series, the aromatic or benzene[2] series C_nH_{2n-6}, with equally complex molecules, is also present in some oils, mostly in the lighter liquid portions; and still another series, the olefin series, which has the same C_nH_{2n} formula as the naphthenes, is found in some crudes and more frequently in certain refinery products. Its chief distinction is that its complex molecules are *unsaturated*, which means capable of taking in more hydrogen atoms.

benzene

Other series closely related to these—diolefins, acetylenes, terpenes, anthracenes, and others—have not been found in crude oils but some of their members have been produced in oil refining and it is probable that most of them could be produced from crude oil if there were reason to do so.

hexene (olefin)

[2] Note the spelling, and don't confuse it with *benzine*, which is a rather nondescript catch-all term for gasoline and the lighter compounds related to kerosene.

The various asphaltic series abound in snappy names. The lightest member now known is cyclopropane, which is a complex molecule that adds up to C_3H_6; one of the heavy members is diocaryophyllane ($C_{30}H_{50}$). If you don't want to pronounce diocaryophyllane, try a somewhat lighter member of the series called piceneikosihydride ($C_{22}H_{34}$).

H   H
 \ /
  C
 / \
H—C———C—H
  | |
  H   H

cyclopropane

Paraffin, Asphalt, and Mixed-Base Oils

The U. S. Bureau of Mines petroleum-base classification defines nine bases according to the proportions of paraffin and naphthene compounds present in an oil. The bases run from *Paraffin, Paraffin Intermediate, Intermediate,* et cetera, to *Naphthene.* The more general terms, *paraffin-base, asphalt-base,* and *mixed-base,* are used more often, however. When an oil man calls an oil asphaltic, he implies that its naphthenic compounds predominate and that the heavy paraffin content is small or absent, although light paraffins are probably present.

An oil composed wholly of members of the paraffin series would be called a pure paraffin-base oil. Such an oil might be a gas (we are still using the term oil in its broad family sense) or it might be solid paraffin or it might be almost anything between, depending on the relative proportions of the lighter and heavier constituents present. Even though minor amounts of asphaltic constituents were present, oil men would still call it a paraffin-base oil.

A pure asphalt-base oil would be composed wholly of members of the naphthene, naphthalene, benzene, olefin, and related series, with no member of the paraffin series present. Such oils are not common, but oils in which the heavier constituents are nearly all members of the naphthene and related series are called asphalt-base oils, even though the gaseous and some of the lighter liquid constituents and a minor part of the heavier constituents belong to the paraffin series. If the heavy members of the paraffin series predominate in the oil, it is known in the industry as a mixed-base crude.

Naphthenes make up by far the largest part of asphaltic oils, with members of the other series present in comparatively few oils and in only minor proportions. Logically, refiners use the terms *asphaltic* and *naphthenic* interchangeably. The term asphalt is applied to dark-colored hydrocarbon solids of variable hardness that are comparatively non-volatile, that contain little or no crystalline paraffins, and that are often associated with mineral matter. It follows that many asphalts contain compounds of nearly all of the hydrocarbon series and that many asphalt-base crudes do also. The composition of a typical asphalt-base oil might be: Dry and wet gases of the paraffin series, the gasoline fractions being mostly paraffinic but with some aromatic compounds, the kerosene fractions about equally divided between paraffinic and naphthenic members, the gas oil, the furnace and heavy fuel oils mainly naphthenic, and the solid constituents almost wholly naphthenic, with only minor amounts of paraffin waxes.

Remember that in the asphaltic series as in the paraffin series the heavier constituents are soluble in the lighter ones, so that what by itself would be a solid asphalt might be in liquid form in an oil. Many of the heavier members of the asphaltic series are also soluble in the lighter members of the paraffin series, and paraffinic and asphaltic constituents are often found in intersolution.

Black Oils, Green Oils, Sweet Oils, and Sour

Often asphalt-base oils are called black oils and the paraffin-base oils called green oils due to their colors. Pennsylvania crudes have been famous for a century because of their green hue and high lube-oil contents. Oils with high sulphur content are *sour oils*; most of them smell unmistakably like rotten eggs because of their hydrogen sulfide. Oils with little or no sulphur are *sweet*. The green oils are usually sweet, the black, sour.

So Many Oils

After studying the various series of petroleum compounds it is easy to see how there can be so many different kinds of oil. Hundreds of different hydrocarbons—gaseous, liquid, and solid, paraffinic, naphthenic, and benzenic—have already been recognized in oils, and further study is continually finding more. Almost any combination of these hydrocarbons may make up an oil. You can figure for yourself how many different and differing crude oils there may be, and can understand why every refinery necessarily needs to be designed for the crude or crudes it is to process.

Vaporization and Boiling

Now let's turn to physical chemistry for a moment, and consider the everyday phenomena that we call *boiling* and *vaporization*.

The molecules of a substance are in constant motion. They oscillate through minute arcs, collide with one another, bounce back to another collision with other molecules, and so on, endlessly. The life of a molecule is an infinite number of infinitesimal oscillations and collisions.

The molecules of a solid move more slowly and through shorter distances than those of a liquid. Those of a gas move faster and farther than those of a liquid. Heat speeds up motion, whether the substance is in the solid, liquid, or gaseous state.

The constant motion of the molecules of a fluid and their impact on one another sets up a force tending to make the molecules fly apart. This is known as the fluid's *vapor pressure*. Tending to oppose the vapor pressure and to hold the molecules together are the attraction of the particles of the fluid for one another, known as cohesion, and whatever force may be imposed on the fluid from without, which may be either the pressure of the atmosphere or some other imposed pressure. When the temperature of a liquid is increased, the molecules move faster and faster and their impact on one another becomes greater and greater. Finally, when the temperature becomes great enough, the vapor pressure exceeds cohesion plus imposed pressure and the liquid *boils* or *vaporizes*, which is to say it becomes a vapor or gas.

Different liquids have different vapor pressures, which is why they have different boiling points. Obviously the liquid with the highest vapor pressure will boil or vaporize at the lowest temperature. Pentane, for example, boils at about 100 degrees, water at 212 degrees at sea level atmospheric pressure. The lightest members in a chemical series as a rule have the highest vapor pressures and therefore boil at the lowest temperatures. They are often referred to as the "low-boiling fractions" of the series.

The vapor pressure of a mixture of liquids is determined by the vapor pressure of the components of the mixture. If, for instance, a high vapor pressure liquid is mixed with one of low vapor pressure, the vapor pressure of the mixture will be somewhere between those of the two at a point determined by the proportions of the two liquids in the mixture.

Now we are ready to consider some of the principles on which refinery design is based.

Distillation

The simplest principle in refining, and the one on which all early refining was based, is this: When heat is applied to an oil, the lightest constituents of the oil boil first, the next lightest boil next, and so on. Boiling, as we have seen, is nothing more than the changing of a liquid into a vapor or gas under the influence of heat. If the vapor is cooled it will condense into a liquid again and is then called "condensate" or "distillate." The process of boiling off vapors and then condensing them is known as *distillation*. It is not peculiar to the oil industry; some people have a fondness for a certain distillate-and-soda!

If we could put a sample of mixed-base oil and gas into a tea kettle at 300 degrees below zero and then gradually warm up the kettle, the methane would vaporize and pass out the spout at about 259 degrees below; the other dry gases would vaporize by the time the melting point of ice was reached. The wet gases would successively vaporize below body temperature; the gasoline would begin to boil at about body temperature and would all be evaporated at 437 degrees. The kerosene would distill over next, followed by the gas oil, lubricants, paraffin wax oils, and fuel oils, until at last, at about 1,000 degrees, most of the asphalt would have decomposed into lighter products and evaporated. The remainder of the asphalt would have been converted into coke.

If, while our tea kettle is boiling, we should condense and collect in separate glasses the successive vapors that pass out of the spout, we would have separated the oil into *fixed* or dry gases (which we would not be able to condense), natural gasoline, gasoline, kerosene, gas oil, *lube stock*, wax distillates, fuel oils, and asphalts, and, in the bottom of the kettle, coke. The common term for such a separation is *fractionation*, and each condensate is called a "fraction" or "cut."[3]

---

[3] In such a simple boiling-off or distillation process, a sharp fractionation of the various products is not obtained because each cut brings over with it some of the lighter and heavier cuts. At about 800 degrees, moreover, *cracking* would begin to take place, and thereafter the process would not be a true distillation. Nevertheless, the description epitomizes the refining process as it was carried on years ago.

Flash Vaporization

If the light fractions are the first to vaporize when the oil is heated, they are obviously the last to condense if the entire oil is vaporized and then condensed. This principle is used in the process *flash vaporization*, in which the oil is heated to a high temperature and allowed to flash into vapor. The vapor is then cooled and the various fractions or cuts are recovered as they condense at various lower degrees of temperature. The method lends itself to more accurate fractionation, and to more efficient use of heat, than does fractionation by successive distillation.

Vacuum Distillation

We have seen that the temperature at which a liquid boils or vaporizes is lower if the pressure on the liquid is low than if the pressure is high. For example, water, which boils at 212 degrees under the atmospheric pressure of 14.7 pounds, boils at 102 degrees at a pressure of one pound. In modern refining practice, advantage is often taken of this fact and both distillation and flash vaporization are done in a partial vacuum. The method, under certain circumstances, may have one or both of two advantages: It may lower fuel cost by reducing the operating temperature, and it may preserve some of the molecules that would break down at higher temperatures.

One method of producing a partial vacuum is to introduce steam into the condensing chamber, which is sealed against the influence of outside pressures. As the steam condenses, it shrinks to a fraction of its former volume, and the reduction in volume creates a partial vacuum. Under the lowered pressure thus produced, the oil fractions flash into vapors at lower temperatures than they would at atmospheric pressure.

Steam Distillation

Steam is also used, by taking advantage of the loss of vapor pressure, to obtain a result similar to vacuum distillation but without creating a vacuum.

If two liquids of different vapor pressures are introduced into the same chamber, even though they are of such a nature that they will not form an intimate mixture, the effect on the boiling point of the low vapor pressure liquid is the same as though the two liquids constituted a true mixture. The boiling point of the mixture will be some-where between the respective boiling points of its components. Water (or steam) has a higher vapor pressure, and therefore vaporizes at a lower temperature, than most oil fractions. It follows that if steam is introduced into a chamber where oil is being vaporized, whether a still or a flash chamber, the various oil fractions will vaporize at lower temperatures than they would if the steam were not present.

This action of steam was recognized by refiners long before most of them knew the theoretical principle involved, and steam distillation has played an important part in refining practice for many years. It is in use in most refineries today.

Topping, Skimming, and Straight Run

The refining methods so far described, you will notice, are intended merely to take out of the oil the constituents already in it, without changing the character of those constituents in any material way. As a matter of fact, some chemical changes take place in the heavier con-stituents due to the high temperatures necessary for some distillations, but these are incidental and are not an object of the methods em-ployed. So far as the objectives of the operation are concerned, the resulting products are chemical compounds that were present in the crude oil before it was refined. Products so produced are called "straight-run" products, and the refining methods by which they are produced are *topping* or *skimming*.

The term *skimming* is usually confined to a refining operation in which only the lighter products are taken off, leaving all the heavier products in an unfractioned residuum. The term *topping* is sometimes used in the same restricted sense, but is often applied to complete frac-tional distillation of the whole oil except for a heavy residual fraction.

Cracking (Breaking Molecules)

For more than half a century after the oil business got under way straight-run methods were used by practically all refiners, and the gasoline recovered from a crude was confined to the gasoline com-pounds originally contained in it. Meanwhile it had been discovered that the larger and more complex molecules of the heavier compounds would break down or *crack* if subjected to enough heat.

Suppose we should again fill our tea kettle with the same oil as before, and this time should tie down the lid and seal up the spout and should heat the kettle to 1,000 degrees, and suppose (which is most improbable) that the kettle did not blow up from the great pressure that would be developed in it. If we should then let the kettle cook, open up the spout and attach it to our condensing apparatus, and should then heat up the kettle gradually as before, the percentage of the different fractions we would obtain would be markedly different from those we formerly obtained from the same oil. We would have more gas, more asphalt, and doubtless more coke.

When oils are subjected to such temperatures the groups of atoms of some of the larger molecules crack and re-form themselves into other molecules, some of which are smaller, simpler, and lighter than the original molecule and some of which are heavier and may be more complex. For example, molecules of pentadecane, which is in the kerosene range, can be cracked into molecules of octane, which is in the gasoline range, and molecules of carbon, which is coke. In addition, some fixed gases will be formed. The equation, neglecting the fixed gases, may be something like $9C_{15}H_{32} = 16C_8H_{18} + 7C$. It is obvious that by this method the gasoline yield of an oil can be increased at the expense of some of the heavier fractions. In practice, every effort is made to maintain such operating conditions that the gasoline yield will be increased as much as possible with minimum production of fixed gases and heavier fractions.

The heat and pressure necessary for cracking may be applied to an oil while it is in liquid form, which is known as *liquid-phase* cracking, or the oil may first be vaporized and then cracked, which is *vapor-phase* cracking.

In many refineries the crude is first topped of its gasoline, part or all of the remainder is then cracked, and the cracked material is then topped to recover the additional gasoline, but in practically all modern installations, the whole job of cracking and fractionation is carried out as a single operation.

Cracking is said to have been discovered accidentally, about 1860, by a stillman who forgot to turn off the fire under a still when the lighter fractions had distilled off. He discovered, to his surprise, that more light fractions were distilling over. Whether or not that story is true, numerous patents were taken out from 1860 on, but no important commercial application was made until after 1910, the main reason being that in those days the desired product was kerosene, and no refiner wanted to produce more gasoline than he had to. Gasoline was merely a nuisance that had to be disposed of, and more than one refiner was arrested for dumping it in the creek, to the danger of the folks living downstream. By 1914, however, commercial cracking installations were in use, and for the first time, the production of cracked gasoline exceeded that of straight-run gasoline.

Before cracking began, the gasoline recovery from each barrel of crude averaged about 18 per cent; today it averages about 60. If it were not for cracking, the wells of the United States would not produce enough oil to meet current gasoline requirements, and you would be paying considerably more for your gasoline than you do.

Catalytic Cracking (Breaking Molecules Better)

The effectiveness of simple cracking depended almost solely on the temperatures, pressures, and times employed, and the efficiency with which they were utilized. Just before World War II a new and revolutionary factor was introduced; refinery engineers found that greater yields of higher grade products are obtained by cracking in the presence of a *catalyst*. A catalyst is a foreign substance that affects a chemical reaction without being itself affected.

When two or more chemical compounds react on one another so that their molecules break down and the atoms rearrange themselves into new and different molecules, the character of the reaction may be modified by the presence of a catalyst. The nature of the new molecules formed may be different from that formed if the catalyst were absent, or the reaction may take place at a different temperature, or the reaction may be speeded up or retarded. Remaining unchanged itself, the catalyst may produce almost any combination of these effects. This is known as *catalytic action* and the process is called *catalysis*. It is a major industrial process.

When a certain oil is subjected to a certain temperature at a certain pressure a certain cracking reaction takes place, resulting in certain products. When the same oil is subjected to a higher temperature, all other factors being the same, different products will result. Subjecting the oil to the lower temperature in the presence of a catalyst, at the same or even a lower pressure, may give the same results as if it were subjected to the higher temperature without the presence of the catalyst.

Knowledge of the principle of catalysis is not new. In its commercial application to the cracking of oil its general effect is to lower the temperatures and pressures at which certain desired results can be obtained. With its advent, simple cracking had to be given a distinguishing name, and is now known as *thermal cracking* or *pyrolytic cracking* to distinguish it from *catalytic cracking*.

Catalytic cracking is divided into three main types. The first two, *stationary-bed* and *circulating*, are alike in passing hot oil feed stock through beds or reaction chambers of catalyst pellets. The catalyst becomes coated with carbon after use and must be rejuvenated, so each catalyst-containing vessel operates by batches—reacting with feed stock for a time, then being rejuvenated so the catalyst can be used over and over again.

In the once-through or *fluid-cracking* process, the oil feed and finely-divided catalyst are brought together, heated in a pipe still, and flowed together into the reaction chamber. Inexpensive catalysts are used once and discarded after separation from the cracked oil.

Cracking catalysts are very porous, clay-like materials. The earliest ones used were composed of spent filter clays from certain natural clays. They were so variable in composition that it was impossible to control the cracking operations properly; this led to the development of synthetic silica-alumina catalysts of controlled composition. These are very porous, hence they provide large surface areas, important in cracking.

Other catalysts may be used in cracking pure feed stocks in such processes as reforming and hydrogenation. The more important of these are metals such as molybdenum, nickel, magnesium, platinum, and zirconium.

The refining industry did the world a big favor when it developed catalytic cracking. By this process it has produced more gasoline of higher quality and at a lower price than ever before. Gasoline prices today are about the same as they were in the late 1920's despite the enormous increases in costs of labor, equipment, taxes, and the fickleness of currencies. No other commodity of wide use can boast such a record.

Polymerization (Combining Molecules)

Breaking down large molecules into smaller ones is not the only thing that heat will do; it will also cause small hydrocarbon molecules to combine into larger ones. This is *polymerization*.

In any form of refining, and particularly in cracking, fixed (or dry) gases are formed. You probably remember them; whether found in nature or formed in refining operations, they cannot be liquefied at ordinary temperatures and pressures. So long as they remain fixed gases they are of value only as fuel, and when all fuel requirements are met they are a waste product. Their molecules, you remember, are small and simple.

If such gases are subjected to appropriate heat and pressure their molecules can be induced to combine into larger ones, and gases composed of such large molecules can be condensed into liquids. It is possible, theoretically at least, to polymerize a group of light gases into heavy solid hydrocarbons, but the present use of the process is to transform refinery gases and gases from gas wells into gasoline and, with the addition of hydrogen or oxygen, to produce anesthetics, alcohols, and other products.

Knowledge of polymerization, like knowledge of catalysts, is by no means new. It has been a recognized chemical principle for many years, and some polymerization has always taken place, no doubt, in cracking operations. Its development to major commercial importance, however, is about contemporaneous with Mussolini's conquest of Ethiopia. The reasons behind its use are increased gasoline yield, utilization of gases with previously only a limited market, and the production of gasoline exceptionally high in anti-knock value, as well as of products that are new to the oil industry and that have high sale values.

Hydrogenation (Making New Molecules)

Straight-run refining separates the oil into its constituents without disturbing the structures of their molecules. Cracking and polymerization break down some of the molecules and use their atoms to build different molecules. In none of these methods are any outside atoms added. Now we come to another method, one in which new atoms are introduced.

You may have noticed from the formulae given that the heavier hydrocarbon molecules have less hydrogen in relation to carbon than do the lighter molecules. If the heavier hydrocarbons are heated to a high temperature and free hydrogen is simultaneously introduced, the atomic groupings resulting from the cracking that takes place will unite with some of the free hydrogen atoms and combine into the smaller molecules of the lighter hydrocarbons. This is *hydrogenation*.

It differs from cracking, you will notice, in that, thanks to the addition of the hydrogen, all of the heavy fraction treated is transformed into lighter fractions, with no part transformed into still heavier fractions. By this method all of a gas oil or a heavy fuel or even of a coke may be converted into gasoline. It is even possible to make more than 100 per cent of gasoline from an oil, the extra percentage being the hydrogen added.

The process had its first important development in Germany during World War I as a means of making motor fuel from coal. Plants now operating in the United States have shown that practically any desired product can be made: aviation gasoline, high-grade diesel fuel, or any desired quality of lubricants, as well as alcohols, formaldehydes, and a host of other products.

A Chemical Manufacturing Industry

It is obvious that, with the development of these newer processes, the term *refining* falls far short of telling what the modern refiner does. Once the industry took crude oil and *refined* out of it the various chemical compounds already in it. It still does a great deal of this, but more and more it takes the carbon and hydrogen atoms in the crude oil and rearranges them into the molecules of the products most desired, and even adds other atoms to increase the yield or improve the quality of the desired products or to produce new ones. Refining has become a broad and versatile chemical manufacturing business.

CHAPTER 12

Refining the Oil—Specifications and Design

The Refinery Designer's Job

THE engineer who designs a refinery starts by making a thorough study of the crude the refinery is to process, paying particular attention to the products that can be made from it. Next he compares the products with the specifications that the market demands of them and chooses the refining methods that will give the most salable products at the lowest cost. Then he designs the equipment. The embryo refinery may spend weeks or months in the laboratory and the drafting room before construction begins.

Specific Gravities

One of the first things the engineer measures, as an index to the character of the crude, is its weight. He knows, of course, that the lightest constituents of an oil are gases, the heavier ones are liquids, and the heaviest would be solids were they not in solution. The gasoline fraction is made up of the lighter liquids, and the more light liquids there are in the mixture, the lighter the mixture will be. He knows, therefore, that the lighter the oil, the more gasoline he is likely to get from it.

This rule is not absolute since the members of the asphalt series boil at higher temperatures than their counterparts of the paraffin series; also the presence or absence of a large proportion of heavy fractions affects the weight of the oil. An oil composed largely of the kerosene and gas-oil fractions, with little or no gasoline and no heavy ends, might be as light as an oil with a large gasoline cut and a large proportion of heavy asphalt. Oils containing only middle-weight constituents are exceptional, however, and for the majority of oils, weight is a good general index of gasoline content.

Weight (more accurately termed "density") may be measured in pounds per gallon or grams per liter, or in any system of weights and measures. To have a uniform standard that will avoid questions as to what system is being used, the universal practice is to use *specific gravity*. In case you have forgotten, specific gravity is the weight of a given volume of a substance compared to the weight of the same volume of water. Because most oils are lighter than water their specific gravities are less than 1.0 and are expressed in decimal fractions. To cite two California examples, the specific gravity of light crude from Ventura is 0.769, and the specific gravity of heavy crude from Huntington Beach is 0.951.

Baume and API Gravities

Decimal fractions are cumbersome and hard to remember, and to express a fairly important difference in the specific gravity of two oils might require the use of four or five decimal places. This is true with many other liquids. To overcome these difficulties a man named Beaumé devised an easier method, universally used in the oil business, for measuring and recording the relative weights of liquids.

The measuring instrument is a hydrometer similar to the one with which the garage man tests the winter mixture in your radiator. It is a slender tube with a weighted bulb on the lower end, and numbered graduations, termed degrees, on its stem. In pure water at a temperature of 60 degrees the hydrometer sinks to the 10 mark. In oil from the East Texas field it sinks to about the 40 mark, and in a good grade of gasoline it sinks to 62 or perhaps 66.

Early-day refiners designated the *gravity* of an oil in *degrees Beaumé* or just *Baume*. Later a committee of the American Petroleum Institute revised the scale slightly, and the oils are now designated by *degrees API*, or simply *API*. Oil men say, for example, that East Texas oil is 40 API, Ventura light oil 52.5 API, Huntington Beach heavy oil 17.3 API, and gasoline 50 to 70 API. A good idea of the vast difference between crude oils is given by the fact that they vary from 5 to about 70 API.[1]

Notice that the lighter the oil the higher its API or Baume gravity,

---

[1] Other American Petroleum Institute standards besides those for liquid densities are used throughout the oil industry. Hence the oil man's colloquial use of the term API to mean standard or normal or just what he expected. Thus a standard reaction or attitude or result is "strictly API" where some of the rest of us might say in the same vein "according to Hoyle," or "strictly GI."

which is topsy-turvy, of course, according to specific gravity. The lighter an oil the less it weighs and therefore the lower its weight with reference to water (specific gravity) but the higher its API number. A 5-API oil has a specific gravity of 1.037 and weighs 8.636 pounds per gallon; a 65-API oil has a specific gravity of 0.7201 and weighs 5.994 pounds per gallon.

Molecular Weights

There is a sound scientific reason why *gravity* is an index to the character of an oil. The lightest compounds in the oil are those that contain the greatest proportion of hydrogen; the heaviest are those that contain the greatest proportion of carbon. Hydrogen is the lightest known element; carbon is nearly twelve times as heavy. The numbers of atoms of the two in the molecules of a compound determine the weight of the molecules and therefore the *molecular weight* of the compound. Note that methane (CH_4)—in which the hydrogen-carbon ratio is four to one—is a light gas with the molecular weight of 16, while diocaryophyllane ($C_{30}H_{50}$)—in which the hydrogen-carbon ratio is only five to three—is much heavier and a solid; its molecular weight is 411.

The molecular weight of a compound in an oil is a clue to its other characteristics as well as to its hydrogen-carbon ratio and its specific gravity. For gases it is even better than a clue; it is directly related to density. For liquids and gases the molecular weight is valuable to the refining engineer; he never lets it out of his mind in designing refinery equipment.

Distillation Characteristics

The refinery engineer is not content, however, with knowing merely the gravity of the crude; he strives to know a lot of other things about it, and he works in the laboratory until he can be reasonably sure of his facts.

About the first thing he explores is the crude's distillation behavior. He puts a hundred cubic centimeters of the oil into a flask (not the kind carried on the hip, but a large round one of clear glass), puts a cork in the flask's neck, puts a thermometer through a hole in the cork, and from another hole conducts a tube through cold water or ice to a graduated cylinder. Then he heats the flask carefully at a regulated rate. Beginning with the lowest boiling constituent, each is driven off as a vapor. The vapors pass through the tube, are cooled and condensed into liquids, and collect in the graduate.

When the first drop distills over, our engineer reads the thermometer and records the temperature, the *initial boiling point*, as *IBP*. When the first 5 per cent of the sample has distilled over he again notes the temperature, and so on until all of the distillable fractions have gone over the *hump*, leaving a solid residue in the flask. All of this is known as *ASTM Distillation*, because it has been standardized by the American Society for Testing Materials. A similar procedure, differing only in minor particulars, is known as an *Engler Distillation*. Refinery men often refer to either as *running an Engler*.

When the engineer or chemist has finished the distillation, he measures the API gravity of each 5 per cent fraction he has collected and studies each of them as to its distillation range, sulphur content, gum content, wax content, viscosity, and other characteristics. He also studies the solid residue. When he is through, he has the *distillation properties* or *distillation range* or *boiling range* of the crude, and the individual properties of each of its fractions.

The Engler distillation is the simplest inspection made of a crude; it makes no attempt to break down the components as they distill. It is used as an exploratory procedure to be followed by more exhaustive methods, usually with larger samples, if its results warrant. The apparatus used for more exhaustive studies varies from a tube full of glass beads, known as a *Hempel column*, to elaborate devices with electrical controls throughout, which so efficiently separate the distilling fractions that pure compounds can be detected, estimated, and even isolated.

Flash Point and Fire Point

Another thing the engineer or chemist needs to know is the temperature at which the oil starts to explode or catch fire, the *flash point* and the *fire point*.

When the oil is heated enough, as we have seen, vapor begins to form, collecting just above the surface of the oil. At the flash point, enough vapor will form to ignite from the flame, causing a small explosion or flash. At higher temperatures vapor is evolved rapidly enough to burn continuously. The lowest temperature at which this takes place is the fire point. Flash and fire points have wide ranges. Gasoline and gasoline-bearing crude will flash and fire at room temperatures; heavy oils may not flash below five hundred degrees.

The flash points and the fire points of the crude and of each product to be made are important to the design engineer; he determines them, particularly on products heavier than gasoline, by carefully standardized tests.

Viscosity

A viscous liquid is one that flows slowly; usually it is thick or sticky or both. Viscosity, in other words, is the measure of a liquid's internal friction or resistance to flowing. Molasses has a high viscosity; liquids like water or gasoline have much lower viscosities. All fluids are more viscous when cold than when hot; the higher the temperature the greater the fluidity and the lower the viscosity.[2] The engineer needs to know the viscosities of the crude and its products; knowing viscosities he can calculate through what sized pipes, with what sized pumps, at what pressures and temperatures, and how fast he can move his fluids from unit to unit about the refinery.

Viscosities may be determined in various ways and designated in various units, but the viscosity of an oil is usually stated in *seconds Saybolt Universal* or *seconds Saybolt Furol* at a given temperature—which means the number of seconds required for sixty cubic centimeters, placed in an instrument known as a *Saybolt viscosimeter*, to flow through the standardized orifice in the bottom at the temperature specified. The Universal orifice is smaller than the Furol and is used for the lighter oils, the Furol for the heavier and more viscous oils.

Further Distillation

Having determined these and perhaps some other fundamental data, the engineer makes further studies with more elaborate apparatus in which refinery conditions can be more closely simulated.

The first thing he does is to make a *true boiling point* study. In a *Peters-type* or other distillation apparatus in which accurate fractionations can be made under vacuum if necessary, he comes as close as he can to separating the oil into its pure hydrocarbons. Then he plots the quantity of each against the temperature at which he has obtained it. The result is a *true boiling point curve*. In other apparatus he may

---

[2] The housewife, who often boils a liquid to thicken it, may think this is not true. But when she thickens gravy by boiling she evaporates some of the water, leaving the fats and other heavy, more viscous constituents. She thus distills out the low-boiling fractions, leaving the viscous, heavier fractions.

take the oil apart by flash vaporization and plot the *flash-vaporization curve*. He may then distill a sample of the oil in a small topping still, to get a line on the results he may obtain by distillation under atmospheric pressure. If he wants cracking data he may use laboratory cracking apparatus and thus determine how the oil behaves when heated under pressure.

The results obtained in these studies will not be quite the same as those that will later be obtained in the refinery. Refinery conditions can never be reproduced exactly in small-scale apparatus, but the experienced engineer correlates many laboratory data with those obtained in refinery practice. When he has completed his laboratory studies he can predict with reasonable accuracy what the oil will do in a refinery.

Characteristics of Products

Next, the engineer is likely to study the products. His object is to adopt the processes and equipment that will yield the most products having the greatest sale value. Because sale value is a function of quality, he is keenly interested in the qualities of the products his refinery will make.

Let's watch him; many of the characteristics he studies are those the consumer should consider in buying the products.

Gasoline Criteria

The points observed in gasoline are gravity, sulphur and gum content, corrosiveness, vapor pressure, octane number, and distillation range. Some are more important than others.

Gravity

Years ago gasolines were judged mostly by their gravities, the idea being that lighter gasolines are better. In time people awoke to the fact that though a light gasoline gives a quick start, it has less power than a more dense one. It also has a greater tendency to vapor lock—a factor that will be considered a little later. For these and other reasons gravity is no longer considered an index to quality, but within rather wide limits it is a useful guide. If a man buys too light a gasoline, his car is likely to develop vapor lock in the fuel line;[3] if he buys

---

[3] Vapor lock is caused by a vapor bubble developing in the fuel line at a point in the line where the temperature reaches the initial boiling point.

too dense a gasoline he may have trouble starting his motor.[4] The gravity of motor gasolines sold in the United States averages 60 API.

Sulphur Content

Some gasoline, as it comes from the stills, is high in sulphur, some is low. Some sulphur is volatile and easily removed; some is stable and can only be removed at considerable expense and with notable loss of gasoline octane number. Some sulphur compounds may corrode valves and other parts of a motor. Whether the more stable forms of sulphur have a corrosive effect in a modern motor once provoked a hot and prolonged argument.

When low-sulphur Pennsylvania and Mid-Continent products dominated the industry, sulphur was a demon to be cast out, and gasoline containing more than one-tenth per cent sulphur was taboo in the United States. Then for a few years, particularly in California and Europe, the sulphur phobia declined in importance and gasoline containing nearly a half per cent sulphur was sold and used and no one seemed to suffer.

An unfortunate angle of the matter is that in removing the sulphur from a high-sulphur gasoline the octane is lowered, but high-sulphur gasolines are much less susceptible than sulphur-free gasolines to octane-raisers like Ethyl fluid, so most refiners now reduce the sulphur content to about one-fourth per cent as a matter of course.

Vapor Pressure

We have seen that the vapor pressure of a liquid is the force with which its molecules try to disperse, and the more volatile the lightest parts of the liquid are, the greater its vapor pressure. If a gasoline containing volatile fractions with a high vapor pressure is put into the gas tank, some of it turns to vapor, filling the feed lines or the carburetor with bubbles, thus blocking the passage of the liquid gasoline. When this happens the motor starves for gasoline and is said to be vapor locked. Reasonably then, wise refiners watch the vapor pressures of the gasolines they manufacture. The commercial range in the United States is from about 6 to 13 pounds per square inch, measured

---

[4] Starting trouble occurs with too dense a gasoline because of the fact that the gasoline will not vaporize rapidly enough, at carburetor temperature, to make a combustible mixture in the cylinders.

by a standard procedure known as the *Reid Vapor-Pressure Test*.

Incidentally, the higher the elevation at which you are driving the lower should be the vapor pressure of your gasoline. You need a very low vapor-pressure gasoline to drive up Pike's Peak without danger of vapor lock. Refiners advertise—*Mountain Made for Mountain Driving*, or *Western Refined for Western Climates*, when they are really saying, "This is a low vapor-pressure fuel."

Anti-Knock Qualities

Place a heavy block of wood on your work-bench and give it a sharp, quick blow with the end of a stick. You may move the block a few inches. Now push the block with the stick, using the same total amount of energy you previously put into the quick blow. The push will move the block several times as far as the blow.

The poke or push of burning gasoline on the pistons in the engine is what makes your car go. The gasoline expands tremendously on burning and forces the pistons down. A slow-burning gasoline gives the piston a smooth and effective push; a quick-exploding gasoline gives it a swift kick, far less effective. If you are using a quick-exploding gasoline and are pulling up a hill, you can hear your engine knocking, the knocks or "pings" being the explosions in the engine.

The more the gasoline (mixed with air) is compressed before it is ignited, the more readily it explodes. The modern motor—with its high compression ratio—is smoother and more efficient than its low-compression forerunners, but it also knocks more readily. Gasoline that drove a 1919 Buick without a murmur would knock in a 1960 Buick loafing along a level road. The great demand these days is for *anti-knock* gasoline.

In general, gasoline from asphaltic oils knocks less than that from paraffinic oils. This fact tends to compensate for the greater gasoline content of most paraffinic oils. Chemists have found that gasoline produced by cracking knocks less than straight-run gasoline. Straight-run gasolines have such poor anti-knock qualities that they are usually cracked or *re-formed* to improve them. Gasolines from hydrogenation or polymerization can have excellent anti-knock qualities.

The tendency to explode or detonate is retarded by the addition of certain chemicals. The one in common use is *ethyl fluid*; its chief ingredient is tetra-ethyl lead—$Pb(C_2H_5)_4$. Almost all motor gasolines now are *leaded*; the premium automotive grades may contain three cubic centimeters of ethyl fluid per gallon, and some aviation gasolines have five cubic centimeters. About the only gasoline you can buy these days which is free of an ethyl addition is the white gas you order for your camp stove and lantern, and now the camp equipment manufacturers are making their products so that they will burn "regular leaded auto gas."

Octane Rating

Along with the demand for greater and greater anti-knock qualities came the need for some standard of measuring them. After a few passes in other directions, the industry adopted the *octane rating* or *octane number*.

Iso-octane (C_8H_{18}) is a member of the paraffin series, and heptane (C_7H_{16}) is another. Heptane knocks more readily than iso-octane. The octane number or octane rating of a gasoline is the percentage of iso-octane in a mixture with heptane that knocks the same as the gasoline. If a gasoline has the same knocking characteristics in a test engine as a mixture of 65 per cent iso-octane and 35 per cent heptane, then the gasoline's octane number is sixty-five.

Octane ratings of gasolines are determined in two ways; they are called the "Motor Method" and the "Research Method," although the gasoline to be rated is burned in a test engine in each. Both methods are used to rate aviation gasolines; the Research Method is generally used for automotive grades. Aviation gasoline octanes range from 80 to 145, the highest octane gasolines being used primarily in military airplanes. Most automotive gasoline is from 80 to 98 octane.

In making motor gasoline, the refiner obtains the desired octane number by blending various gasolines—topped, cracked, or re-formed— to as high an octane number as is economically feasible. Then if necessary he adds ethyl fluid to bring up the blend to the desired octane number.

Each cubic centimeter of ethyl fluid in a gallon of gasoline increases the cost more than a cent. Hence the refiner's careful blending and sulphur removal before adding ethyl fluid.

Distillation Range

Now we come to the most important thing of all, the temperature at which the successive fractions of the gasoline boil or vaporize. Consider the initial boiling point and the 10 per cent point (the temperature at which the first 10 per cent will vaporize). If they are too high, you will have trouble starting your car. If they are too low, your car may vapor lock in warm weather. The *end point* (*ep*) is the temperature at which all of the gasoline vaporizes. If it is too low your mileage may be low also, for the heavier fractions deliver more power than the lighter fractions. If the end point is too high some of the gasoline may not burn in the cylinders; instead it may work down past the piston rings into the crankcase, dilute the crankcase oil, and lessen its lubricating qualities. If the temperatures at which successive fractions vaporize are not properly spaced between the initial boiling point and the end point, your motor will perform erractically under certain operating conditions. The smooth and efficient performance of your engine depends in large measure on a proper distillation range.

Initial boiling points must vary according to climate; for summer gasolines the average ranges from 90° to 110°. In winter they are lowered to 75° to 85° for cold climates by adding natural gasoline. End points run from 340° to 440°; aviation gasoline has the lowest, usually below 360°. Automotive gasoline end points are high in summer and lower in winter, somewhere around 400°.

The 10 per cent and 90 per cent boiling points are also important. The temperatures between these boiling percentages govern the ease of starting, rate of acceleration, and loss to crankcase dilution. The 10 per cent temperatures are from 110° to 145°; those of the 90 per cent vary from 330° to 380°. These also must change with the seasons and the climate for the automotive gasolines.

Kerosene

The criteria for kerosene are of much the same nature as those for gasoline, with color, burning test, flash point, and fire point added, and low sulphur content being of real importance. Once upon a time, when kerosene was the principal product and gasoline an undesired by-product, the authorities had to watch the flash point closely to keep refiners from putting gasoline into the kerosene.[5] Needless to say, that watchfulness is no longer necessary.

Fuel Oils

The specifications for the fuel oils, such as tractor fuel, diesel fuel, furnace oil, and heavy fuel, concern mainly gravity, flash point, fire

---

[5] Domestic kerosene is acceptable with the flash point as low as 130 degrees, although the specification may be 110 degrees in kerosene for a very cold climate.

point, pour point, viscosity, and distillation range, and include another important quality known as *fuel value.*

Some oils develop more heat or energy than others. The amount of heat that a pound of the oil will develop under certain standard burning conditions is called its "fuel value." It is usually measured in *British thermal units,* but in countries that use the metric system it is measured in *calories.* A British thermal unit, usually referred to as a *Btu,* is the amount of heat required to raise the temperature of a pound of water one degree Fahrenheit. A calorie is the amount of heat required to raise the temperature of a gram of water one degree Centigrade. The fuel values of oils range from 18,000 to 20,000 Btu per pound, or 130,000 to 160,000 Btu per gallon.

Lubricants

The specifications for lubricating oils include the familiar factors—gravity, color, flash point, and fire point. Freedom from gums and other impurities is important, so is the *cold test* or *pour point,* and most important of all is the viscosity.

The function of a lubricant is to permit one hard surface to move easily on another. A banana peel on a sidewalk, for example, is an excellent lubricant; it permits the hard surface of your heel to move easily on the hard surface of the sidewalk. The lubricant's job is to keep the two solid surfaces from impinging directly on each other, which it does by maintaining a slippery film. If it is too thick, it will be sticky and will tend to retard the movement. A lubricating oil must therefore have enough *body* to keep the two surfaces apart and yet not be so thick and heavy as to be sticky, stickiness being a handy word for internal friction. The ideal lubricant for any given job is the thinnest one that will maintain an effective film between the bearing surfaces.

It is obvious that the lubricant should be chosen for the particular duty it is to perform. If the surfaces are moving rapidly and the pressure between them is slight, a light lubricant is called for; it will have the minimum internal friction and film strength great enough to keep the surfaces apart. If the movement between the surfaces is slow and the pressure between them is great the internal friction of the lubricant is not so important but it must have greater film strength.

Lubricant Viscosity

The most convenient indicator of film strength and internal friction is viscosity. By definition it is the measure of internal friction. It is not an exact measure of film strength, because two oils of equal viscosity may differ in their film strengths, but speaking broadly with reference to oils (to which no non-petroleum constituent has been added), film strength varies with viscosity. The more viscous the oil, the greater is its film strength, so viscosity is an index of both qualities. In choosing a lubricating oil, pick one with a viscosity that indicates the necessary film strength without excessive internal friction.

This is simple enough where the oil is to be used at a constant temperature, but consider what happens in your motor, and remember that the viscosity of an oil varies with the temperature. When you start your engine on a winter morning the oil temperature may be 10 degrees below zero. After you have driven a few miles the oil temperature may be up to 175 or 200 degrees. An oil that is thin and fluid at 150 degrees may be too viscous to flow at minus 10 degrees; an oil that flows readily at ten below may have no film strength at 150.

If you don't want to burn out your bearings or score your cylinders you must have an oil that flows readily on the coldest morning and that has adequate film strength even when the antifreeze in your radiator is about to boil. There you have one of the refiner's nice little problems.

Viscosity Index

Not all oils behave alike. As the temperature increases, some oils lose viscosity much more rapidly than others. Two oils may have the same viscosity at zero, and yet one may have several times the viscosity of the other at 150 degrees. To compare the differences in behavior a rough scale, known as the *viscosity index*, has been adopted, on which lubricants from Gulf Coast crudes are taken as zero, and those from average Pennsylvania grade crudes are taken as 100. Some oils have viscosity indices as low as minus 300, which means that they are no good at all as motor oils; some have viscosity indices as high as 140, which means that they are very good indeed.

Cold Test

Akin to viscosity is *pour point*, which is the temperature at which a liquid lubricant begins to congeal and will no longer flow or pour. Do you know what was wrong that morning in camp when the starter couldn't turn your engine until nearly noon? The oil in the crankcase was semi-solid instead of liquid. Next time you go hunting up north in late fall be sure your motor oil has a low pour point. Axle grease has a pour point above a hundred degrees and who cares; it is not required or expected to flow; but an oil intended to lubricate high-speed machinery must flow readily at the lowest temperature at which the machinery is to operate. Some oils, otherwise good, start congealing at twenty degrees above or even higher, and are all right for use in Florida; but Quebec winter drivers want an oil that is still liquid at twenty below. The car manufacturer has these facts in mind when he recommends either one grade of oil for summer use and another for winter use, or a *multi-grade* oil of high viscosity index; and the refiner has them in mind and makes oils with the necessary specifications.

The test for pour point is known as the *Cold Test,* and the two terms are used interchangeably and are practically synonyms.

Color and Purity

A lubricating oil should of course be pure. If it were to contain solid particles they might scratch and wear the moving parts. If it were to contain free carbon or gums they might clog up the apertures through which the oil must move. Most specifications, therefore, provide against gum and resin-forming tendencies, excessive carbon formation in use, and similar undesirable characteristics.

Color is usually specified, for two reasons: First, it is something of an index of freedom from carbon and other impurities, and, second, oil with good color sells better than oil that looks dull and muddy. Many refiners actually "trade-mark" their oils by adding a coloring agent, but they are all sparkling clear and clean-looking.

Addition Reagents

In recent years most lubricating oils have been greatly improved by the addition of non-petroleum substances, loosely grouped under the name *addition reagents.* Every refiner who uses these reagents has a list of his own and keeps it pretty much to himself. They are added to the lube oils in relatively small proportions to alter one or more of the properties of the oils.

One group is used to increase film strength. Another is used to lower the pour point without affecting other physical characteristics. Another is used to prevent the oxidation that sometimes results in the formation of gums and resins; it usually includes a detergent to keep impurities from settling as *varnish* or *sludge.* A related type inhibits the corrosion of the metal parts of the motor that might otherwise take place after the oil has been in use for a long time. There is even a re-agent designed to improve the oil's viscosity index.

Not all refiners use addition reagents of any type, and few refiners use all types. Some refiners decry the use of many of them, while other refiners consider most of them beneficial. It is probably safe to say that most multi-grade oils are now detergent oils as well. Some motor mechanics "swear by" the detergent or addition reagent oils, whereas others swear at them.

Asphalts

The desiderata for asphalt products are of a different nature, naturally, than those for gasoline or fuel oil or lubricants. We need not discuss them exhaustively, but a few deserve mention.

An asphalt, whether it is to be used for paving a street or binding gravel on a highway or coating a roof, should be heavier than water; otherwise it might float off in a rain. It should be free from paraffin or similar waxes, for these ruin its binding qualities. These two requirements pretty well rule out products made from paraffin-base or mixed-base oils.

The ratio of an asphalt's melting point and ductility to its hardness is important. The melting point is the lowest temperature at which it will flow. The ductility is the number of centimeters that a filament of it will stretch without breaking. Hardness is expressed as *penetration*, or the number of *degrees* (1/10 of a millimeter or 1/250 of an inch) that the point of a No. 2 sewing needle, weighted with a hundred grams, will penetrate the asphalt in five seconds at a temperature of 77°. The harder the asphalt the less the penetration. A ten-penetration asphalt is about as hard as a soft pine board; paving asphalts run from 40 to 70, and road binder may have a penetration as high as 90.

The paving engineer naturally would like to use a hard asphalt which will not squeeze into trenches or rolls under heavy traffic in warm weather. But he knows that as penetration becomes less, ductility also is less and the melting point is higher. The harder the asphalt the less ductile it is and the more heat it takes to melt it. This makes it more difficult to handle and to keep fluid. If the paving engineer picked an asphalt for hardness alone he would probably find himself with a product so lacking in ductility that it would break up in the expansion and contraction caused by changes in temperature. He may not be able to keep it fluid long enough to lay it. He therefore specifies an asphalt that has a satisfactory ratio between hardness on the one hand and ductility and melting point on the other.

One other point might be noted. If an asphalt is heated too much, either in producing it or laying it, it loses much of its stickiness or binding quality, due mainly to the breaking down of some of the molecules and the formation of fixed carbon. Overheating must be avoided.

Most people think that practically all asphalt products are used for paving and highways. The fact is that nearly half of the asphalt produced in oil refineries finds other uses—roofing, paint, rubber filler, battery seals, wood preservatives, waterproofing, and caulking. Each of these uses has its own specifications, in most of which the points already mentioned are included.

The Consumer's Choice of Products

By this time, having watched the intensive study the refinery engineer gives to the specifications of the products he manufactures, you are probably wondering how the ordinary consumer can be assured of selecting the grade of products he should use. Should you install a chemical laboratory in the basement and test every tankful of gasoline you buy? Obviously, no. You may safely leave the tests to the refinery engineer; the studies he makes are performed for your benefit.

The reputable refiners and distributors know that you will buy their gasoline only so long as it gives you good results. If you buy a certain brand of gas and are troubled with vapor lock or undue knocking or erratic performance, you will buy some other brand next time. Every refiner and distributor is competing to sell every gallon he can, and no responsible refiner is going to risk the loss of business by turning out an inferior product. The desire to build gallonage through quality is so strong that much of today's third-grade gasoline is better than the best gasoline available a few years back. You may get stuck if you buy from some cut-rate bootlegger on the edge of town, but you are safe if you buy from reliable dealers handling a recognized brand made by a reputable company, large or small.

What grade of gasoline should you use? The answer depends on your engine and how you operate it. If you drive a modern car with a motor of moderate-compression ratio, buy a good regular grade. If you have a high-compression motor, or if your driving requirements are severe, or if you like the last little finishing touch in performance, you may well consider using premium gasoline.

Motor oil can safely be selected on the same principle. Buy a good quality of oil, made by a reliable manufacturer with a modern plant and modern methods, and be sure to use the grade recommended by the manufacturer of your car.

The Society of Automotive Engineers has simplified the matter for you; it has laid down specifications for the various oil grades, which are given numbers such as SAE 20, SAE 30, SAE 10W-30, and so on. These grades are based only on viscosity and viscosity index, but you are safe as to the other characteristics such as color, flash point, carbon residue, and pour point, if you buy oil made by a reputable refiner.

The life of your motor depends far more on the oil you use than on the gasoline, so don't count pennies too closely when it comes to quality.

And don't think, from what has been said, that all lubricating oils go into automobiles. There are a thousand and one other uses for lubricants, some of them more complex than filling crankcases. The refiners make lubricants—liquid or solid—to fit them all. If you have unusual lubricating problems, call in the lubricating engineer of some reliable manufacturer; he will be glad to advise you and will save you money, too.

Now for the Design

Having studied his crude and its behavior under various methods of treatment, and having studied the products the various methods have given him, the engineer is at last ready to draw conclusions and design the plant.

He decides on the initial treatment to which the oil is to be subjected, and on the successive treatment, step by step, to be accorded each *cut* or fraction, until he has a *flow sheet* outlining every step from the pre-heating of the incoming crude to the finishing of the last product.

Then comes the design. Every still and tower and tube and condenser must be so designed that the proper amount of material will move at the required velocities. The proper temperatures must be maintained and the right amount of heat supplied here or subtracted there. Hot materials must interchange their heat with colder ones and the equipment must be strong enough to contain them. All operations should be as nearly automatic as possible, and the danger of fire or explosion must be considered throughout all this planning. Every pump and valve and fitting has to be selected for its special purpose. A steam plant, an electric plant, a fire-protection system, tankage for storing products, railroad sidings and loading racks, water supply— all these must be carefully figured from fundamental data, and presented to the draftsmen.

When the engineer is through, he has dozens—oftentimes hundreds —of blueprints showing in minute detail every structure and every piece of equipment in the refinery. There on paper is the refinery. It remains only to convert it into fabricated iron, steel, and copper, bricks, mortar, and cement.

CHAPTER 13

Refining the Oil—The Refinery

Now for a trip through a refinery or two. We shan't stop to study everything, lest we never get through, but we will try to see the most important units in operation.

While we are waiting for our passes we might pick up a bit of refinery vernacular.

A Little Terminology

Stock is a stock word around refineries, almost as much so as *location* and *string* around an oil field. It is applied to any material that is to be processed. *Charging stock* for some particular unit is oil that is to be fed into or treated in that unit, *cracking stock* is material that is to be cracked, *lube stock* is material from which lubricants are to be made, and so ad infinitum.

Nobody in the oil business bothers to say "lubricant"; *lube* (plural *lubes*) is shorter and sufficiently elegant. *Overhead* is not something the harassed executive is trying to keep down; it's the product or products taken from a scrubber or absorber in the form of vapor or gas. The *tops* in any operation are the products distilled or flashed off, not teenage slang for a "cool cat." The *bottoms* are the parts that remain. *Light ends* are the lighter volatile fractions of an oil, such as gasoline and kerosene; *heavy ends* are the heavier fractions such as fuel oils, lubes, waxes, and asphalts.

When a refinery unit is in operation it is *on stream*; the rest of the time it is *shut down* or *cold*.

Here are our passes. Let's go.

237

Shell Stills

The grizzled veteran of oil refining is the shell still. In one form or another it has been stewing products out of petroleum since before the first American oil well was drilled. Almost every conceivable form of closed vessel has been used. Anything in which oil could be heated and from which the resulting vapors could be conducted away to be condensed has seen its day in refining. For nearly sixty years the shell still distilled the world's oil; for a decade and a half it dominated cracking. Today it is crowded into the background by the more efficient pipe still.

A typical *shell still* is a horizontal steel cylinder, six to fourteen feet in diameter and twenty to forty feet long, held four to ten feet above the ground by solid rectangular brick or concrete walls. These walls are lined with fire brick and form the fire box or heating chamber. The lower half of the circumference of the still receives the heat generated in the heating chamber by the burning of gas or oil. Some stills are heated with steam and are called "steam stills"; those that are heated by direct firing are called "fire stills."

Oil is charged into the still through a pipe, and the vapors are taken off through a pipe in the top.

Batch and Continuous Operation

Early shell stills were run on the *batch* principle; only a very few are so operated at present. The still is charged with a batch of oil and heated until all of the desired products are distilled off. If only a partial distillation is desired, enough heat is applied to vaporize the required fractions; then the bottoms are drawn off for further treatment or for use in other processes.

Batch operation has some obvious disadvantages. To overcome them, refiners many years ago worked out methods of continuous shell-still operation. A battery of stills is erected side by side, with pipes leading from each to its neighbor. A moderate heat is maintained under the first, a somewhat higher heat under the second, and so on. Oil (feed) is charged continuously into the first still. The feed rate, the capacity of the still, and the amount of heat supplied are so interrelated that by the time the oil has lost the first fraction desired, the remainder has flowed through the connecting pipeline into the next still. There it loses the next desired fraction and passes into the third

SHELL STILLS

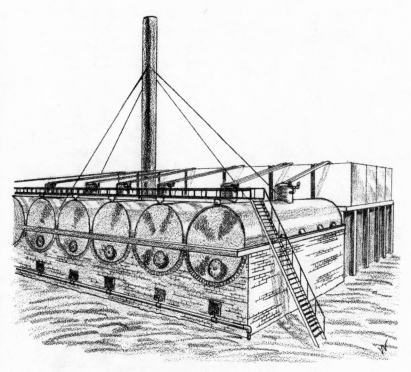

In former days all refining, including a great deal of cracking, was done in "shell stills." Back of the stills are condensers, where vapors from the stills are condensed into gasoline and other products.

BOX-TYPE FURNACE

Much refining depends chiefly upon furnaces or heaters in which the oil is heated while passing through tubes still, and so on into successive stills until all of the desired fractions have been distilled out.

Steam is often introduced into shell stills to lower the feed vapor pressure and thus lessen the heat required for distillation. (The stills are not called steam stills since the main heat is from direct firing.)

One of the most famous of the shell stills was the *Burton still*. Now obsolete, for years it was the major cracking unit. In a still such as this, oil was heated to temperatures of seven hundred to nine hundred degrees, under pressures of seventy to one hundred pounds, and the resulting distillate was recovered by condensation. The Burton process operated originally on a batch basis, but later improvements permitted the use of higher temperatures and pressures, and afforded continuous operation.

Whether batch or continuous, shell stills are only used now for re-running small amounts of special naphthas, recovering solvents from various products, reprocessing asphalts and *slops*, and for other special separations.

Pipe Stills

In the shell still, oil is heated while more or less stationary. With the development of flash vaporization and vapor-phase cracking, a different type of still—one in which the oil is heated while passing through tubes—came into general use.

Every pipe still or *tube still* is specially designed for the service it is to render. Around the walls of the heating chamber are vertical tiers of horizontal pipes, each connected at one end to the pipe next below it, and at the other end to the pipe next above it. These tiers are known as *coils*, but they are flat, back and forth coils, not spirals. The U-shaped connections between the pipes fit over the pipe ends and can be removed to permit cleaning the pipes. The coils on the different walls may be connected to form a continuous system, so that oil entering at one corner of the chamber passes through all the coils before leaving the chamber at the farthest corner, or different connections may be made so that different streams of oil pass through the different coils.

Heat is usually supplied to the chamber by gas or oil flames, and the radiation from the flames, with a smaller amount of heat from contact with the combustion gases, heats the coils. This first chamber is known as the *radiant section* of the still.

Beyond or beside or above the radiant section, and usually separated from it by a checker-work of fire brick built to intercept the radiant-heat rays, is another chamber containing other coils. This is called the *convection section* and its usual purpose is to raise the temperature of the oil on its way to the radiant section, thus reducing the amount of heating that the radiant section has to do. It is heated by the flue gases on their way from the radiant section.

As a rule the charging stock or oil to be treated, which may or may not have been preheated, passes first through the coils in the convection section and then through part or all of those in the radiant section. By the time it has completed the journey it has been heated to the desired temperature.

Notice that no distillation or fractionation takes place in a pipe still. The sole purpose of the still is to heat the oil; the fractionation takes place in another unit. Notice also that pipe stills are used for topping, cracking, polymerization, and hydrogenation, each unit being designed for the particular temperatures and pressures required for the service it is to render.

In almost any modern refinery, you see several distinctly different pipe stills. If you visit other refineries you will note more pipe stills, different from those you saw before. Engineers have developed about a dozen basic types, differing from each other in the distribution of heat in the radiant and convection sections, the temperature levels at which the heat is absorbed, and the kind of fuel used—gas or oil. There are other factors of differentiation, but these are the important ones.

Bubble Towers or Fractionating Columns

The hot oil from the pipe still goes to a *fractionating column* or *bubble tower* in which most of it flashes into vapor.

A typical bubble tower is a vertical steel cylinder ten or more feet in diameter[1] and forty to a hundred feet high, with from eight to eighty horizontal steel partitions or *trays* spaced at intervals from near the bottom to near the top. Each tray is pierced by a large number of holes through which short lengths of pipe, *chimneys*, are inserted, each projecting a few inches above the tray. Each chimney is covered, but not sealed, by a *bubble cap*. A *down pipe* extends from each tray nearly to the tray below. The upper end of the down pipe extends two to

---

[1] If the amount of light fractions in the charging stock considerably exceeds the amount of heavy fractions, the upper part of the tower may be considerably larger in diameter than the lower part. This is especially true in vacuum towers.

four inches above the tray, so that a corresponding depth of liquid may accumulate on the tray before the liquid begins to run down the pipe. This serves the same purpose as the overflow pipe in an aquarium equipped with a water circulating system. Vapors from below the tray rise into the chimneys, pass through slits in the bubble caps below the liquid level on the tray, and bubble up into the chamber above.

At the top of the tower may be a *reflux condenser,* a cylinder two feet or more in diameter with a series of parallel tubes inside arranged like matches in a pill bottle. A stream of oil, usually the charging stock for the pipe still, passes through the tubes and cools the vapors circulating around them in the cylinder. This exchange of heat, from the vapors to the incoming oil, preheats the oil for the still and so the unit may also be known as a *reflux exchanger.*

From the top of the reflux condenser or from the tower a *vapor line* runs to a condenser and then to a gas separator similar to those used for separating gas from oil at an oil well. From the gas separator or *gas trap* a *reflux line* leads back to the tower, entering just above the top tray.

In the lower part of the tower (a few trays above bottom) is an exceptionally tall space between trays. This is the *vaporizer section.*

Superheated steam may be fed in at the bottom of the tower, usually below the lowermost tray, to lower the vapor pressure.

Hot tower feed comes from the pipe still through a *transfer line* and enters the tower in the vaporizer section. It is hotter than the boiling points of most of its constituents; therefore most of it flashes into vapor instantly. The vapor starts upward through the tower, passing through the chimneys in the successive trays.

The heat in the tower, except for that supplied by the steam, is maintained by the oil itself. Consequently, the vaporizer section and the lower trays are the hottest parts of the tower. Each tray above the vaporizer section is a little cooler than the one next below; the top tray is the coolest.

On each tray a liquid fraction accumulates that is condensible at the temperature of that tray. The liquid cools the vapor from below as it bubbles through, and the cooling causes the vapor to leave most of its content of the same liquid on the tray.

The vapor that does not condense, on the other hand, strips out of the liquid any lighter fractions and carries them to the higher trays where they condense.

When enough liquid accumulates on a tray it spills into the down

BUBBLE TOWER

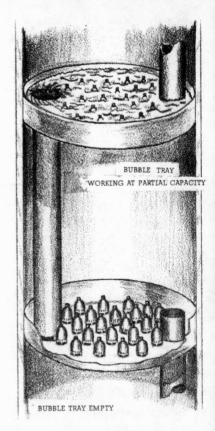

BUBBLE TRAY
WORKING AT PARTIAL CAPACITY

BUBBLE TRAY EMPTY

BUBBLE CAP AND CHIMNEY

The hot oil from the pipe still goes to a "bubble tower" or "fractionating column," where by flashing into vapors, condensing, revaporizing, and recondensing it is separated into liquid "cuts" or "fractions."

pipe to the tray below. There most of it is again vaporized and joins the other rising vapors, to bubble up through and be condensed again on the upper tray—a continuous merry-go-round called "refluxing."

Thus each particle of oil is flashed into vapor and so on over and over, perhaps twenty or more times. Each cycle tends to segregate on each tray a more homogeneous fraction, more and more free from lighter or heavier fractions. From certain trays the fractions particularly desired are allowed to get off the merry-go-round and flow through pipes. These fractions are known as *side draws* or *side streams*. One side draw may be naphtha, the next lower a light kerosene fraction, the next a heavy kerosene fraction, the next a gas oil fraction, and so on.

The lightest vapors do not condense in the tower but pass up through even the highest trays. If there is a reflux condenser on the tower, part of the vapors condense in it and fall back into the top tray. The remainder pass out the vapor line at the top of the column and go to a condenser. In the condenser the part that will condense at ordinary temperatures drops out as a liquid, usually gasoline. The remainder is fixed gas. It goes next through a separator or "gas plant" where any liquefiable fractions are removed like natural gasoline is removed from oil-field gas. The liquid thus obtained is added to the plant's gasoline output. The dry gas remaining is either used as fuel somewhere in the plant or goes to the *polymer* section where part of it may be polymerized into gasoline.

Part of the gasoline from the separator, well cooled by now, goes through the reflux line back into the bubble tower onto the topmost tray, where it cools the rising vapors and is itself again flashed into a vapor and taken off through the vapor line. Thus the gasoline, like the heavier fractions, is refluxed until it attains a high degree of homogeneity. If, as it collects on the top tray, it has retained some heavier materials, as it usually has, these eventually flow through the down pipe to the next lower tray where they belong.

As each tray is slightly cooler than the one below it, the liquid flowing through the down pipe tends to cool the lower tray and counteract the heating effect of the rising vapors, the net effect being to keep each tray at the desired temperature.

The part of the oil that does not flash or vaporize at the temperature and vapor pressure of the vaporizer section collects on the tray at the bottom of the section and overflows (tray by tray) until it reaches the lowest chamber and is led off as the *bottom draw* of the operation. It is the heaviest product of the fractionation. As it starts its descent it may contain constituents that would vaporize in the vaporizer section if they were not mixed with heavier ones. So it is subjected to steam or oil vapors bubbling up through it on each tray in its descent. These strip out the lighter fractions and they rise as vapor to the vaporizer section and on higher in the tower.

Thus from a single fractionating column may come a wide range of distinct products—from fixed gas and aviation gasoline to heavy fuel oil or heavy lube stock.

Because of the refluxing, the products or cuts from a fractionating column are more homogeneous than those from the simple vaporization and condensation of the shell still and condenser. For this reason and because their fuel consumption is notably less, the pipe still and bubble tower have almost displaced the shell still.

Strippers

Even the number of successive vaporizations and condensations possible in the bubble tower may not satisfy the refiner. In that case he places a *stripper* or *stripping tower* alongside the bubble tower. The stripper is a smaller tower like the bubble tower but divided by horizontal partitions into compartments. Each side draw from the bubble tower enters a stripper compartment above the topmost tray. Steam is fed into the compartment near the bottom. It bubbles up through the liquid on the trays and strips out the lightest fractions by vaporizing them. These vapors are returned to the bubble tower above the topmost tray from which that particular side draw is collecting. The liquid from the lowest tray in the stripper compartment is the true side draw; it goes to a cooler and thence to the stock tanks or the treating units.

Condensers and Coolers

We have been talking a lot about condensation and cooling of this and that; it's time we learned something of how it is done. Condensing and cooling are important parts of refinery operation; the efficiency of a refinery depends as much on the efficiency of its cooling units as on that of its heating units.

Much cooling of liquid or vapor products is done in *heat exchangers*, in which the liquid or vapor to be cooled is passed through pipes surrounded by colder liquid that the refiner wishes to heat. Remember our reflux exchanger on the bubble tower? One such heat exchange saves fuel for preheating the charging stock and saves water for cooling the side stream.

The big cooling job, however, is the condensation of overhead vapors from the shell stills and fractionating columns. It is done by passing the vapors through coils of pipe immersed in water. The water takes up part of the heat of the vapors and the resultant cooling condenses the vapors into liquids. If plenty of cold water is not available, the water, after it has become hot in the condenser, may be cooled by spraying it into the air—the fountain we sometimes see as we pass a refinery. If the water is heated to the boiling point, it too may first pass through a condenser to cool it.

These are present-day methods. In earlier refining days, vapors were cooled by passing them through pipes suspended in the air, and many a refinery, on hot summer days, could operate only in the hours before dawn when there might be a cool breeze.

Vacuum Distillation

We long ago considered the theoretical advantages of vacuum distillation for some operations. Now we can see how it is done. Large volumes of steam, condensed at the vapor outlet of the bubble tower, create a vacuum in the condenser, bubble tower, and stripper. The vacuum is highest in the condenser, slightly lower at the top tray of the tower, and increasingly low at each lower tray. If extreme vacuum is being used, the pressure in the vaporizer section may be a small fraction of atmospheric pressure, and the reduced pressure may extend back into the transfer line toward the pipe still.

The fact that the partial vacuum in the tower and stripper causes the various fractions to vaporize at lower than normal temperatures is a big fuel saver in many operations. Remember, however, not every bubble tower operates under vacuum. Many function at atmospheric pressure, with or without the use of steam to lower pressures, and cracking may be conducted at several hundred pounds per square inch.

Recycling

After the gasoline has been topped off the crude, part or all of the remainder may be cracked. Recall that cracking makes small molecules out of big ones, and that small ones are typical of gasoline. Then, more newly created gasoline can be either topped or flashed off the cracked material. Part or all of the remaining large molecules of the cracked material may then be cracked again. This is known as *recycling*, and under some circumstances and with some crudes it may be repeated a number of times until the maximum amount of gasoline has been obtained, leaving nothing but a small amount of fuel oil and coke. This, like any other treatment that is continued until coke is formed, is called *running to coke*.

Making Lubes

If lubricating oils are to be made from a paraffin-base oil, the crude is usually topped into six fractions: Gasoline, kerosene, *straw distillate* or *furnace distillate* (from which diesel oil, domestic heating oil and similar products can be made), *wax distillate, cylinder stock*, and *tar*, which may be converted into an asphalt product or burned for plant fuel.

The wax distillate and cylinder stock are the *lube stocks* from which the lubricating oils will be made. Both contain paraffin, but in different forms. That in the wax distillate is crystalline; if the distillate is chilled long enough the wax will separate out of the oil in large crystals. The wax in the cylinder stock is amorphous; it will stay in suspension indefinitely unless removed by special processing.

The wax distillate is chilled and put through a *filter press* under high pressure. The filters retain the wax and the oil passes through. The wax-free oil, called *pressed distillate* or *neutral stock*, is treated with acid and clay to remove impurities and stabilize the color. It may also be given a solvent treatment to clarify impurities, stabilize the color, and improve its viscosity index. It is then cut by redistillation into fractions of the desired viscosities and specific gravities, known as *neutral oils*. These are ready for marketing or for blending with *bright stocks* into motor oils.

The cylinder stock is also treated, and naphtha is added in removing the amorphous wax. The naphtha lowers the viscosity and the specific gravity of the oil, and thus increases the specific gravity difference of the wax and the liquid. Then the oil is chilled and put through a centrifuge which works on the same principle as a cream separator. The wax is separated and discharged as a *cream*. The wax-free, diluted oil is redistilled to remove the naphtha and to cut the dewaxed cylinder stock into the desired fractions. These fractions are *bright stocks*;

they are blended with appropriate neutral oils to make the lubricating oils that are sold to the public.

Pressing and centrifuging are costly operations and they are giving way to two newer and distinctly different methods of crystallizing and separating the waxes. In one, solvents are mixed with the feed stock and the mixture is chilled to a low enough temperature to crystallize the amount of wax to leave the oil with the desired pour point when filtered. The refiner has his choice of many solvents.

In the other method the chemical urea is mixed with the feed stock at room temperature. The wax molecules combine with the urea to form large crystals; these are easily filtered. Oils thus treated have pour points as low as minus 70 degrees.

Treating

Lubricating oils are not the only things that need treatment before they are ready for sale. The first step in the manufacture of oil products is to make them; the next step is to make them fit for use. Each product requires its own special treatment, depending on the use for which it is intended and on the crude from which it is produced. The kinds of treatment are many and varied. In general, the chief objects of treatment are to improve color, odor, and stability, to reduce the amount of sulphur, to remove gums, resins, and other impurities, and to create the desired properties of the product.

Kerosene does not always come water-clear from the stills, nor does gasoline unless made from *straight run* stocks, but kerosene must be colorless to sell, and so must gasoline unless it is to be colored before it is sold. Water-clear gasoline or kerosene must stay clear when exposed to light and air. A malodorous product will not sell well. Kerosene and gasoline must be low in sulphur. Gums and resins are harmful in gasoline or lubricating oils. A lubricating oil must have a clear, stable color and a good viscosity index. All these requirements mean treatment.

Chemical Treatment

The most useful chemical treating agent is sulphuric acid. It is useful in improving color, odor, and stability, in removing sulphur, and in reducing gums and resins. After the treatment, the excess acid is neutralized with sodium or ammonium hydroxide or a neutralizing clay.

Sour products (those high in certain forms of sulphur) are usually given the *doctor treatment*, which consists of agitating the oil with an alkaline solution of sodium plumbite and a little free sulphur. A satisfactory product is said to be "sweet to the doctor test."

Copper chloride, lead sulfide, sodium hypochlorite, calcium hypochlorite, and other chemicals are also used for various purposes in some refineries, but none of them is as important as sulphuric acid and the doctor treatment.

Various solvents are used to improve the viscosity index and to remove impurities from lubricating oils. This is a relatively new development that has already brought notable improvement in quality. Nearly every important refiner of lube oils has its own solvent process.

Most of the chemical treatment is done in *agitators* (not members of a labor union) in which the chemical is mixed and agitated with the product being treated. The agitation may be mechanical or may be done by compressed air. You will see plenty of agitators scattered about any large refinery. Some chemical treatment is done in centrifuges, the reagent and the product to be treated being first mixed violently in a mechanical mixing device.

Clay Treatment

If an oil product is filtered through or otherwise brought into intimate contact with certain clays such as Fuller's Earth, the surfaces of the clay particles will adsorb the resins and asphaltic residues in the oil and improve both color and quality. Clay towers are as common in refineries as agitators. The liquid oil may be allowed to percolate downward through a column of clay, or it may be mixed into a slurry with the clay and pumped through a filter, or it may be passed as vapor through a bed of clay fragments. When the clay has reached its limit of adsorption, it is removed from the tower and either discarded or rejuvenated by being heated to burn away the adsorbed materials.

Some Refineries at Work

Now that we are acquainted with the principal refinery units, let's follow the oil through a couple of refineries and watch what happens to it. We will see the manufacture of only a few of the many hundred different petroleum products, but we will see enough to illustrate the main principles of operation. Remember always, as we go, that no two refineries are just alike, and that the same products can be made from the same crude by more than one method. Our refineries will represent good modern practice but the same results might be obtained by other practices equally good.

Topping for Light Ends and Fuel Oil

Suppose our crude stock is East Texas oil and we want to top it to make gasoline, naphtha, kerosene, furnace oil or diesel fuel, and undistilled fuel oil.

The crude goes to a pipe still where it is heated to somewhere between 500 and 600 degrees, and thence to a bubble tower into which steam is introduced. Both pipe still and bubble tower are operated at atmospheric pressure; that is, no imposed back pressure is applied to either. The fixed gases and the gasoline are taken off through the vapor line at the top of the tower. The naphtha, kerosene, and furnace distillate are drawn off as side streams from trays above the vaporizer section, and the fuel oil is taken off as bottoms from the bottom of the tower.

The fuel oil is sold without further treatment. The furnace distillate is sold as furnace oil or diesel fuel, also without further treatment. The kerosene and naphtha may require doctor treatment before being sold. If the gasoline is high enough in octane number and sweet it can be sold without treatment (remember, straight-run gasoline is water-clear and color stable). Of course, if we are running a sour crude the gasoline may require both acid and doctor treatment.

Straight-run gasoline from such a paraffinic or mixed-base crude will be relatively low in octane number. It can be sold as third-grade gasoline, or some tetra-ethyl lead can be added to bring it up to the octane standards of standard-grade gasoline. Not many straight-run gasolines from paraffin-base crudes can be sufficiently *ethylized* to bring them up to premium grade because they may require more ethyl fluid than the rules of the Ethyl Corporation and economics permit.

The gasoline may even have too low an octane number to permit its sale as third-grade gasoline. Perhaps the lighter half would pass muster, but the heavier half is far below octane number standards. In that case the part below standard may be *re-formed* by cracking it.

It goes through the vapor line to the condenser and thence to another pipe still where it is heated to 900 to 1,100 degrees temperature at a pressure of 500 to 1,000 pounds. Thence it goes to a very simple bubble tower, having no side draws. Perhaps 2 per cent is polymerized into "tar" and drawn off as bottoms to be used as plant fuel. The remainder goes through a condenser to a gas separator. Something like 20 per cent will be fixed gas and the remainder is re-formed gasoline with a satisfactory octane number.

Thus we have done our job of making gasoline, naphtha, kerosene, furnace oil, diesel fuel, and fuel oil from a 40-API mixed-base crude, predominantly paraffinic.

Topping and Cracking for Light Ends and Fuels

Another refiner, using the same East Texas crude and with a properly equipped plant, wants to make more gasoline at the expense of the fuel-oil cuts. This is how he may do it:

The crude here also goes at atmospheric pressure to a pipe still and a bubble tower into which steam is injected. The bubble tower overhead is straight-run gasoline and side draws of naphtha and kerosene are taken off as before. No furnace distillate side stream is drawn off, however; instead everything below the kerosene is drawn off the bottom at about 500 degrees and is known as *reduced crude*, or *topped crude*.

The topped crude, under perhaps fifty pounds pressure, next mixes in a *quenching valve* with the discharge stream from a cracking pipe still, or *pressure still*, of which more anon. The discharge stream, at about 975 degrees and a thousand pounds, goes from the quenching valve to an *outside vaporizer*, a vertical cylinder like a bubble tower without trays. Here the pressure and temperature drop to about thirty-five pounds and 775 degrees and the stream flashes into vapor except for a small amount of heavy, viscous material formed in the cracking still. The heavy *pressure-still tar* is drawn off the bottom to be used for fuel.

The vapor from the outside vaporizer goes to a bubble tower. Cracked gasoline is taken off through the vapor line, furnace distillate is taken off as a side stream, and the remainder, known as *recycle stock*, goes back into the pressure still at about fifteen hundred pounds.

Note that part of the recycle stock or *hot oil charge* comes from the reduced crude from the topping unit. It has not been through the pressure still; the balance comes from a previous trip or trips through the pressure still. Eventually the reduced crude goes through the cracking unit—quenching valve, outside vaporizer, bubble tower, and pres-

The fundamental units of a modern refinery are the pipe stills, bubble towers, and condensers. There are many of each in a large refinery. In most refineries there are also many strippers, clay towers, agitators, and gas separators.

sure still—until all of it has been cracked into gasoline, furnace distillate, and tar.

The cracked gasoline is high in octane rating, but it probably needs to be acid and clay treated, followed by doctor treatment for sweetening, color, odor, and stability. It may be sold direct or may be blended with the straight-run gasoline to impart a satisfactory octane number.

The furnace distillate is sold without treatment as furnace oil or diesel fuel.

Thus our second refiner, from the same crude we formerly processed, makes straight-run gasoline, cracked gasoline, naphtha, kerosene, some furnace distillate, and fuel oil. He may have doubled the gasoline yield of the former operation.

Topping for Light Ends and Lubes

Still another refiner of East Texas crude wants to make lubricants from the bottoms that the first refiner sold as fuel oil and the second refiner cracked to make more light ends. Here the reduced crude from the bubble tower is heated to about 775 degrees in a pipe still and goes thence to a vacuum-type bubble tower. In the vaporizer section the pressure may be less than a pound absolute, nearly fourteen pounds less than atmospheric pressure. The overhead, except for the fixed gases, is condensed into a furnace distillate. Several side streams of wax distillate and cylinder stock may be drawn off from the bubble tower through strippers and pumped through a cooler to storage. The bottom-stream tar is used as plant fuel.

Each wax distillate cut is chilled and filtered. The crude wax is *sweated* or deoiled in big, open pans by first being chilled solid and then heated with steam. It is filtered through clay and sold as paraffin.

Solvent deoiling is now replacing sweating. The crude wax is dissolved in a solvent and chilled to the temperature required to make a wax of the desired melting point. The wax is separated from the mixture on a rotary filter. The crude wax is yellow; it is clarified by filtering through Fuller's Earth or by acid treating and caustic washing. Wax melting points vary from 90 to 190 degrees; the melting point of the paraffin used in covering jelly jars is 130.

A small special cut may be taken from the neutral stock or *pressed distillate* from the wax filter, treated for color and stability, and sold as mineral oil. However, because a medicinal product of this nature is used by the tablespoonful instead of by the gallon, and because the treatment cost is high, only a few refiners bother to make it.

Most if not all of the neutral stock is treated and is then ready for blending with bright stocks for motor oils and other lubricants.

The cylinder stock from the bubble tower is also treated, diluted with naphtha, and then chilled and centrifuged. The heavy, waxy material from the centrifuge is *crude petrolatum*. Part of this is further refined and purified to make such products as Vaseline; the remainder is charged back to the cracking stills.

Dewaxed solution from the dewaxing filters or centrifuges is reheated and distilled to knock out the diluent; this is condensed from the vapor line of the tower and reused in diluting more cylinder stock. The tower bottoms are bright stock, ready for blending with neutral oil to make motor oils and other lubricants.

Thus our third refiner, from the same East Texas crude, may make gasoline, naphtha, kerosene, furnace oil, diesel oil, motor oils, paraffin waxes, mineral oil, and Vaseline.

Topping for Light Ends, Lubes, and Asphalts

The three refineries we have just studied have all operated on East Texas crude. Their operations on any other paraffin-base oil would have followed the same general lines, though with minor variations to fit individual crudes, available markets, and engineering preferences. Let's consider the refining of an asphalt-base crude such as that from Signal Hill, California, to make gasoline, naphtha, kerosene, furnace distillate, lubes, and asphalt.

The gasoline, naphtha, and kerosene are topped-off as before.[2] The straight-run gasoline is much higher octane than that from paraffinic crude; it can be made standard grade or premium grade by proper ethylizing, and blending gasoline.

A side stream of furnace distillate may also be sold as furnace oil or diesel fuel.

The reduced crude bottoms, heated to about 775 degrees, go to a vacuum tower with steam injection. The overhead is *light-lube distillate*; the side streams are *medium-lube distillate* and *heavy-lube distillate*. The treatment so far does not differ greatly from that for a paraffin-base oil, but a major difference appears in the further processing of these lube stocks. They contain no wax, hence require no filtering, dilution, or centrifuging. They are given chemical and clay and perhaps solvent treatment and are then ready for sale or for blending

---

[2] In any refinery, bubble towers may be supplemented by strippers.

into lubricants. They are known as *pale oils* and are all classed as neutral oils, though the heaviest ones may have the viscosities of bright stocks made from paraffinic oils.

The bottom stream from this bubble tower is asphalt. It can be made to rigid specifications as to flash point, melting point, ductility, and penetration—all regulated by the temperature in the pipe still and the amount and character of the side draws. The product may be a light road oil or a paving asphalt or any asphalt between. If further modification is desired it may be run to a *converter*, virtually a batch shell still into which steam and air are introduced in such amounts as to cause controlled oxidation. The oxidation helps alter asphalt characteristics to the desired specifications.

Thus from Signal Hill crude we have made gasoline, naphtha, kerosene, furnace oil, diesel fuel, motor oils, and other lubricants, and specification road oils and other asphalts.

Refining Control

Perhaps you've noticed how few people we have seen among the stills and towers and pipes in the refineries we have visited. Almost all refining processes are controlled from a single room somewhere on the premises. So let's visit a control room on our way out.

As we step in the door we are confronted with more gauges and meters and pushbuttons and switches than there are on the flight deck of an airliner. Most of them are on a panel as long as the room, tied together by the colored lines of the flow diagram that represents the refinery piping. The meters and gauges tell the operating engineer the important pressures and temperatures and flow rates at a glance; he adjusts and checks and keeps his records of the plant's operation from the hundreds of bits of data the meters and gauges present him. Through telemetering and electronic computers and remote controls he maintains his rate of crude input into the refinery and constant outputs of specification products.

Farewell to Refining

We could study almost any number of other refineries and refinery processes, but enough of refining for a while. Most refineries have a pungent and pervasive odor that only refiners appreciate. If you would like to fill your lungs with unpolluted air, suppose we take to the open road and watch the marketing of the products we have seen made.

CHAPTER 14

Disposing of the Products

The Magnitude of the Task

GREAT as may be the twin jobs of producing two and a half billion barrels of oil a year in the United States and of transforming it into finished products, the task of distributing and selling the products is even more staggering. The thousands of producers have only to operate 600,000 wells located in thirty-two states; the five-hundred-odd manufacturers can concentrate on 900 refineries[1] in forty states, but the distributors, many thousands of them, operate something like one and three-fourths million retail outlets, serving every city, town, and hamlet in the United States, and many spots that are not even in hamlets.

In 1958 they sold nearly sixty billion gallons of gasoline, to run 57 million cars, more than eleven and a half million trucks and buses, four and three-fourths million tractors, and more than five million motor boats, and more airplanes than I have figures for, not to mention a surprising number of light plants and other stationary units.

Nor is this all. These one and three-fourths million gasoline retailers, augmented by a few thousand dealers in other products, sold more than one billion barrels of fuel oils, more than four and a half billion gallons of kerosene, nearly two and a half billion gallons of lubricants, over a million tons of road oils, nearly eighteen million tons of asphalt, more than 31 million barrels of coke, four and a half million barrels of wax and over nine million barrels of miscellaneous products.

You will have to estimate for yourself how many diesel engines were

---

[1] This includes 583 natural gasoline plants not connected with refineries of crude, and includes the plants that were in operation, building, or shut-in at the end of 1958.

run, how many multi-engine jet planes were flown, how many steam boilers were fired, how many homes and factories and offices were heated, how many lamps were lighted, how many miles of road and street were surfaced, how many miles of intestines were lubricated, how many fields were fertilized, how many plastic toys were made for Santa Claus, and how many floors were waxed and jelly glasses were sealed by those products.

The Consumption of Gasoline

You may be interested to know who uses gasoline and how much. Automotive vehicles use about 87 per cent; the rest is used by airplanes, tractors, motor boats, stationary engines, light plants, and others. Of the amount used by vehicles, about 29 per cent is consumed by trucks and buses and the other 71 per cent by passenger cars.

Airplane consumption has increased rapidly over the twenty years since the first edition of this book went to press; but the advent of turbine and jet engines in heavy aircraft use is beginning to pare away at aircraft gasoline figures. Kerosene and jet fuel consumption seem to be headed for an all-time record. Jet fuel consumption is already pushing the 500,000 barrels per day mark. The percentage change in gasoline used by trucks and buses is starting to show the same trend because of the shift to diesel fuel.

Total consumption increases with amazing persistence. Through 1930 every year since the invention of the automobile saw greater United States gasoline consumption[2] than the year before. Beginning with 1931, consumption dropped for three years; then the upward march began again. By 1937, consumption was 12 per cent above that for 1929, that for 1938 was still higher, and that for 1958 was three times greater than 1938. How many other articles of commerce can show so consistent a record?

How are these billions of gallons and millions of tons of products distributed from the point of manufacture to the point of use? The methods differ for different classes of products.

Some products—many hundreds in number but constituting only a small fraction of the total volume—are boxed or canned or bottled or barreled at the refinery and follow the regular channels of trade. You buy them at the drug or grocery or hardware store. Others go in tank cars or in barrels to the factories of other industries and come to you in

---

[2] Includes exports as well as domestic consumption.

tires or paint or building paper or roofing or batteries or candy or some other article of commerce. The distribution of these two classes is no different from that of other manufactured products.

Lubricating oils are in a different class. As a rule they are canned or barreled at the refinery, and thence are distributed by rail or truck to bulk stations or to retailers. When your car needs a quart of oil, the service station attendant is likely to sell it to you in a sealed can that he opens in your presence, thereby assuring you of getting the grade of oil you think you are getting.

Road oils and paving asphalts usually go directly from the refinery to the users. As a rule they are shipped in tank cars to the railroad siding nearest the point of use, and then by tank truck to the job. The tank cars contain steam coils, with which, when they reach their destination, they are heated until the material is liquid enough to pump into the truck. Some of the heavier asphalts are shipped in light steel drums which are cut open and discarded or saved as scrap steel at the point of use.

Gasoline, Kerosene, and Fuel Oil

It is gasoline, and to a lesser extent fuel oil and kerosene, that has a unique distribution system.

The next time you take a train trip, notice that in every town you pass, somewhere beside the tracks, there is a collection of tanks and a building or two, all neatly painted, and blazoned with the name of some oil company or jobber. These are known as bulk stations, and they are the foci of distribution to the retail outlets of the surrounding territory. Their size and number vary with the size of the town and the density of the nearby population.

Most gasoline goes by tank car directly from the refinery to the bulk station. Some goes by tank truck instead of tank car. Some goes by pipeline or steamer from the refinery to large distributing centers and thence by barge, tank car, or tank truck to the bulk stations. From the bulk stations it goes by tank truck to the retail outlets, or to the tanks of large commercial consumers. Some of it goes in barrels from the bulk station to the farm trade, for in many areas farmers now buy at a special price and get free delivery.

Kerosene and fuel oils follow the same procedure, except that there is little pipeline transportation of these products, and that fuel oil—instead of being distributed from the bulk stations to retail outlets—is usually distributed direct from the bulk station to the consumer.

The tank truck that fills the tank buried in your yard doubtless came from your local bulk station.

The kerosene and fuel oil bulk stations are not necessarily the same as those that handle gasoline and lubricants. Much of the kerosene and fuel oil business is done by jobbers who own their own bulk stations and are independent of the dealers in gasoline and lubricants.

Who Sells Gasoline?

In 1954 (no later figures are available), there were about one and three-fourths million places in the United States to which a man could drive up and say, "Fill 'er up."

Not all of these outlets were service stations. About one million five hundred thousand of these were garages, country general stores, groceries, lumber yards, restaurants, hardware stores, and other businesses to which gasoline sales were probably a minor incidental. The remaining 250,000 outlets were service stations.

Who operated all these outlets? The figures furnish a surprise for anyone who thinks the major oil companies ran most of them. Of the 250,000 service stations in 1954 only 10 per cent were chain operated. This included all the stations operated by the oil companies, major and minor, and also most of those operated by the many independent jobbers and wholesale distributors. The remaining 225,000 were presumably operated by independents, each running a station or two, or at most five or ten. It would appear, therefore, that of one million seven hundred fifty thousand retail outlets, about 25,000, or slightly more than 1 per cent, were run by the big oil companies in 1954. The figure is even smaller today, as we shall see.

Why, then, does a person get the impression that four out of five service stations are operated by Standard of Texas or Continental or some other major company? The answer is simple: The major companies make most of the gasoline; it follows that most stations sell major-company gasoline. The stations are designed and labeled to call attention, not to the identity of the operator, but to the identity of the gasoline being sold. If you look carefully, however, you will usually see somewhere about the place a sign that says JOHN SMITH, OWNER, or EVAN JONES, LESSEE.

The Iowa Plan

That EVAN JONES, LESSEE, sign is significant. In most cases it means that the manufacturer of the gasoline built and owns the station and is now leasing it to Jones. For years the major companies plunged deeper and deeper into the retail business; now they are practically out of it.

Each company, in order to sell as much gasoline as possible, bought or leased every favorable location it could find, built thereon a *service station*, and employed attendants to run it. By the middle 1930's, they owned nearly 135,000 stations. In time, the administrative burden of hiring and supervising men in hundreds of scattered stations became a major problem. Eventually the executives began to realize that almost any man, no matter how conscientious, will put a little more heart into his own business than into a salaried job. They began to lease their stations. In those days the principal activity of a station was to fill your car with gasoline, therefore they were logically called "filling stations."

About that time chain-store taxes started to crop up in various states, and some made no exception of service stations. This materially accelerated the movement toward leasing.

Hand in hand with the leasing movement went an even more important change in practice. There are four stages in the gasoline price: the *refinery price* which the gasoline brings at the refinery, the *tank car price* at which it is delivered to the bulk station, the *tank wagon price* at which it is delivered to the retailer, and the retail or *pump price* at which it is delivered to the motorist. In the retail stations operated by the manufacturer he naturally fixed the retail price as well as the tank wagon price, which is to say that for such stations he fixed the *spread* between the two prices. The retail price thus fixed was the *posted price* of gasoline. As a rule other retailers took this price and adhered to it rather rigidly. To some extent, then, the manufacturer was determining not only what the service station operator must pay for gasoline, but also what he charged for it.

With the substitution of Evan Jones, Lessee, for company employees, the situation has changed. The manufacturer has practically ceased to exercise control over the retail price by the example of company-operated stations. If John Smith, Owner, on the next corner, shades the price to quantity buyers such as the owners of fleets of trucks, Jones has to choose between losing business or meeting Smith's price. He has no one to consult but himself and nothing to consider but the tank wagon price at which he buys the gasoline, his cost of doing business, his craving for a profit, and last but not least, the price his competitors are charging.

These two trends—toward the retirement of the majors from the operation of service stations and the cessation of their influence over the retail price—have developed into a policy with most of the major companies and make up what is known as the *Iowa Plan*.

The Fight for Gallonage

The multiplication of stations until there is one for every 445 licensed drivers (1958), the ceaseless struggle for more and more retail outlets—all of this is part of one of the greatest commercial battles the world has ever seen, the fight for gasoline gallonage.

A refiner, like any other manufacturer, has certain expenses that are more or less fixed. Administration, depreciation, and interest on investment remain about the same whether he puts out fifty thousand gallons a day or a hundred thousand gallons. Total sales expense may increase somewhat with increased volume, but it by no means increases in proportion to the volume. Every refinery operating cost does not increase in direct proportion to throughput (a term used by the refiner when he expresses how much he can handle or put through his refinery in a given time); refining ten thousand barrels a day does not cost twice as much—far from it—as refining five thousand barrels a day in the same plant. It is obvious, therefore, that the more gallons per day a refiner produces and sells, up to his most efficient operating capacity, the less is his cost per gallon.

For this reason every refiner wants to operate his plant or plants at capacity, but it does him no good to produce more gasoline unless he can sell it. Under these circumstances the desire for distribution of output, for *gallonage* of sales, becomes an insatiable appetite; lack of gallonage may terminate a refiner's profits or even his existence. Gallonage is vital to a refiner, whether the refiner is the Standard Oil Company of New Jersey, the world's largest, or a new thousand-barrel plant in Illinois.

The Inflexible Market

When a refiner, big or little, seeks additional gallonage he is confronted by the stark fact that the total gasoline market cannot be expanded by his efforts. The prize for which he struggles is not a new market for his product, but only a part of an existing and satisfied market.

The manufacturer of an entirely new product, such as epoxy cement,

SUPER SERVICE

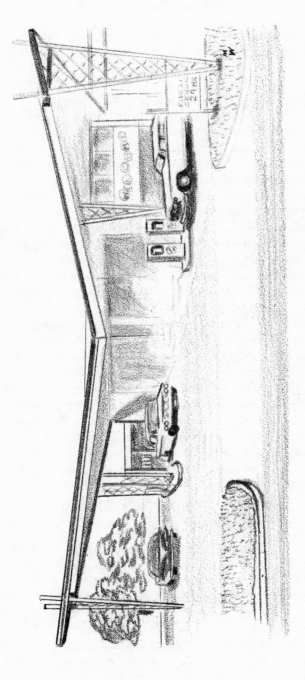

The gasoline retailer uses all the weapons in the salesman's arsenal in the fight for gallonage. Attractive service station decor, customer conveniences, and a good location help sell a few more gallons of even the best known gasoline.

sends his salesmen to people who have never before used such a thing. The business he gets is new business for the most part; no one else loses the business and no one opposes his getting it. Until he reaches the saturation point for his product, the volume of his sales is limited only by the extent and efficiency of his sales campaign.

The manufacturer of a new but competitive product such as nylon or orlon has a more difficult task; he must take a part of his market away from weavers of rayon, silk, and cotton fabrics. The competition between him and his rivals, however, is likely to increase the total market; their combined advertising, sales efforts, and price adjustments will make two cocktail dresses hang in many a home where only one would have hung otherwise.

The manufacturer of gasoline has no such possibility. Motor fuel is not sold to people who never used motor fuel before. Only in the rare and negligible case does it displace some other product. If a refiner sent ten thousand salesmen into Ohio and backed them with a million dollars' worth of advertising, he probably would not increase Ohio's total gasoline consumption by one per cent.

True, the total market grows. The increase in population, the increasing horsepower, comfort and efficiency of the automobile, and above all, the increasing mileage of super-highways—all these are steadily increasing total gasoline consumption. Relative prosperity has much to do with it; every time an additional dollar goes into the American pocketbook a part of it seems to come out for gasoline.

However, this growth offers little opportunity for the refiner striving to increase his gallonage. The increase in the market is fully satisfied by the time it comes into existence. The only way a refiner can get a greater share of the market is to take it away from someone else, or to keep someone else from getting a certain part of the increasing demand and sales.

The Weapons of Competition

The manufacturer competing for a market has four weapons in his arsenal: price, quality, service to retailer, and sales efficiency. The refiner uses all of them, though he wins few battles with the first two because the arms of his adversaries are likely to be equally good.

Price

If he has a low production cost or a location that gives him a freight-rate advantage, he may be able to shade the existing price. But in all probability, every competitor will meet his price within twenty-four hours; they have to or lose their gallonage. Every refiner, big or little, pounces on a chance to pick up some gallonage by a price cut, of course, and there is constant jockeying for price advantage, with frequent price adjustments. All of this results in a thin margin of manufacturing profit and in lowering prices to the retailer and the public, but it seldom materially increases any particular refiner's gallonage for any length of time.

Quality

As to the quality of his gasoline, the refiner can seldom claim any striking superiority over the principal competitive brands. Gasoline quality has been greatly improved over the years, as we have seen, and many of the improvements have been effected by the use of patented processes and additives, but licenses are obtainable for the use of the more important patents, and those that are not available for general use are not broad enough to prevent the attainment of similar results by other methods. This healthy competition has greatly improved the quality of petroleum products. For example, the octane rating for regular gas has risen from 55 octane in 1925 to 92 octane by the end of 1959—an advance of 37 octane numbers in 37 years. Now and then some major refiner develops a new manufacturing wrinkle that may for a time give his product superiority in some respect; but the ingenuity of his competitors soon enables them to duplicate his results and all are again on a par. Speaking generally, any responsible refiner can obtain the right to make as good gasoline as his competitors are making. Only rarely can one of them make a gasoline that is enough better to increase his gallonage materially for any extended period.

The refiner can, of course, stress the superiority of his product over those of irresponsible manufacturers and *prairie-dog* skimming plants, but the volume of business he can take away from these is small for the obvious reason that the volume of business they do is small. To gain material gallonage, a refiner must take business from other refiners as good as he is, making essentially as good a product.

This does not mean that all gasolines are alike, nor that the claims made for their respective merits are without foundation. Different refiners, equally good, make different gasolines and advertise their respective advantages. Some stress the quick starting and flashy power that come from a relatively high proportion of light fractions. Some stress the sustained power and mileage that come from a higher percentage of heavier fractions. Some stress a balance between the two; others emphasize special additives, or seasonal changes in blends to match local climates. The motorist can pick the brand that gives him the type of performance he likes, but whatever brand he picks—if made by a responsible manufacturer—will be a good gasoline. A refiner cannot build up much gallonage by talking quality alone.

Service to Retailer

Service is a big builder of gallonage. We are not only talking about service to the motorist, which is primarily the retailer's responsibility, but service to the retailer, who is the customer so far as the refiner is concerned. Even though the retailer may be one of the lingering number of major-company employees operating a station, he still insists on service. It covers a multitude of items.

One of the principal items of service is delivery. The retailer wants to know that at any hour, day or night, a telephone call will bring him a tank truck carrying whatever amount of gasoline he wants. The distributor who lets his bulk station stocks get low, or who maintains an inadequate tank truck fleet, or whose tank-wagon men fail to get the goods to the retailer promptly and pleasantly, will soon find his dealers buying from someone else. The day when the station operators or the local trucker got the gasoline from the bulk station and took it to the retail station is long since past; competition for gallonage forced the distributors into free delivery to the service station, at call—a service that also adds to the cost of each gallon.

Another item is equipment maintenance for the retailer. Theoretically the service station operator is responsible for his equipment, and a pump out of order would be his hard luck. In practice he calls up the distributor's representative and says, "One of my pumps has quit. Send a man down right away to fix it." The man is sent and the pump is fixed; otherwise the station might lose some sales, or, in the case of a station selling more than one refiner's products, the sales might go to the pump delivering another refiner's gasoline.

A third item is advertising. A dealer prefers to handle a gasoline that is widely and persistently advertised in the press, on the billboards, and about his station. He prefers a station of some typical construction or color scheme that identifies it with a well-advertised brand. He likes to be supplied with posters and signs for use about the station. All of these things he classes under service.

Then there is service to the retail customer where each and every gallon is eventually sold. The manufacturer must be concerned with the gallon-by-gallon sale, too, but we will discuss customer service later.

Sales Efficiency

Closely interwoven with service to the dealer is sales efficiency, perhaps the most potent weapon in the fight for gallonage. Like any other sales effort, it depends on good organization and good personnel. Different distributors have different set-ups, all aimed at maintaining the closest and most satisfactory relationship with the dealers, but most of them are built on some such plan as this:

Each locality is in the charge of an agent whose business it is to build and maintain sales, to see that dealers do not desert his company and tie up to some other, and to get additional dealers if he can. He sees his dealers frequently, studies their strengths and weaknesses, and advises, encourages, reproves, and helps them in every way he can. As a rule he works on a commission, so much for each gallon sold.

The agent has one or more tank-truck men who drive his trucks and deliver his products to the dealers. They are in even closer contact with the dealers than he is, for though the agent may see each dealer every day or two, the tank-truck man may see him two or three or four times a day, if he is doing a good business. The tank-truck men, moreover, are the visible embodiment of the service he receives. If they are prompt, efficient, and friendly, they do much to cement the dealer to the distributor.

If the agent encounters something too tough or too hot for him to handle, he calls the district supervisor, who has charge of a group of agencies. Perhaps a dealer is dissatisfied with his contract and wants to change it. Perhaps the agent thinks that with a little help he can land the business of a crossroads store that now handles a rival brand. Perhaps the farmers along Pine Creek claim they can get more prompt delivery from a competitor's agency in the next town to the west. Perhaps, worst of all, some trucker is hauling in gasoline from a skimming plant somewhere and is cutting the price a half cent. The supervisor comes on the run. He and the agent work on the problem until it is whipped, or they are. If they are whipped the supervisor may call in the district manager, and he in turn may call in the division manager, who either may come himself or send one of his able assistants. Before the problem is finally solved—or given up—it may have been all the way up to the head office and back again.

Such an organization, with attention to detail and its ability to concentrate on a problem, begets confidence in the dealer's mind. He is in frequent contact with the truckers and with the agent. They see that he gets the service to which he is entitled and a bit more. If he and the agent fail to agree about something, the district supervisor will be calling by in a few weeks. If he has a chance to land an attractive account he can get the help of the agent or the district supervisor or a crack salesman from the manager's office. He is made to feel a definite member of the organization, and only the assurance of equal service from an equally good and equally friendly organization can induce him to consider changing his allegiance. These are the terms in which a refiner seeking additional gallonage must talk.

Multiplication of Stations

Watching this battle for gallonage, it is easy to see why there are so many service stations. If an agent's town has a corner that is unoccupied and appears desirable, he watches it until he is confident that new business could be gained by building a station there; then he recommends to his company that it buy the corner and build the station, and he locates or brings a man to lease and operate it. While he is still making up his mind about the corner he may learn that a rival distributor is considering it. This will probably remove all doubt at once; his company may not make much money by putting a station there, but it would lose gallonage if a competitor should do so. Then we see that one station begets another at the same intersection, and two attract a third until we find many intersections with four—and thus the battle for gallonage rages.

Curiously enough, the mushrooming of service stations around a single intersection has proven beneficial to all, in most instances. Competition is intense, yes, but experience has proven that the gallonage of each established station jumps up when the new neighbor opens for business. Once again, this is an application of sales psychology.

Selling Lubes

The gasoline sales organization handles motor oil sales as well, and applies its efforts with equal force to the latter. It does not cover all lubricant salesmanship, however, for a goodly proportion of oil and grease goes into industrial use.

Most of the industrial sales are handled by special salesmen from the manager's office who devote themselves solely to this line of work. When they need help they call on the lubricating engineers who also work out of the manager's office. Such an engineer will go into a plant, study every piece of equipment, discuss the needs with the plant engineer, and submit to the owner a detailed report specifying the type of lubricant best adapted to each service and recommending the particular lubricant to be used therefor. If new equipment is added to the plant or its operations are otherwise changed, the report will be brought up to date. If a special lubricant is called for and the order is large enough, the refiner will make it to order. Many an operator of a mine, mill, or factory has been surprised to learn that the recommendations of a lubricating engineer have made important savings in his lubricating bills, and at the same time given his equipment longer life, and thus have given better service to him.

Customer Service

We have said that service to the customer is primarily the retailer's responsibility, but the refiner sees to it that the retailer maintains certain standards, for it is important to the refiner that the motoring public associate high quality of service with the refiner's brand name. Not the least of the responsibilities of agents, supervisors, and managers is to see that dealers handling their brand of gasoline and lubricants maintain attractive premises, good pumps conveniently located, clean rest rooms, well-equipped grease and wash racks, and adequate tire inflating service. They see that the customer is not given a short measure or sold inferior or substitute motor oil, and that such services as radiator filling, windshield washing, battery testing, and tire checking are performed promptly and cheerfully. The retailer who falls below these standards not only loses business; he is also hurting the reputation for service that the refiner insists be associated with the brand name of his gasoline. If the retailer persists, the refiner is likely to refuse to sell him gasoline; the refiner would rather lose the retailer's gallonage than have his gasoline's name associated with inferior service.

Each service station operator pays his price, in cost of services, in order to attract new customers and to raise gallonage. Marketing research runs through a long list of subjects—from the psychology of a child's influence on the buying habits of parents to traffic pattern studies—in an attempt to find new appeals which will boost gallonage. The familiar and lavish "open house" parties, celebrating the opening of a new station, including give-aways of cigars for men, balloons and suckers for the kiddies, flowers for the ladies, free pony rides, rides on the ancient fire truck, bottled soft drinks or color photographs for the whole family, are testimonials to the fight for gallonage. Have you given thought to the tremendous advertising effort which nowadays is directed to promoting Junior's influence on Mom's and Dad's buying habits? It works well on Dad if Junior insists on a stop at Mr. Evan Jones' station because Evan always gives him a raspberry sucker or another cowboy-decorated milk glass. Mom likes the "free" trading stamps Evan gives out too, because his sales make a regular contribution toward her eventually owning that gorgeous jewelled dresser set at the stamp redemption store.

Customer service problems are under constant watch by the oil companies and the National Congress of Petroleum Retailers. A quicker or a cleaner service or a less hazardous or a less smelly technique of dispensing gasoline is always attractive to the buyer.

A customer will trade where he finds these added features. The complete fuel pack, the service-free car, interchangeable and disposable containers of packaged gasoline are all in the picture for petroleum marketing. All are features designed to corner just a little more of the market and to keep a jump ahead of the competition.

Open-Market Sales

By no means is all of the gasoline that refineries make distributed through the sales organizations or by the sales methods we have been describing. Much gasoline is sold to wholesalers, jobbers, and other refiners on the open market.

Gasoline consumption in any particular area is subject to rather sharp fluctuations. A heavy snow or a severe cold snap may cut it in two; a hot spell or some special event like a convention may increase it materially. A refiner may find himself with surplus gasoline on hand or he may find himself facing a shortage. If he is oversupplied, he either puts the gasoline into storage against the day of need, or, if his storage supply is already ample, he offers it on the open market where it may be bought by jobbers or refiners in whose territory business is good and who need additional supplies. If he is short, he buys in the open market or buys from some refiner who is oversupplied. Some refiners sell most of their output on the open market, with little or no distribution through sales organizations of their own.

This constant offering and buying of gasoline creates what is known as the *spot-market* price, and the spot market governs the price at which practically all gasoline is sold. No refiner could hold his dealers if he tried to sell them gasoline for more than they could buy it on the open market. No refiner is big enough or strong enough to set the spot-market price, because any small refiner with a few cars of gasoline for sale can knock the spot market down to whatever price he is willing to take.

There is no oil exchange or board of trade at which bid and asked prices are recorded, but spot-market quotations are published daily in a sheet called "Platt's Oilgram," and these prices are generally accepted as gospel. Most contracts for future gasoline deliveries are based on them.

Jobbing

We have been talking as though the distribution of gasoline and lubes, from the refinery down to but not including the retail outlet, were necessarily handled by the manufacturer. Such is not always the case, as we learned when we first met the oil men. A fair proportion of the business is handled by jobbers who buy from the refiner and sell to the retailer. Many jobbers also operate retail outlets and sell to the consumer.

A jobber may distribute gasoline either under the brand name of the refiner or under a brand name of his own, or he may distribute to retailers who sell under their own brand name. He usually pays a higher price for gasoline to be sold under the refiner's brand name than for gasoline to be sold under some other name; the difference in price is created by the refiner's advertising cost. If he sells under the refiner's brand name he is also required to maintain the refiner's standard of service to the retailers, and his retailers are required to maintain the refiner's standard of service to their customers.

The jobber who distributes under his own brand name or under the brand names of his dealers usually buys his gasoline in the open market; that is, he buys it from whatever refiner is at the moment offering gasoline in the market at the best price. Jobbers' purchases furnish a fair part of the spot-market business and materially affect its quotations.

The jobber's success, like the refiner's, depends on gallonage, for only by having adequate gallonage can he keep his fixed costs per gallon low enough to make a profit. So far as he can, he uses all of the weapons used by refiner-distributors to build sales, and thus he contributes his share to the intensity of the battle for gallonage.

The jobber survives and prospers so long, and only so long, as he can distribute products as cheaply as the refiner can distribute them. Thus the jobber, like the middleman in so many other lines, is in jeopardy from the competitive struggle for volume and the tendency to eliminate middleman's profits in order to quote a lower price to the consumer.

The Effect on the Industry

Much that we have seen regarding the battle for gallonage and the activities of manufacturers, jobbers, and retailers could be seen in many other competitive industries, but in few other industries is the struggle for sales so intense. In no other industrial battle are the chief contestants so large and well organized or their smaller antagonists so numerous and active. The intensity of this battle for sales in the marketing department reacts throughout the industry and explains, for example, the eagerness of the search for peak production, the anxiety of the producer to prevent waste and obtain maximum ultimate production, and the rapidity of improvement in refinery technique. These reasons are all founded on the necessity of offering gasoline to the consumer at the lowest possible price in order to get gallonage.

Because all these activities reap their rewards in the disposal of the refined products, this is an appropriate place to consider some of the things about which a man is likely to wonder while his gas tank is being filled.

Who Fixes the Price of Gasoline?

One of the questions most often asked is, "Who fixes the price of gasoline?"

To begin with, Uncle Sam fixes the price to the extent of four cents a gallon; you pay him four cents in gasoline tax every time you buy a gallon of gas. Your state is a bigger villain; it collects tax of from 3 to 7 cents a gallon. Your city and county may also collect a tax. The total gasoline tax for 1958 averaged 10.1 cents, which was an average of $68.41 per motor vehicle for the year.[3] This means that federal, state,

---

[3] The American Petroleum Institute estimates that other automotive taxes and license fees raise the total tax bill to $92 per vehicle per year. That's what the average operator of a motor vehicle pays in special taxes for roads and traffic cops and the privilege of driving a car or truck.

county, and city governments fix the price to the extent of 10.1 cents a gallon.

The railroads and the Interstate Commerce Commission fix a substantial amount of the price. If you live in Chicago, for example, you may use Oklahoma gasoline, on which the freight from the Oklahoma refinery, whether it comes by rail or by pipeline, is several cents a gallon. If you use gasoline refined in Whiting or East Chicago, the crude from which it was made has had to pay freight or pipeline charges, which are also subject to supervision by the Interstate Commerce Commission.

Taxes and freight are fixed charges, over which no one in the oil business has any control. Subtract them from what you pay per gallon, and you will have what is left for the producer of the oil, the refiner, the jobber, and the retailer.

The retailer can figure on a spread of 3.5 to 5.5 cents a gallon, but during "gas wars" this is cut in many places. A lot of the boys who leased stations under the Iowa Plan have found that they can't break even, let alone make a profit, with a smaller spread, unless they handle a locally popular brand.

Most jobbers find that they must have about 2 cents spread between the tank-car and tank-truck prices to cover their costs and give them a reasonable profit. If the manufacturer, instead of a jobber, operates the bulk stations and distributes to the retailer, the cost of the operation is still there, though he may forego a profit on this part of the business in order to maintain or increase his gallonage.

What's left of the gasoline price goes to the refiner. At the end of 1960 he was selling at a gross price of 12.0 cents a gallon in Oklahoma, 10.2 cents on the Gulf Coast, and 11.3 cents in California. Out of that he had to buy his crude, pay his operating costs and interest on his investment, and make a profit if he could.

The Refinery Price

As we have seen, the refinery price is mainly determined from day to day by the spot-market quotations. The spot market, in turn, is determined by what purchasers are willing to pay and by what refiners are willing to take. If there is an oversupply of gasoline, the buyers will rule the market; if there is an undersupply, the refiners will name the price. The price the refiner makes on a seller's market—and tries to get on a buyer's market—is of course his cost plus a profit; as much

of a profit as his competitors will let him make, for if he tries to make
too much they will undersell him.

He sells his other products at a profit, too, for his profit is what
keeps him in business, but the income from gasoline sales is greater
than that from all other sources for most refiners. Gasoline sales and
prices are the most important by far in determining who survives and
who loses out.

The refiner's cost is fixed by his manufacturing and sales expense,
plus the cost of his crude and its transportation to his plant. How,
then, is the price of crude determined?

The Price of Crude

In general, and with many exceptions, the price paid for crude in a
field is posted[4] from time to time by the principal purchaser in the
field, usually (again with many exceptions) a pipeline company which
may be affiliated with a refining company. It is an open, public price,
available to all producers, large and small, integrated or independent.
Under the purchase contract known as a *division order* (in use in practically all fields) either the producer or the purchaser may terminate
or modify the agreement without notice; the purchaser may refuse to
take more oil from the producer or the producer may refuse to sell
more oil to the purchaser. If the producer can get a better price elsewhere he is free to do so; if the purchaser can buy his supply cheaper
from another source he has the same freedom.

If the purchaser is oversupplied, he is forced to make reductions
in the amount of oil he will buy each day from each well, instead of
dropping the price of crude. This is *pipeline proration* and is usually
only a temporary restriction on production. Although the pipeline
takes the brunt of the blame for this reduction it is more commonly
a response to seasonal oversupply at the refinery. Continued oversupply
would force down the price of crude. In such a free, open, and flexible
market the law of supply and demand has full sway.

There are times when a dominant purchaser might, for a limited
period, arbitrarily depress the price of crude in some field, but such

---

[4] The term *posted* originated in the early days when the buyers posted announcements in the field or in the oil exchanges stating the price they would pay
for crude. Nowadays the announcements are made through the press, but the
term has persisted.

occasions are rare because of the fact that, though a single purchaser may dominate one field, no purchaser dominates all the fields in a territory, and every field has its effect on the price. Moreover, most refiners are producers also, and are therefore interested in maintaining profitable crude prices.

The main factor in determining the price is competition for supplies among the refiners. If the price posted by one purchaser is lower than its true market value, some purchasing company affiliated with another refiner will soon be offering a higher price, or independent refiners will start paying a premium, and eventually the price will be equalized at its proper level. Conversely, if the posted price is too high producers will begin to offer distress crude below the posted price, and eventually the posted price will come down.

Moreover, the results will not be confined to a single field, for pipelines so link up the country that a refiner in Chicago, for example, may draw his supply from almost anywhere between the Alleghenies and the Rockies. Thus an oversupply of crude in Illinois affects the price of crude in Louisiana and Wyoming.

The net result of these interacting forces is a crude price that is fairly sensitive to the supply of crude available and the refinery demand for it.

The Effect of State-Controlled Proration

Many states fix the producing rate of each well and each field. If the crude supply fluctuated more rapidly the crude price would also fluctuate more violently. It follows that proration, which smooths out the hills and hollows of the production curve, smooths out the price curve as well.

In theory, the sole purpose of the various state conservation or proration laws is to prevent waste of oil and gas and to maintain equity between competing producers in the same field and competing fields in the same state. That they serve this purpose is beyond question, and that this purpose is uppermost in the minds of most proration officials few people will doubt. Proration officers are human, however, and no state official likes to see the producers of his state suffering from a price lower than he thinks they might have. It is not too surprising, therefore, that proration laws are sometimes administered with an eye to their effect on the price structure.

If a number of states did not derive a large revenue from per-barrel taxes on crude production, and if there were no rivalry among the states over their respective shares of the crude market, and if every pro-rated producer were not constantly anxious to produce more oil, and if, moreover, there were no foreign fields from which oil could be imported, there might be danger that production would be so restricted as to create an artificial shortage and maintain an artificial high price.

Every state wants its own producers to be able to produce as much as possible without forcing a ruinous price. In consequence each state presses for a higher production than the other states think it should have. Even if all the states had and enforced proration laws and if none of them wanted high production for taxation reasons, this inter-state rivalry would prevent inordinate restriction and unduly high crude prices.

The interstate rivalry is caused by the natural desire of every producer to make as much profit as he can as fast as he can. We have seen that as an operator's production per diem is decreased, his cost per barrel increases. The more barrels he produces daily, the greater his profit is on each barrel. Still more obviously, the more barrels he produces each day, the more dollars he receives. In the long run he may make a greater total amount under restricted than under less-restricted production. But he has to wait longer to receive it, and neither oil producers nor their bankers are of a breed that likes to wait. It follows that no conservation commission can unduly restrict the production of its state without bringing the state's producers in a swarm about its ears—and the ears of state officials are usually sensitive.

If, however, the thousands of producers in the United States should cease to clamor to be allowed to produce more oil, and should persuade the conservation authorities to impose restrictions that would materially raise the price, there would still be a safeguard against such an action. A barrel of oil from Mediterranean ports can be delivered in New York at less cost than a barrel from Oklahoma or Texas. A modest excise tax on foreign crude gives the American producer some protection, but any large increase in the domestic crude price would bring a flood of foreign oil over the excise barrier. So long as there is unused low-cost production in nearby countries, no combination of producers and conservation authorities can unduly raise the price of crude. No flood of foreign oil will come, however, while there is a potential over-supply of crude in the United States; the constant pressure of domestic producers on the conservation bodies will keep the price low enough to prevent it, or, conversely, excise tax high enough.

Proration has at times prevented the price from falling to ruinous levels, but by conserving oil, conserving the gas that helps to produce the oil, and thus increasing ultimate recovery, it defers the day of shortage and consequent higher prices. Proration's effect, therefore, is to prevent violent fluctuations in price. It does not and cannot, however, divorce price from potential production. The price of crude is still determined by the available supply, which is determined by proration under the dominance of potential production.

The Law of Supply and Demand

Thus we have the factors that determine the retail price of gasoline: (a) the posted market price of crude, determined by the amount of crude available and the needs of competing refiners for supplies; (b) the spot-market price for refinery gasoline, determined by the refiner's manufacturing cost, including the cost of his crude, and by the amount of gasoline and products on the market and the demand therefor; (c) the distribution cost, which includes the refiner's or jobber's distribution and sales expense, and the retailer's cost of doing business; (d) whatever profits the producers, refiners, jobbers, and retailers make, which in the long run must be at least sufficient to keep them in business; (e) transportation costs, including pipeline charges on the crude and freight on the refined products; and last, but by no means least, (f) taxes.

Through all of this runs the interplay of familiar economic forces, the forces that determine all commercial activity and all commodity prices. What it all comes to is this: Leaving aside the major items of taxes and freight, the price of gasoline is fixed by the same law that fixes the price of boots and bricks and brandy, the law of supply and demand.

That is why the price of gasoline is the same at practically every service station in town, just as the price of potatoes is the same at practically every cash-and-carry store in town. The price you pay for potatoes at the supermarket is determined by what the farmers are willing to take in order to dispose of their crop, plus whatever shipping cost there may be, plus the wholesalers', jobbers', and retailers' cost of doing business, plus whatever profits the competition for sales will permit them to make. The law of supply and demand determines what the farmer gets, what profits the wholesaler, jobber, and retailer make, and therefore what you have to pay. It's a tenacious law. Despite all efforts, no one has repealed it yet—for either potatoes or gasoline.

The Participants

The law of supply and demand is kept busy in the oil industry by the large number of competing units in every branch of the business, and by the constant struggle for advantage between the branches.

There are thousands of oil and gas producers in the United States. Each, naturally, wants the highest obtainable price for his crude. Some of them are integrated companies owning pipelines and refineries in addition to their wells; some non-integrated producers have only one pipeline outlet or are tied up by long-term contracts; but several thousand of them are free to sell their crude wherever they can get the most for it. If they have more crude than the market will take at the posted price, they are free to sell below the posted price, and if enough of them do so, the posted price will come down.

The producers have about 200 petrochemical plants for customers, 317 refineries, and 583 natural gasoline plants. They like to buy their crude low and sell their products high; but the moment one of them has an excess of products that he cannot sell at the going price, he is almost certain to offer it for less—and down comes the general price level. If it does not, and the refinery profit remains high, before long there will be producers, jobbers, and outsiders building refineries. Anyone with the necessary capital who can obtain a supply of crude can go into the refining business—and plenty of them do whenever a local situation arises that warrants it.

The manufacturers sell, either directly or through jobbers, to the retailers, who are always struggling to buy as cheaply as they can, and who cut prices on one another whenever they can do so and still make enough profit to stay in business.

Thus the producer's, manufacturer's, refiner's, jobber's, and retailer's continuous jockeying for a better position keeps the law of supply and demand polished up and running at high speed; such high speed that it stabilizes the net profits of the industry. In 1958, the latest year for which an authentic figure is available, the industry's net profit on its net worth was 10.2 per cent.[5]

The Result

The result is a happy one for the consumer, which means practically everybody in the United States. In 1920 the average retail price of gaso-

---

[5] First National City Bank of New York.

line, less tax, in fifty representative cities was 29.73 cents; in 1959 it was 21.5 cents. Think that over. In forty years the price of gasoline (less tax) dropped! During the same period the average octane rating of the standard grade of gasoline has been raised from 55 to above 92, and the quality has been even more improved in other respects. What other important commodity has had its quality so much improved since 1920 and yet has enjoyed a drop in cost?

Back of the Result

This notable result has not come about by accident, or by the altruism of the oil men; the price gets whittled away, as we have seen in the merry and merciless *free-for-all* for additional business. That such a continual whittling is possible is due to an activity of which so far we have had only glimpses, one that underlies and runs through the other activities of the industry—the ceaseless campaign of research. We have been living in the known world that is fairly well explored and occupied; let's try to spy out some of the new worlds that have been explored only partly, if at all.

CHAPTER 15

New Worlds To Conquer[1]

Research and Researchers

SOMEONE has estimated that the oil industry spends about $50 million a year on refinery research alone. Whether the estimate is high or low I do not know, though I think it is probably low, but in any case refinery research is only one of the many fields in which oil research workers are busy.

We can glance at only a few of the lines of work that are under way. We shall make no attempt to distinguish between the work done by the industry itself and that which is done, with or without financial aid from the industry, by such agencies as the American Petroleum Institute, the Bureau of Mines, the Bureau of Standards, the United States Geological Survey, the various state organizations, the American Association of Petroleum Geologists, the American Society for Testing Materials, the universities and colleges, and a number of others. All of the results sooner or later find their way into the common reservoir of scientific knowledge.

The objectives of all this expenditure of time and money and effort are the finding of more oil, the development of new useful products, the improvement of present products, and the reduction of costs. The relationship between some of the more fundamental types of research and these objectives is sometimes hard for the layman to see, but it is there. Even the most abstract scientist would lose enthusiasm for his work if he did not believe that in some way, perhaps yet unseen, it would ultimately benefit mankind.

---

[1] In this chapter we shall continue to refer to the whole oil-gas family as oil, without attempting to distinguish between products made from gaseous and the liquid forms of the family.

The man who questions the practical value of research in the realm of pure science forgets an important historical fact. Many of the most valuable discoveries, the beneficial use of which we now take for granted, were the result of scientific research carried on with little or no thought of its practical application. When Benjamin Franklin experimented with electricity he was not planning today's television and telephones in his beloved Philadelphia, much less an electrically operated railroad passing through it.

The Composition of Oil

After more than sixty centuries of acquaintance with oil, man is still trying to determine what it is made of. He knows that it is composed of atoms of carbon and hydrogen combined into an amazing variety of hydrocarbons, but as to the number and nature of those hydrocarbons and each one's individual physical and chemical characteristics, he has only the beginnings of exact knowledge.

Some thirty-five years ago Dr. Edward W. Washburn and his associates in the Bureau of Standards, backed by the American Petroleum Institute, began an attempt to find out. Dr. Washburn died in 1934, but the work continued. The research men selected a Mid-Continent oil and started to take it apart. By distillation with minutely-controlled temperatures, by freezing or solidifying a hydrocarbon in a solvent in which its companions are not soluble, by adsorbing a hydrocarbon on silica gel without adsorbing its companions—by one or another or all of these methods they worked steadily away, isolating one hydrocarbon after another.

When they obtained a pure hydrocarbon they proceeded to study it. They boiled it, froze it, weighed it, photographed it under the microscope, dissolved it in aniline, subjected it to polarized light, and broke it down into its carbon and hydrogen components. By the time they finished with it, the hydrocarbon had few secrets.

After years and years of work by multitudes of chemists, they have made their way down through the dry gases, the wet gases, the gasoline range, and the kerosene fraction. They have isolated and studied scores of pure hydrocarbons. If you remember that there are thousands of possible hydrocarbons in the gas-oil family, that the light end is the simple end, and that the number of hydrocarbons increases rapidly down the scale, you can understand that they have made only a fair start toward determining and studying the hydrocarbons present even in a single oil. The number of man days of work that will be necessary to complete the job is almost as staggering as the complexity of the oil itself.

Now and then someone asks, "What is the practical value of all this? Who cares how much dimethycyclohexane there is in an oil?" To this the researcher replies that he doesn't know, that he is pursuing a purely scientific study without regard to its practical aspects, and that his job is to determine the facts and let others determine what practical use they may make of those facts. If the chemist were so inclined, however, he could suggest many possible practical applications. With the oil industry being called upon to furnish an ever-increasing number of chemicals to other industries as well as to medicine and agriculture, it requires little imagination to see the practical value of knowing the number, quantities, and characteristics of all the chemical compounds that an oil contains. The industry and the public may well hope that the chemists will devote many more thousands of man days to their studies of what oil is made of.

Origin of Oil

Important as the work on composition may be to those engaged in the utilization of oil, the work that a Dr. E. Berl and a number of collaborators started at Carnegie Institute of Technology is equally important to those engaged in finding oil, and thus to those who use oil products. Dr. Berl and his associates showed that the whole oil-gas range can be made from organic material, and demonstrated the general truth of the theory held by geologists as to the origin of oil. They applied heat and pressure to organic material, in the presence of mineral substances and with an alkaline reaction, and transformed it, through an intermediate stage that Dr. Berl called "protoproduct," into asphalt. Then, continuing the application of heat and pressure, they showed that the asphalt could be transformed into asphaltic oils, these into mixed-base oils, these into paraffinic oils, and these into wet and, finally, into dry gases.

In addition to upholding the general organic theory of the origin of oil, this work has thrown new light on the order of the successive steps in the transformation of it. Many geologists and chemists have thought that the paraffinic oils were the primary product of the source material and that the asphaltic oils and asphalt itself were formed by the oxidation or sulfation of the paraffinic oils. Dr. Berl's work indicates that asphalt is the primary product and that all the oils, including the paraffinic ones, are derived from it, although minor amounts of secondary asphalt and asphaltic oils may have been formed by the action of oxygen or sulphur on paraffinic material. This conclusion was foreshadowed by a Dr. Bergius of Germany, who applied heat and pressure to organic material and produced a substance more closely allied to asphalt than to paraffin. As Dr. Berl put it, "I have no doubt that the asphalts are the parent materials of oil."

Dr. Berl's conclusions have been important to the refinery chemist, the geologist, and the consumer. An oil that has been subjected to long-continued heat and pressure and subsequent oxidation is less likely to be sensitive to further heat and pressure. Therefore it may be less amenable to refinery processing than an oil that Mother Nature has treated more gently. If the chemist knows that nature has not subjected asphaltic oils to such heat, pressure, and oxidation as was formerly supposed, new possibilities are opened for new products research and for increased refinery efficiency.

Geologists have a long-standing question: Why are comparatively young oils—such as those of the Gulf Coast—as a rule more asphaltic than much older oils such as those of the Mid-Continent? Dr. Berl's work suggests that young oils, not having been subjected to heat and pressure for so long a time, have been less altered from the parent material, and therefore are more asphaltic than older oils. This may not be of immediate help to the geologist in finding new fields, but he knows that every ray of light he can get on the successive steps in the origin of oil will ultimately help him in his search for it. In a science so new as his, he needs every bit of help he can get. He would like to relate the kind of oil he finds to the source rocks whence it comes, and to the reservoir where he finds the oil pool.

Source Sediments

It is only a step in thought from the nature of the ancient source material and its transformation into oil to the nature and environment of the modern sediments in which source materials are laid down. This is of direct importance to the geologist, for it bears directly on the nature of the source rocks for which he seeks.

A number of projects are under way. The Geological Society of America, the American Petroleum Institute, the United States Geological Survey, the Coast and Geodetic Survey, the hydrographic office of the Navy, the American Association of Petroleum Geologists, ocean-ographers of many universities, and numerous major oil companies are all active. The topography of ocean bottoms is being mapped by depth soundings, their sediments are being sampled, and their under-lying beds being cored. Men are learning about how waves and currents and sea-floor configuration affect the distribution of the materials washed into the sea, and in what types of sediments organic materials tend to concentrate. In a few years we shall probably know much more than we know now about what are, and what are not, source rocks; and what a help that will be in exploring new territory!

The Time of Migration

After studying the nature of the source material, the nature of the containing sediments, and the nature of the transformation from source material to oil, the next natural question is, at what stage in the compaction of the sediments into rocks and the folding of the rocks into structures did the oil migrate out of its source sediments and accu-mulate in its present traps? The bearing of this question on the finding of oil is obvious, and many research groups all over the country are trying to contribute their bit to the jig-saw puzzle in an attempt to find the answer. They have performed many experiments, but they recognize that the conditions of rock formation and folding cannot be completely duplicated in the laboratory. They are therefore, with-out abandoning their laboratory studies, attempting to focus the at-tention of the scientific world on the field problem, in the belief that if all the facts observed by geologists, chemists, and physicists are known and analyzed, the answers will begin to appear. Imagine the task of collecting and analyzing such a mass of facts and opinions, and you will realize what they have tackled.

This is only one of many geological and geophysical research proj-ects under way in the never-ending effort to learn more of how and why and where oil accumulates, with a view to finding more of it at less cost. All of them are interesting, but we cannot stop for more, lest we become so fascinated we spend the rest of our lives with these research projects.

Reservoir Conditions

When the oil has come to rest in its structural or stratigraphic trap, how does it exist there? To what extent is it free, and to what extent adsorbed on the surfaces of the sand grains? What are its relations to the water surrounding it, and to the gas dissolved in it or in a free state above it? What happens when a well is drilled into a reservoir? What changes take place in the character of the oil and what solids are precipitated from it? What fluid movements are set up? How are they affected by the porosity and permeability of the reservoir rock?

Every one of these questions and a host of similar ones have bearing on the best way to handle a well and obtain the most oil from it at the least cost. To find the answers engineers and technicians are measuring reservoir pressures and taking fluid samples from the bottoms of thousands of wells. Barrels of midnight oil are burned while other engineers delve into the solutions of gases in oils under different conditions of pressure and temperature and into the analyses of oilfield waters and thousands of cores of reservoir rocks. The Bureau of Mines is a leader in this general field of research, the American Petroleum Institute is sponsoring projects, and in company, university, and private laboratories scores of individual investigators are busy on studies of individual fields to learn such things as the function of water in the production of oil, and of the retention of oil by sand. Only the most studious production engineer can keep abreast of the new knowledge that is gained to guide him.

Drilling and Production Problems

The alert production engineer must also keep abreast of other research besides that devoted to reservoir conditions. Professional magazines such as *Petroleum Technology* and *The Petroleum Engineer*, for example, contain the results of studies of radioactive cement in cased wells, the behavior of fluids in consolidated sands, the interrelations in oil-natural gas mixtures, drilling muds, turbo drilling, and the discovery of new features in waterflooding. Here we have many papers of interest to the production engineer, representing the research work of hundreds of men. The next publication that comes to hand may carry a number of papers on the choice and use of drilling and production equipment, or on studies of the surveying of wells to determine the amount and direction of deviations from the vertical. The sonic logging of wells, the relation of drilling speed and weight to the character of the rocks penetrated, or any of a dozen other subjects on which the production engineer must keep posted, may be in each new publication.

You will notice that in these lines of research the direct value of the results is more evident than in those we have examined previously. They are sometimes referred to as practical or applied research as distinguished from pure research, but the distinction is more apparent than real. Practical research is based on sound science, and scientific research yields results of practical value.

Refinery Research

When we leave the field and enter the refinery, we find research at its fullest flower. Thousands of chemists, physicists, and engineers work day after day to discover ways of reducing cost, improving quality, and making new products. We already know something of one result; we have seen such rapid improvement in refinery technique that no refinery can keep up with it. There is little doubt that the improvement will continue. The men who in a few short years have perfected the application of hydrogenation, polymerization, and catalytic cracking have not locked up their laboratories and gone home. They are still on the job, and month by month they spring new discoveries on a perspiring industry. The refining business is kept panting, trying to keep up with its own progress.

Improvement in Quality

One consequence of this improvement in manufacturing technique has been a consistent lowering of manufacturing costs. Another consequence has been the striking improvement in the quality of the products. You and I can remember when we changed motor oil every one thousand miles or else; now we change every five thousand miles, and most of us know someone who may trade-in his car first. Part of the change has been due to improvements in the motor; most of it is due to improvements in the oil. In a few years we shall probably drive for the life of the car without either changing oil or needing a lubrication job.

The improvements in diesel fuels, furnace oils, road oils, and other products has been equally notable, and we have already seen some of the improvement in motor fuel. The most obvious difference between the car of today and its predecessors of twenty, thirty, or forty years ago, aside from appearance, is the greatly increased power per unit of weight, an improvement that would have been impossible without an equally striking improvement in motor gasoline.

The most notable recent advance is in aviation fuel. Piston-type airplane engines require gasoline with an exceptionally high octane value; if the motor is designed for high anti-knock fuel, the higher the octane number, the better the performance. In 1934, the best gasoline available, except for small amounts made especially for the Army, had octane ratings of 80 to 85. Today, thanks to hydrogenation, polymerization, and catalysis, gasoline is commercially available in large quantities with octane values of 110 and better.

The increase enables a properly designed airplane to take-off in half the distance, to carry 20 to 30 per cent more payload, and to fly and climb at least a quarter faster. Think what halving the take-off distance and increasing the speed by 20 to 30 per cent means to commercial aviation, or, if you are war-minded, think what a 25 per cent advantage in speed and climb meant in a dog-fight during World War II.

New Products and New Uses

The most spectacular success for the research worker, however, is in the development of new products and of new uses for old products. The field is almost unlimited. The refinery chemist is convinced that he can derive from oil every hydrocarbon that can be derived from coal, many products now or formerly derived from vegetable material or animal fat, and many products that cannot be obtained from any of these sources. What's more, to a large and growing extent he is actually doing it mainly from by-products that formerly went to waste or were burned as refinery fuel.

The rate at which an oil derivative displaces a product from some other source depends, of course, on price, quality, and dependability of supply. Many oil and gas products have already displaced products from other sources and have brought prices down to a fraction of their former levels. Others have tougher competition to meet. Glycerine, for example, is a by-product of soap making, and as such, its price has fluctuated between ten cents and seventy cents a pound since World War I. A large amount of it is used in the cosmetic, soap, explosive, chemical, and other industries. The oil industry is now making glycerine commercially and could supply the glycerine needs of the world at steady prices in the neighborhood of fifteen or twenty cents a pound.

The industry looks gleefully at competition with coal-tar products. The majority of producers of a compound called pthalic anhydride use naphthalene as the raw material. Naphthalene was once manufactured mainly from the coal-tar by-product of coke ovens, but, with a falling demand for coke by the steel companies, the industry had to find more reliable sources for its demand of nearly six billion pounds per year. They have switched to petroleum as the raw material. Benzine, which was also largely supplied from by-products of the coke plants, is required for the manufacture of aniline dyes, plastics, and synthetic fibers. The supply of benzine was once controlled by the coke market, not the benzine market. But now in this age of plastics, petroleum has been called upon for over half of the source material.

Jet Fuels and Lubricants

The piston engine is only one of the power units used in planes; another, the gas turbine or jet engine, has become the major source of power in the fast commercial and military planes of today. The fuels used are a special distilled cut composed of heavy naphthas and light distillates. The amount of jet fuels used daily is greater than that of aviation gasoline.

Ten miles up, where the jets cruise faster than sound, temperatures from 70 to 100 degrees below zero make old-fashioned lubricants thick and sticky. A few feet away, inside the engines, jet parts must be protected against 600° temperatures. New oils conquer withering jet heat which would burn up ordinary lubricants, and at the same time provide free-flowing lubrication for the cold parts.

Alcohols and Antifreezes

About 300 million gallons a year of ethyl alcohol, which used to be called grain alcohol, is now being made from oil. Refinery chemists claim that the many millions of gallons used yearly in the United States could be made from cracked-oil gases at a cost lower than that of fermenting grain. The oil industry also makes methyl alcohol, which used to be called wood alcohol, isopropyl alcohol, and other alcohols in an amount not made public, but probably equal or nearly equal to its production of ethyl.

The isopropyl alcohol is unpalatable and only slightly toxic or intoxicating, which gives it a big advantage over ethyl alcohol for industrial purposes, for it can safely be sold without denaturing. Isopropyl alcohol also makes a better winter radiator mixture than grain alcohol because its boiling point is somewhat higher; it therefore boils at a higher temperature and is slower in boiling or evaporating away.

Perhaps, however, you are one of the ninety per cent of people who prefer a permanent antifreeze, such as Prestone. The scientific name is ethylene glycol, and it is an oil product made from cracking natural gas or refinery gases. This is one reason why you can buy antifreeze from your service station which bears the oil company name as the manufacturer, just as the lubricating oil and gasoline carries his label.

Lacquers, Paints, Varnishes, and Solvents

Service as an antifreeze is only one of ethylene glycol's many uses. Perhaps of equal importance is its use in the manufacture of lacquers. Some of the alcohols made from oil find a similar use, and so do large quantities of acetone, a product whose price has been materially reduced since it has been produced from oil. Ethylene glycol, the alcohols, and acetones enter the lacquer fields as solvents, but a great many new lacquer bases are also oil products. Other oil derivatives have largely displaced Stoddard's solvent, a coal-tar product used in the manufacturing of lacquers, paints, and varnishes. The fact is that varnishes, paints, and lacquers depend on oil products for more and more of their ingredients. And speaking of lacquers and the increasing use of oil derivatives in their manufacture, you may not have heard that 40 years ago a car used to require a new paint job every year or two.

In cleaners' fluids, one oil product is displacing another. Cleaners' naphtha, an oil product, displaced Stoddard's solvent; now there is virtually no cleaners' naphtha used, and flame-proof chlorinated solvents—which are also made from oil—have taken their place. Carbon tetrachloride, another solvent and cleaning fluid often sold for household use and fire extinguishers, is commercially produced from oil.

Nylons, Rayons, and Plastics

The use of acetone is not confined to lacquers. It and various ketones and acetates are used in the production of some—not all—rayons and plastics. Ketones and acetates as well as acetone are made from oil, some of them at a fraction of their former costs from other sources, and this reduction in cost has been a boon, especially to certain parts of the rayon and nylon industry.

Following World War II, E. I. du Pont de Nemours and Company built their mammoth chemical plant at Orange, Texas, principally designed for the production of nylon intermediates. Starting with only 375 employees, it has grown to over 2,000 employees and stands as a magnificent symbol of the progress of industry in the nylon and plastics field. The du Pont process utilizes not only petroleum but butane and natural gas, which are other products of the petroleum industry, to make ethylene and then polyethylene.

Some plastic products are rubbery and pliable like polyethylene. Others are so tough and hard that they are replacing brass, copper, and aluminum in some applications. They are taking the place of glass or pottery or wood in many uses, such as on casseroles, baking dishes, the cabinets of small radios, automobile instrument panels, lunchroom table tops, toys, utensils, and modernistic furniture. Plastics are produced by the chemical reaction of liquids with one another. By varying the chemical constituents they can be given almost any quality desired. As a matter of fact the chemists keep on developing new combinations at such a pace that by the time this book goes to press it will already be out-of-date in the plastics and synthetic fiber field. Some new petroleum-base compounds developed in 1959 were already in large scale production in 1960. Spectacular growth stories can be told about the acetal resins, the polycarbonate resins, and the flexible polyurethane foam used in mattresses and cushions. Its rigid twin is becoming popular for insulation and for construction panels.

Plastics are playing an increasing part in modern life, and though not all plastics are dependent on oil products, oil products are playing an increasing part in the production of many of them. Mr. Johnson's oil well helps to keep us well dressed, comfortable, and on the move.

Dyestuffs, Textile Oils, and Leather Oils

We were speaking a while ago of ethylene glycol. It finds another use as a solvent for dyes in the textile industry. Secondary butyl alcohol and other products that may be made from oil are also used in the manufacture of dyestuffs. The dyes may still come from coal tar, but oil derivatives are used in their manufacture and application.

That reminds us also of textile oils. The modern spinning machine spins so hard and so fast that unless the wool or silk or rayon fibers are lubricated they will be eroded. The oil used to lubricate them and prevent wear must of course wash out readily. Olive oil and castor oil are the traditional textile oils, but the textile oils from petroleum show signs of displacing them.

In the same way, petroleum oils are competing with neatsfoot oil and turkey red, which is sulphonated castor oil, for oiling leather, without which your shoes and your belt would be of little use to you. Oil derivatives are also used in the manufacture of artificial leather, of rubber-coated fabrics, and the synthetic resin and silicone-resin-coated fabrics. World War II research developed many new products for the coating of special weatherproof clothing for our armed forces. Most of these grew into major industries from the complex products of oil.

Synthetic Rubbers

Leather leads naturally to thoughts of rubber, and here important developments are taking place.

Before World War II the rubber industry was almost entirely dependent upon supplies of natural latex from the forests of the tropics. The curtailment of seaborne imports and the increased demands of war-time and normal variations in supply forced the industry to turn to synthetic rubber made from oil. During the war, some 85 per cent of all our needs for rubber were produced by a mushrooming synthetic rubber industry making a product called buna-S or GRS rubber. Most of this general-purpose rubber is made from an oil product known as butadiene. By the end of the war, in 1945, GRS rubber production was demanding 500,000 tons of butadiene. By this time the door stood wide open to the research chemist in the field of synthetic rubber. Many new products given such odd-sounding names as SBR, Cis-4, Diene, and EPR have rocketed into prominence as better-than-rubber substitutes. Two of the best-known "gasolineproof" rubbers are neoprene and thiokol. They do not deteriorate in contact with gasoline and oil as do natural rubbers, and they therefore have displaced natural rubber for gasoline hose, conveyor belts for carrying bituminous materials, gaskets, and liners for oilfield pumps, hydraulic lines, and similar uses. Some of them resist the destructive action of electric currents and are consequently being used with natural rubber in the insulation of electrical cables.

Neoprene and thiokol do not lend themselves to ordinary vulcanization, but the German buna rubbers vulcanize like natural rubber. The bunas are said to be even more resistant to abrasion than natural rubber. One of them, known as buna-N or perbunan, is as resistant to gasoline and other organic solvents as is neoprene.

American synthetic rubbers are fast replacing natural rubber in the more rugged uses such as truck tires. Shell Chemical Company was one of the early producers of cis-polyisoprene rubber for tires. Firestone Tire and Rubber Company built a 30,000-ton-per-year plant near Orange, Texas, in 1960 to enter into the competition with Shell. Their product is called Diene. Phillips Petroleum Company has joined the race for the production of a polybutadiene, called Cis-4 rubber, at their 25,000-ton-per-year plant in Borger, Texas. And so the battle for the tough and rugged truck-tire market rages. Butyl rubbers, once used only for innertubes in the automotive field, are now sold in tires. The introduction of the tubeless tire probably had much to do with that conversion.

Butadiene is also valuable for the production of other plastics besides synthetic rubber.

Medicines, Poisons, and Toiletries

Turning to an altogether different field, we have already noticed the widespread use of highly refined oils as intestinal lubricants such as Nujol and liquid petrolatum, and as a base for nasal drops and sprays such as Mistol and Vick's drops. Petroleum jellies, of course, have long been used, as in Vaseline, and in the host of toilet and medicinal products of which they are the base. Increasing amounts are used in cold creams, vanishing creams, and other cosmetics. Lighter oil derivatives are used in making perfumes, toilet waters, hand lotions, shampoos, shaving lotions, wave lotions, mouth washes, nail polish removers, and many similar articles. Oil products are also used as the siccative or drying agent in brushless shaving creams.

Large quantities of oil derivatives are now being used in soaps, both the conventional types we have always known and the detergents which lather in any water.

These are all benign preparations, except perhaps to germs. A number of powerful germicides, disinfectants, and surgical antiseptics are made from oil derivatives. Insects fare no better when subjected to some of the powerful insecticides made with oil derivatives. Oils of the benign type serve as poison carriers in certain preparations not at all benign to insects. The advantage of these preparations is that they are harmless to warm-blooded animals, which you and I are supposed to be, and deadly to cold-blooded animals such as insects. "Quick, Henry; the Flit!"

The propellant in your pressurized Flit can and the well-known

DDT which we find listed on the label of many bug-spray cans and in many other "trademark" components of commercial poisons are derived from oil. Some manufacturers add the glamor of secrecy to their advertising of poisons by using technical-sounding symbols or code names. We may be completely convinced that "Product T2" in Brand X is far superior to the competitor's "Product TdT," because Brand X sponsors our favorite television program; yet both poisons contain the same ingredient—toluidine trichloroethylene, a product of oil. Who wouldn't search for an abbreviated name to describe that tongue twister?

Explosives and Poison Gases

Not all oil derivatives are harmless to warm-blooded animals. Some of them are used in the preparation of poisonous gases and explosives. An oil derivative is used in the making of tear gas, and any one of several may be used in the manufacture of smokeless powder, cordite, TNT, and nitroglycerine.

If we were not trying to keep our minds off the possibilities of war we could spend hours studying the uses and possibilities of oil products in these fields; as it is, we merely note in passing that oil products can be, and have been, used in the preparation of some of the most powerful explosives and some of the most deadly poison gases.

Anesthetics

Chloroform and ether are made from oil. Even more interesting, the oil-derived gases ethylene and cyclopropane are reported to be superior to ether and nitrous oxide or laughing gas for general anesthesia because they are free from the dangers of nausea and pneumonia. Deep surgical anesthesia is readily induced, insensibility to pain comes quickly, there are no harmful after-effects, and there is a general feeling of comfort and well-being during anesthesia. Who knows, the operating table may become positively attractive.

Even a Vitamin

Everyone nowadays must have his vitamins, and food chemists are diligently seeking ways to make them synthetically. It is often easier to give a child a tablet than to induce him to eat his dinner. Synthetic vitamin E uses trimethyl phenol, an oil derivative, as the key starting material. Most of the synthetic vitamin B-complex tablets find at least a part of their starter materials in petrochemical factories. Still others of the vitamins require petroleum products somewhere in the manufacturing process. We might say that the health and vigor and stature of our nation's people can be attributed, at least partially, to Mr. Johnson's oil well and over a half million others like it.

Food Fats

There has been no occasion in the United States or Canada to commercially produce the fats for soap or produce from oil the fats for food, but before and during World Wars I and II Germany was short of animal and vegetable fats and produced edible fats—perhaps no worse for the digestion than any others—from oil derivatives.

The Germans were even compelled to produce fatty acids for soaps and food fat from other than the oil-derived chemicals. They were forced to resort to combining the fundamental elements—hydrogen, oxygen, and carbon—uniting hydrogen with carbon monoxide under the Fischer-Tropsch process to produce the necessary hydrocarbons. From these they made the required fats. Seldom do we stop to think how dependent the world is upon the petroleum industry, even for food products, until the oil supply is cut off.

Fruit and Vegetable Uses

Even though Americans are not eating oil products, except a small amount of candy oils in good candies and perhaps some paraffin in cheap candies and cheap ice cream, the oil industry is lending a hand with foodstuffs. Ethylene and butylene accelerate the ripening of fruits. If a late spring makes the growing season short, the grower of such fruits as tomatoes, peaches, apples, pears, apricots, plums, prunes, cherries, or even walnuts may enclose his trees in tents, fill the tents with ethylene or butylene, and ripen his fruit before the frost comes. If, however, he is compelled to pick his fruit before it is ripe he may dip it in certain liquid oil derivatives and the fruit will continue to ripen while in storage or on the way to market.

And here's something to think about. Dormant potatoes have been induced to germinate by dipping or soaking them in a solution of ethylene chlorhydrin, a petroleum derivative.

Looking back at paraffin again, although we don't eat very much of it, we find it in many places in our kitchen. Next time you buy a rutabaga at the supermarket, just see if it isn't coated with paraffin to prevent drying and shrivelling. The dairy industry is rapidly converting most of its milk bottling plants to handle paraffin-coated or plastic-coated cartons. Maybe Mother doesn't can as much jam and jelly as Grandma did, but when she does she probably uses paraffin to seal the glasses, and wax (paraffin) paper to cover the tops. Almost every fresh vegetable item on the produce counter at the supermarket is now packaged in plastic bags—made from oil or gas.

Advertising specialists have found a great "eye appeal" factor in displaying products of all kinds behind a plastic-covered window in the food package.

Fertilizers

Not long ago we thought of fertilizers only as manure mixed with straw which we put on our fields and gardens in a very smelly operation. Mixing of various kinds of fertilizers began in this country in about 1849 and has now developed into a real science as well as a major industry (mostly free of horrible odors, by the way).

I am sure many of us have wondered why our favorite service station is also in the fertilizer business in the spring of every year. Possibly it seems a bit odd that the bag he sold last spring also carried the name of his major brand of gasoline. No, I'm sure he wasn't trying to diversify his services so much as he was handling an important product of his company's petrochemical plant. Mr. Evan Jones probably sells ammonium sulphate and ammonium nitrate, and others, too, in large paper sacks. These are reasonably pure, crystalline petroleum chemicals which we can spread lightly on our lawn, but they are only a small part of the list of chemical fertilizers which at least in part are derived from petroleum.

On an assortment of bag labels in Evan's station we can find several combinations of numbers like 10-6-6, standing for the formulae of the contents in terms of nitrogen (N), phosphorus (P), and potash (K). Most of the nitrogen is probably provided by compounds derived from oil, such as ammonia and several of the nitrates. The potash is combined with sulphur compounds in one variety of fertilizer; the sulphur was once a part of crude oil. You might say that most of the active ingredients of commercial fertilizer today came from petroleum. The label on the bag also refers to a large percentage of inert matter which is nothing but the carrier for the chemicals. These may be pulverized minerals, sand or peat moss or black soil, or factory wastes which help to loosen and condition the soil.

In the days before the fertilizer business had grown into a major industry, most of these now valuable ingredients were burned off as waste at the refinery or carried to the factory dump. The research chemists have let no grass grow under their feet in finding uses for petroleum products, even if the grass was fertilized "by petroleum"!

Petroleum Mulch

Gardeners and farmers sometimes spread a mulch of straw, leaves, sawdust, paper, dead moss, or similar substance to protect the roots of young plants from cold and drought. Now the *Esso Research* firm has developed a mulch of inexpensive petroleum resins that does the job far more effectively than traditional materials, and aids in producing large increases in crop yields!

This material performs four functions. Because it is dark, it absorbs heat from the sun and warms the soil. It reduces evaporation of water from the soil. It holds the soil, thereby decreasing erosion; and, finally, it prevents water and other agents from dispersing mineral and chemical fertility ingredients from the soil.

The mulch has been tested in various parts of the world for the past three years.

Odds and Ends

The oil industry is one of the largest consumers of sulphuric acid as well as one of the largest producers. Oil could easily produce the world's requirements despite its own demand. Petroleum derivatives are displacing pine oils and other vegetable oils for floating ores in metallurgical practice.

Oil derivatives are coming into increasing use in the manufacture of photographic films, and in the preparations of solutions for drying such films. Some of the same products are used in making certain types of safety glass. Detergents, emulsifiers, wetting agents, and a long list of other products and derivatives have come out of the research laboratory and have taken their places among everyday commercial products. We cannot stop to look at all of them; an hour would be required merely to read the list.

Competitive Sources

I trust I have not given you the idea that oil products have all these fields to themselves. A dozen other industries make products with which oil products must compete, and their research men are striving as hard as those in the oil businesss to make better products as well as less expensive products. In some of the fields, such as the rayons and the plastics, oil derivatives are comparative newcomers and have dislodged their opponents from only a small part of their entrenched positions. A position won today may be lost tomorrow and regained the day after. The only constant winner is the consumer, to whom all this competition gives an endless supply of new and better products at less cost.

This much may be said: In the manufacture of its ordinary products the oil industry perforce makes a great volume of by-products that would be wasted or used for fuel were they not converted into more valuable derivatives. In these by-products, there is a larger volume of valuable hydrocarbons than is readily available from any other source, and the products of the oil industry are gaining and will doubtless continue to gain new ground.

To suggest how much ground has been and may be gained, let's run through an ordinary day of an ordinary man and see how much he owes, and may come to owe, to the oil business and the refinery research chemist.

Your Day

You wake up with a headache and take an aspirin. It's made from coal tar or from oil. You notice that the house is unduly warm and realize how much the tar-coated insulation, in which the tar is an asphalt product, cuts your fuel bills. By the time you have made a quick trip to the thermostat to turn down the oil or gas furnace you think you feel a cold coming on, so you use a nasal spray made from oil. Then you wash with a soap in which the fatty acid may be an oil product, shave with a brushless cream in which the drying agent comes from oil, follow with a face lotion containing a refined oil product, and rub into your scalp a hair tonic with an oil base.

Back from the bathroom you proceed to dress. In the bathroom you stood on linoleum made in part of an oil product; in the bedroom your nylon slippers rest on varnish made with oil products. The bedroom furniture is shiny with oil-derived lacquers. The wool in your suit was lubricated with oil before being spun. The dyestuff may have been an oil product in whole or in part, and a wetting agent made from oil was used in its application. Your socks and tie are nylon into the making of which oil derivatives may have entered. Your shoes have been treated with leather oil, probably a petroleum product. Your belt, which you thought came from a pig, may have come in part from an oil refinery, or it may be vinyl made entirely from oil. You pull your clean shirt from a plastic bag, remove a plastic collar stiffener and button the plastic buttons.

Down to breakfast, to drink orange juice from oranges whose ripening may have been hastened by an oil product, to eat breakfast food kept crisp because the box was wrapped in wax paper, and eggs kept fresh by being dipped in wax, and to spread on your bread some jelly from a jar that was sealed with paraffin. The jelly was persuaded to jell by the use of pectin, in the making of which isopropyl alcohol was used. The percolator handle and switch are plastics into which oil products may have entered. The paper you read while your wife tries to talk to you is printed with ink into which an oil product has gone. After the paper was printed the press was cleaned with an oil derivative. You gulp down a daily supplement vitamin, pull on a pair of synthetic rubber overshoes and kiss the wife goodby, taking with you some of her petroleum-derived lipstick.

Then to the garage, to crawl into a car with a plastic instrument panel, plastic coil, battery, and distributor housings, a plastic steering wheel, and some two hundred other plastic parts. Its body is resplendent with a paint job of lacquer dependent on oil products. It is run by recently improved gasoline, lubricated with new-type solvent-treated oil, and kept from freezing-up by ethylene glycol. You are protected by safety glass in the making of which acetone may have been used, and by hydraulic brake fluid containing another oil product. The tires are made of synthetic rubber. Your windshield is dirty so you clean it with a solvent and put an anti-freeze and solvent solution in the automatic windshield washer bottle under the hood. As you drive away you admire the pattern of the asphalt composition shingles on your roof and feel a sense of security in their fire-resisting quality.

Rolling along the asphalt street on your way downtown, you decide to stop for gas. The service station attendant fills your tank through a neoprene hose, and polishes your windshield with a new oil product for making glass shine.

After you are gone your wife washes the dishes with a detergent and counteracts the effects of dishwater with a hand lotion with an oil base. She gets your other suit out of the chest where it has been protected by an oil-derived moth discourager in a plastic bag, and sends it and her orlon dinner dress to the cleaners, where they will be dry cleaned with flame-proof oil products. Then she shines the furniture with a polish containing a special oil product, and waxes the floor with another combination of oil products. Later in the morning she takes some snapshots of the youngsters on films made with oil products, and takes them to be developed, during which process they will be dried with an isopropyl alcohol solution. Then her favorite beautician gives her a shampoo and a facial, resets her permanent, and winds up with a manicure, using at every stage a cream or solution containing oil derivatives. When she comes home she finds that Junior has cut his finger, and she disinfects it with an oil product.

At the office, you make calls through a plastic telephone set, talk to a dictating machine with a plastic mouthpiece and a plastic record belt or disc, light your cigar with a lighter filled with high-test gasoline, and drop your ashes into plastic ash trays. When the dictating machine gets logy your secretary gives it a few drops of a new, highly-refined machine oil. Out in the factory, you inspect the use of large quantities of industrial alcohol recently bought from the nearest petrochemical plant. The machinery is lubricated with oil tailored especially for it, of course, and is probably run by improved diesel fuel or by electricity made by turbines fueled with oil or natural gas.

You keep an appointment at the dentist's, during which he uses a petroleum wax to take an impression. While you are helpless he tells you how the safe of the bank on the corner was blown up last night with TNT, which may have been made with toluol from oil, and how the police caught the yegg by using tear gas, in the making of which acetone was used. The dentist has you wash your mouth with an oral antiseptic containing isopropyl alcohol and hands you a bottle of it to use at home.

When you have escaped the dentist you decide on a light lunch, stop in at a lunchroom and sit at a table with a plastic top. You take your secretary a box of high-priced candy, and it never occurs to her, when she bites into a cream center, that the rich smoothness is due to a little candy oil. As she eats it you admire her stockings and wonder whether they are rayon or nylon.

You interrupt your busy afternoon with a pause to refresh yourself with a beverage carbonated with carbon dioxide made by burning oil, and flavored with the help of oil derivatives. At the end of the day you are carried home again by 86-octane gas and sludge-free motor oil over another asphalt street, and find that your shirts are back from the laundry where they were given a preliminary cleansing with an oil-born detergent and a final wash with an oil-born *soapless* soap.

The dripless paraffin candles on the table and your wife's newest orlon dress remind you that friends are coming for dinner. Both meal and company are good, and you particularly enjoy a salad that has been kept cold by the oil-product refrigerant in the electric refrigerator. Because your wife is reducing, the dressing is made with mineral oil. Dessert is a prune pie, the prunes in which were ripened by dipping in an oil product. It was baked in a plastic baking dish made by the help of an oil derivative. You admire the ladies' beauty and their plastic beads but do not realize how much of their beauty they owe to face creams, lipsticks, and mascara, in all of which there are probably oil derivatives. You are particularly intrigued by the perfume of the lady on your right, which neither you nor she knows came in part from an oil refinery; it certainly does not smell like one! After dinner you and your friends discuss the current investigation of the oil business, opine that oil men are a poor lot, and agree that a hundred years ago there was no oil business and people got along all right then, didn't they?

When the guests have gone and your wife is preparing for the night with cold cream containing an oil-product base you remember the cold you thought was coming on. You take a hot tub and follow with aspirin and an alcohol rub using rubbing alcohol made from oil. Then, for safety's sake, you take a swig of mineral oil. And so to bed, to dream of a time when the oil business may, if you wish, supply you with every requirement of food, clothing, shelter, transportation, livelihood, and recreation.

When that time comes the refinery research chemist may be content to call it a day and go fishing, but he probably will not. More likely he will keep developing new products and better ways to make and use old ones, 'til death do us part, amen.

Prepare for a Tour

Now we have watched the operation of this unique combination of Man and Nature known as the Oil Business, from the laying down of the source material millions of years ago to the fueling of a car that just drove in and the invention of tomorrow's products. How would you like to visit briefly the oil fields of the world and hear a little oil history?

CHAPTER 16

Oil in the Old World

The Span of Oil History

WE Americans like to think that we invented the earth, but if we claim the oil business as an American creation we are in error by at least six thousand years. This fascinating oil business is older than recorded history.

The earliest civilizations known to archeologists were developed in the valleys of the Nile, the Tigris and Euphrates, and the Indus, with a little-known but perhaps equally ancient civilization in China. In Egypt, Mesopotamia (Iraq), and India, eager excavators keep unearthing relics of earlier and earlier cultures. Each culture apparently had its oil men, already doing business. Asphalt was in common use, called by terms variously translated as slime, bitumen, pitch, asphaltum, and tar. Liquid crude oils (which appear in later inscriptions and records as naphtha, nepthar, and maltha) were probably used as early as asphalt, but, being burned instead of used as mortar and waterproofing, have left fewer evidences of their use. Gas seepages, ignited by accident and burning for years or centuries, were among the earliest objects of man's awe and adoration. The fire-breathing Chimera of earliest Greek mythology, on the south shore of Asia Minor, seems to have been such a seep. The modern Turks call it *"Yanar tash,"* meaning "stone that burns," and one of the great religions of the ancient world centered about the "eternal fires" of gas seepages.

Oil was burned for light; it was also used as a medicine, as it is today by primitive peoples who have access to it, and by some not so primitive, including ourselves. Some if not all of the uses prescribed about A.D. 60 by Pliny the Elder, the Roman naturalist, had been common among the Mesopotamians and possibly the Egyptians for thousands of years. Among other things, Pliny recommended oil or bitumen for

301

bleeding, cataracts, leprosy, skin eruptions, gout, diarrhea, rheumatism, coughs, shortness of breath, and toothache, and for straightening eye-lashes, hastening menstruation, driving away snakes, and the detection of epileptics.

Asphalt was a common material for caulking boats and ships. Once a great flood inundated the Tigris-Euphrates Valley, probably somewhere between 4000 and 5000 B.C. Judea, Chaldea, Persia, Greece, and India had legends of it, and river deposits in Lower Mesopotamia confirm it. The directions to Noah concerning it were: "Make thee an ark of gopher wood; rooms shalt thou make in the ark, and shalt pitch it within and without with pitch" (Genesis 6:14). In the Babylonian story, as inscribed on a tablet about four thousand years old, the builder of the ark says, "Six sar of bitumen I smeared on the outside; three sar of bitumen I smeared on the inside," from which we may infer that he used two coats on the outside and only one on the inside.

A prophecy of Isaiah (Isaiah 34:9-10) alludes to a devastation of burning asphalt and the resulting pall of black smoke: "And the streams thereof shall be turned into pitch, and the dust thereof into brimstone, and the land thereof shall become burning pitch. It shall not be quenched night nor day: the smoke thereof shall go up forever; . . ."

From these and many other citations we might logically conclude that naturally occurring petroleum materials played a significant part in molding history.

Thus early man was mystified and perhaps warmed by gas seeps. He burned oil and used it as a medicine, and employed asphalt as a seal and binder. His relation to the gas was largely accidental and incidental, but he produced, transported, treated, and traded in oils and asphalt. Such are the beginnings of the oil industry—some speculative, some abundantly authenticated. From these primitive origins the industry has grown, until in 1960 it produced more than a hundred twenty-three billion barrels of oil from more than fifty-seven countries.

Perhaps you would enjoy a leisurely ramble about the world, taking a look at its oil fields, ancient and modern. We'll use post-World War II political boundaries and 1960 figures in the main. Let's start where, to the best of our knowledge, the earliest civilization took root.

Iraq (Mesopotamia)

By 4000 B.C., long before Babylon became the leading city of Mesopotamia, the inhabitants of Kish and of Ur of the Chaldees were using asphalt as a mortar to hold bricks together. The builders of the Tower of Babel, who "had bricks for stone and slime had they for mortar," were following the established practice of Mesopotamia (Genesis 11:3). The artisans of Babylon and Nineveh also used asphalt to waterproof boats, coffins, basements, bathrooms, toilets, drains, cisterns, and silos. Artists used it for paint, to hold mosaics and other decorations, and to repair pottery and statues. Reeds laid in bitumen, overlain by sheets of lead, prevented the Hanging Gardens of Babylon from ruining the luxurious apartments below. Torches were made by dipping bundles of reeds in oil. Bodies were wrapped in bitumen-coated mats and burned on funeral pyres. Sorcerers based their auguries on the shapes assumed by hot oil and asphalt dipped into water, and spells were cast by burning a victim's bitumen effigy or burying it under his front door.

Nebuchadnezzar may not have been the first to use asphalt for road building, but as far as we know he was the first to write about it. One of his inscriptions says, "I . . . placed above the bitumen and burnt brick (road) a mighty superstructure of shining dust, and made them strong within with bitumen and burnt brick as a high-lying road."

He gloried also in other uses. In another inscription he says: "I caused a mighty wall to be built on the east side of Babylon. I dug out its moat and I built a scarp with bitumen and bricks." In another: "I built a palace of brick and bitumen." Of a terrace he says: "With bitumen and brick I made it tall like unto wooded mountains."

Such an extensive use of asphalt argues for the existence of an oil business, the origins of which go back to remote antiquity. At least as long ago as the fourth millennium B.C. men were finding or producing, transporting, and disposing of oil. They heated it to the proper consistency for their purposes (Herodotus says the builders of the great wall of Babylon made "use of hot bitumen in place of mortar") and in so doing drove off some of the constituents, which means that they were refiners as well as producers, transporters, and marketers. Even in those early days the business was doubtless hotly competitive, for the Babylonians seem to have had as keen trading instincts as their Arabic descendants. In Ur, about 2000 B.C., pure bitumen sold for something like the equivalent of thirty dollars a ton, which is about the price of good paving asphalt today.

The ancient industry had its troubles, as its modern counterpart does. A letter written in New-Babylonian time indicates a local shortage of supply and complains about non-delivery of a promised shipment. The industry even had to submit to price fixing, at which modern men shudder. The Code of Hammurabi fixed the price that might be charged for caulking a boat of a certain size.

The supply for Lower Mesopotamia, and perhaps for Upper Mesopotamia also, came from Hit—eight days' journey up the Euphrates from Babylon. The locality abounds in oil seeps and oil pits and was known as the "Fountains of Hit," or, in later days, the "Fountains of Pitch."

Diodorus, with the enthusiasm that oil engenders in most people, wrote, "Whereas many incredible miracles occur in the Babylonian country, there is none such as the great quantity of asphalt found there. Indeed, there is so much of it that it is not only sufficient for so many and such large buildings, but the people who have gathered there collect large quantities of it. And although the multitude is without number, the yield, as with a rich well, remains inexhaustible." Many a man since Diodorus has had that "inexhaustible" idea in sizing up an oil property.

The tankers of the period were probably rafts or the round boats made of hides stretched over osiers that Herodotus described. In these the boatmen-merchants of Armenia floated down to Babylon. After disposing of their cargoes they dismantled the boats, sold the hides, and rode home on the asses that had been their fellow passengers on the boat journey.

Diodorus and Herodotus were not the only Greeks who were struck by the oil occurrences of Mesopotamia. Plutarch says that Alexander the Great, when in the vicinity of Kirkuk, in Mosul, was intrigued by a "gulf of fire, which streamed continually as from an inexhaustible source." He admired also a flood of naphtha not far from the gulf, which flowed in such abundance that it formed a lake. "The naphtha in many respects resembles the bitumen, but it is much more inflammable. Before any fire reaches it, it catches light from a flame at some distance, and often kindles all the intermediate air. The barbarians, to show the king its force and the subtility of its nature, scattered some drops of it in the street which led to his lodgings, and standing at one end they applied their torches to some of the first drops, for it was night: The flame communicated itself quicker than thought, and the street was instantaneously all on fire. . . ."

The name Kirkuk is a contraction of Kirkuk baba, meaning "father of sound." It refers to the roaring of the gas seepage, the "gulf of fire," that so interested Alexander. If Alexander visited the place today

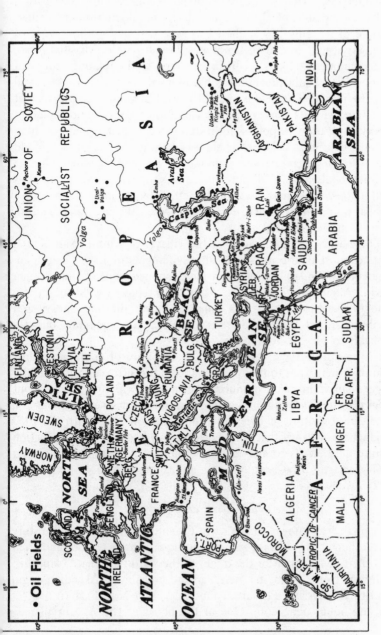

Europe, North Africa and the Middle East

he would see an oil well capable of producing about 90,000 barrels daily, and in lieu of the bull-hide boats of olden times, pipelines capable of carrying 269 million barrels a year across the deserts of Iraq, Syria, and Jordan to Haifa and Tripoli on the Mediterranean, whence modern tankers might carry the oil to the refining centers of Europe. The Haifa pipeline has been closed since 1948, but oil still flows to Tripoli. The field is one of the world's finest examples of unit operation, with its consequent conservation of gas, oil, and capital.

Kirkuk oil must be potent, for it developed a great amount of international heat even before the field was developed. The end of World War I found several nations claiming the oil rights for themselves or their nationals, and the dispute almost blew up the peace negotiations with Turkey. The matter was finally settled by assigning the field to a company in which British, American, Dutch, French, and Turkish interests are represented.

There are some other fields in the villayet of Mosul. Naft Kaneh supplies most of the modest local requirements of Iraq. Quai-yarah-Najmah-Jawan-Qasab, which may have supplied some of the bitumen used in Assur and Nineveh, has many wells reported good for a thousand to five thousand barrels a day each, but the oil is heavy and sulphurous, and the field is shut-in waiting for a pipeline outlet—probably to the Mediterranean.

In the fall of 1960, following the completion of another pipeline from Northern Iraq to the Mediterranean, this little country (slightly larger than California) produced over a million barrels a day, about 5 per cent of the world production. This placed it seventh among the producing countries of the world and about on a par with the 1960 production of Louisiana. Iraq, however, can boast of this production from only 97 wells from 7 fields (56 more wells were shut-in during 1960) whereas Louisiana marked up its production in 1960 from 23,285 wells.

Israel, Syria, and Turkey

Asphalt was known to the early Israelites who recorded the story of the flood. The Hebrews knew the liquid products also, and even had their own version of the origin of the word naphtha. "Naphtha" is derived from "nafta" or "neft," a word probably of Akkadian origin used by the Syrians, Persians, Arabs, and other Near-Eastern peoples for any liquid petroleum. One of the books of the Apochrypha recounts that when the Jews were led captive into Persia the priests "took the fire of the altar privily" and "hid it in the hollow place of a pit without water." When, many years later, Nehemiah sent to fetch it, "they found no fire, but thick water." Then, "when the sacrifices were laid on, Nehemiah commanded the priests to sprinkle the wood and the things laid thereupon with the water. When this was done, and the time came that the sun shown . . . there was a great fire kindled. . . . And Nehemiah called this thing nepthar, which is as much as to say, a cleansing."

The ancient oil industry of Palestine, however, was based on solid and semi-solid asphaltic products rather than on the liquid naphtha. The Jordan Valley, from Lake Tiberias (the Sea of Galilee) through the Dead Sea, is an area of major faults along which oil from deep-seated deposits may find its way to the surface. In Genesis we read that "the vale of Siddim was full of slime pits: and the kings of Sodom and Gomorrah fled, and fell there" (Genesis 14:10).

Since earliest recorded time bitumen from the shores of the Dead Sea has been sold in many parts of the Near East, part of it to the early Egyptians. The Romans called the Dead Sea "Lacus Asphaltites." Apparently part of the bitumen flows into the lake from seeps along the banks and part rises from the lake bottom. As is often the case, it is associated with hot water that gives off sulphurous gases. Strabo, writing about the time of Christ, describes the occurrence as follows: "The lake is full of bitumen which at irregular intervals is thrown up from the depths of the lake. The bubbles burst on the surface of the water, which latter thus appears to be boiling. The mass of bitumen protrudes out of the water and has the appearance of a hill. . . . The natives can tell when the bitumen is about to come to the surface, as then their metal utensils begin to rust. Noticing this, they at once make their preparations for collecting the bitumen by means of rafts made up of a collection of rushes."

Egyptian, Babylonian, Syrian, Assyrian, Persian, Greek, and Roman rulers fought through many centuries for the control of the Dead Sea asphalt "fishery." Mark Anthony captured it and gave it to Cleopatra, and Cleopatra leased it to Malchus, the Nabataean. Malchus, like many an oil lessee since, defaulted on the rental, and Anthony and Cleopatra had Herod punish him.

The Jordan Valley is known for oil seepages and asphalt deposits; some of them have been worked on a small scale.

The centuries of hopeful exploration finally bore fruit for Israel in 1957 with the discovery of the Heletz-Brur field about 50 miles south along the coast from modern Tel Aviv. This was followed in 1960 by the discovery of the Negba oil field, 2½ miles north of Heletz-Brur. At the end of 1960, the two fields were producing 2,500 barrels of oil a day from 25 wells. To the east the Zohar-Kidod gas field, eleven miles from the Dead Sea and 37 miles south of Jerusalem, is now supplying natural gas to the salt industry of the Dead Sea Valley.

Numerous attempts have been made to find the asphalt sources in the Dead Sea Valley. A well drilled along the highway to Sodom produced about twenty barrels of asphalt a day until it was finally abandoned because of production problems.

Syria likewise has mined asphalt. Deposits have been mined near Latakia, a name well-known to pipe smokers, and others are reported near Antioch and Aleppo. Oil seeps are reported near Alexandretta. Drilling for oil near the old oil seeps proved fruitless for many years; then a German company, the Société de Petrole Concordia, found the elusive black gold in the extreme northeastern tip of the country. The Karachok and Souédie fields were developed in the late 1950's, very close to the Tigris River and the Iraq border. In 1960, a 4,000-barrel-a-day refinery was under construction to handle the first production from Syria.

Oil seeps and asphalt deposits have been reported since earliest times from Asia Minor and the upper Tigris and Euphrates valleys in what is now Turkey. Successes in the southeastern part of the country, not far from the Iraq border, have given some encouragement to both the Turkish government and to foreign oil companies. Before the country was opened to foreign exploration in 1954 the small Raman and Garzan oil fields had been discovered by the government. By the end of 1960, there were 51 wells producing about 7,240 barrels of oil a day from these two fields. Two other discoveries, at Kahta and Kayakoy, were being developed and were awaiting an outlet when this book was being readied for the press.

Saudi Arabia

In western Arabia oil seeps are reported at various points in Yemen, but it is the east side of the Peninsula that has jumped into the limelight.

This is an area that has moved forward in discoveries and oil reserves each year since development began in the 1930's. From a meager beginning 30 years ago, Saudi Arabia has become one of the leading oil-producing countries of the world, ranking fifth behind the United States, Russia, Venezuela, and Kuwait. Daily production reached one and a fourth million barrels in 1960, from seven fields with the exotic names Ain Dar, Shedgum, Uthmaniyah, Abqaiq, Safaniya, Dammam, and Qatif. Twelve other fields were shut-in, awaiting outlets or markets.

Exploration for oil on the Arabian Peninsula began along the coast of the Persian Gulf where some world-famed oil fields were discovered. From the early near-coastal fields, Qatif and Dammam, exploration has moved both landward and seaward. A string of oil fields 165 miles long now stretches southwesterly into the Peninsula. The most inland field, Khurais, not fully developed in 1960, was already 31 miles long and 8 miles wide. Exploration successes have moved in Goliath strides northwest along the coast of the Persian Gulf to include the 7-mile-long Khurvaniyah field onshore, and the Manifa and Safaniya fields offshore.

By the end of 1960, the average well in Saudi Arabia was producing more than 5,000 barrels a day, an enviable record in anybody's oil field. Modern methods of pressure maintenance and water flooding are continuously increasing the total producible reserves of many of the fields.

Trucial Coast

The outline of the Persian Gulf south shore resembles the profile of a western riding saddle. The cantle on the west is Qatar, and the Oman Peninsula forms the pommel separating the Persian Gulf from the Gulf of Oman. The Trucial Coast, in the seat between, is composed of seven small states, each claiming independence from Saudi Arabia and the others.

One of the seven sheikdoms, Abu Dhabi, broke into the list of oil-producing countries during 1960 with the discovery of the Umm Shaif field.

No production figures are listed by the operator, Abu Dhabi Marine Areas, Ltd. The structure is 80 miles out in the Persian Gulf, and a marine pipeline company from Texas is laying 20 miles of 18-inch pipe on the bottom to deliver this new oil to the Das Island terminal. Within a year or two, after the pipeline connection, we should see some impressive production statistics for Abu Dhabi and maybe from others of the "saddle countries."

Neutral Zone

A postage-stamp–sized piece of the Arabian desert with about 45 miles of Persian Gulf coast separates Kuwait from Saudi Arabia. This miniature but wealthy buffer ground between the two countries may soon be partitioned. If the partition materializes, we shall see an example of the direct influence of this fascinating oil business on Middle East politics. Kuwait and Saudi Arabia each control undivided half-interests in the oil of the Neutral Zone. The properties are operated for them by foreign oil companies. Natural divisions of each country's oil interests and principal areas of exploration suggest that a political division is in the making which will abolish the identity of the Neutral Zone.

The onshore petroleum rights are owned by Getty Oil Company and a syndicate of ten other American companies. They have the impressively large Wafra field to their credit, where 248 wells produced about 50 million barrels in 1960. Arabian Oil Company, Ltd., a Japanese firm, discovered the offshore field, Ras Al-Khafji, in 1959. It lies at the southern boundary of the concession and is considered by some geologists to be an extension of the Safaniya field, Saudi Arabia's largest offshore bonanza. On last count it looked like each well would be capable of producing 5,000 barrels a day. A large floating flow-station of moored tankers, flow lines, loading lines, and a sea berth for the biggest ocean-going tankers will be running oil to market by the time you read this. As a matter of fact, the Neutral Zone itself may not even exist, but a fabulously productive piece of the Persian Gulf and its shore will be booming the economy of the Arab nations.

Qatar

This independent Arab sheikdom occupies the thumb-like peninsula pointing north into the Persian Gulf toward little Bahrain Island. Bahrain Island has become famous as an oil producer, so we can guess that Qatar should be a good place to look for oil because of its similar geologic setting.

Qatar broke into the list of oil-producing countries in 1940 with the discovery of the Dukhan field on the west coast. Development ceased during the war years and all the wells were shut-in. After the war, exploration moved out to sea; there, Shell Oil Company has spent tens of millions of dollars drilling holes from marine platforms. The latest discovery is 55 miles offshore on the Idd-Ei-Shargi structure under the Persian Gulf.

Qatar doesn't consume much of its own oil—592 barrels a day is pipelined to the capital city of Doha; the balance, 162,000 barrels a day, is exported.

Bahrain Island

Bahrain is an island (the biggest of the group of the same name) in the Persian Gulf just off the coast of Arabia, sheltered in the indentation between El Hasa shore and the Katar Peninsula. It is a British Protectorate and sheikdom and it has been one of the members of the Arab League since 1945. Here, Standard Oil Company of California found a major oil field in which The Texas Company now owns a half interest. The wells are large—a number of them have potentials of 20,000 to 30,000 barrels a day. Only enough has been developed to supply the immediate demand for crude and to maintain the gas-oil ratio calculated to give the maximum ultimate production. The oil field boasts 150 flowing oil wells and 5 gas wells; the gas is used for repressuring. Year after year the Island has held production at about 45,000 barrels a day, feeding it to the local refinery along with about four times as much Saudi Arabian crude oil. Products are sold throughout Europe and the Orient.

Bahrain Petroleum Company's (Cal-Tex) concession, covering all of the Island and the territorial waters (about 1,700 square miles), will not expire until the year 2024—long after this book becomes a dog-eared collector's item.

Kuwait

The little kingdom of Kuwait, near the head of the Persian Gulf, stands in fourth place in world oil production. The entire sheikdom is slightly larger than Connecticut, but it makes its mark in the world by producing more than 32 per cent of all the oil in the Middle East from 350 flowing wells.

The state and the territorial waters out to the six-mile limit in the Gulf comprise the Kuwait Oil Company, Ltd., concession. The company is a 50-50 combination of British Petroleum Company and Gulf Oil Corporation. Kuwait will doubtless produce hundreds of millions of barrels of oil and Kuwait Oil may discover millions more before its license expires in 2026.

Production is mainly from the three flowing fields of Burgan-Magwa and Ahmadi. In April, 1960, the new Raudhatain field was connected by an 80-mile, 30-inch pipeline to the Ahmadi tank farm. Another new field at Minagish, some 22 miles west of Burgan, was discovered in 1959, but was not connected to storage and marketing facilities when this chapter was written. Kuwait oil wields a big stick in world petroleum economics even though Kuwait is small and hidden away at the end of the Persian Gulf.

Iran (Persia)

When Cyrus the Great made himself king of Persia he became ruler of one of the world's most promising oil territories, one that had an oil industry even then.

Though Herodotus is comparatively modern—he wrote about 450 B.C.—the things he describes were probably done in much the same way two or three thousand years earlier; changes in production technique and refinery design were not so rapid then as now. He tells of pits dug about natural springs at Kir Ab in Susiana (who says oil wells are modern inventions?) and how the oil is produced, and how it is refined by separating its heavier and lighter constituents: "At Ardericca is a well which produces three different substances, for asphalt, salt, and oil are drawn up from it in the following manner: It is pumped up by means of a sweep, and, instead of a bucket, half a wine skin is attached to it. Having dipped down with this, a man draws it up and then pours the contents into a reservoir, and, being poured from this into another, it assumes these different forms: The asphalt and the salt immediately become solid, but the oil they collect."

Modern production in Persia began in 1913 and in 1960 it was 385 million barrels. This amount, about 41 per cent as much as that of Texas the same year, placed Iran sixth among the oil-producing countries of the world.

Twelve oil fields dot the country, most of them grouped in an area within 150 miles of the big refinery at Abadan on the Persian Gulf. These have musical names like Lali, Masjid-i-Sulaiman, Naft Safid, Haft Kel, Ahwoz, Agha Jari, Pazanun, Gach Saran and Kuh Binak. Two others not far from Teheran, the Alberz and Sarajeh fields, are only 150 miles from the Caspian Sea.

Naft-i-Shah is close to the border of Iraq, separated by a mountain range from the Mesopotamian Valley and from Baghdad, a hundred miles or so away. Its oil goes to a refinery at Kermanshah, the products from which supply the markets of northwest Iran.

The fields of southwest Iran are in the same region as the primitive production at Kir Ab (Ardericca) and not far from the site of ancient Susa, from which Cyrus, Darius, and Artaxerxes ruled the empire of the Medes and Persians. Their oil goes by pipeline to Abadan or to Kharg Island at the head of the Persian Gulf. Part is refined at Abadan: part goes by tanker (as do the refined products) to Europe, Africa, Australia, and the Far East.

Each of the fields is operated on a unit basis, with wells carefully spaced for maximum ultimate recovery at minimum drilling cost, and with rigid conservation of gas.

Oil has been found at a number of other localities and much promising territory remains to be explored. The development to date, largely by the competent British Iranian Oil and Exploration Company, has been confined to the western part of the country along the Persian Gulf.

British Petroleum Company controls the major interest in a large concession which flanks the eastern border of Iraq and extends all the way to the Gulf of Oman. A dozen or so "big time" companies, both European and American, share the development of the block under a consortium agreement; they call themselves Iranian Oil Participants, Limited. The joint operating license expires in 1979. Concession maps of Iran label the area *The Consortium*.

Offshore concessions in the Persian Gulf give great promise for submarine fields near Kharg Island and Barganshar.

Between the fields already developed and those that will doubtless be found, Iran promises to be an important oil-producing country for many years to come.

Afghanistan

This little country of 250,000 square miles, in area not quite the size of Texas, is a newcomer to the fraternity of oil-producing nations. Not until mid-1960 did 30 years of exploration effort bear the fruits of discovery. By a 1929 treaty, Russian crews participated in the exploration of a 50- to 100-mile-wide strip of Afghanistan on the north side of the Hindu Kush. In April, 1960, Soviet diplomatic sources announced the first discovery, at Aq-Shah about 40 miles south of the Russian border and 75 miles northeast of the only other show of oil, at Sar-i-Pul. Details have not been released. The rest of the country has been opened to exploration by American and other Western interests, but only preliminary mapping and necessary surveying had been completed at the time of the first discovery in the northern border zone.

The town of Kandahar, which has loaned its name to a brand of ski bindings, is one of the exploration headquarters for the western interests.

More oil will probably be found in the southern sector of the country when drilling gets under way. Perhaps some day a man will be able to drive his car through the Khyber Pass and fill up with Afghanistan gas at a street corner in Kabul. How would that strike Alexander the Great and Genghis Khan if they should pass through Kabul again?

Pakistan, India, and Burma

Long before Alexander the Greek and Genghis the Mongol swept through Kabul and the Khyber Pass, even before the first Aryans followed the same route from the highlands of Persia, there flourished in the valley of the Indus a civilization in which the citizen had as many of the comforts and amenities of life as his contemporaries in Mesopotamia and Egypt. Archaeological work at Harappa, Mohenjo Daro, and Nal in the Sind has already traced this civilization back to 3000 B.C.

Here, as in the other early civilizations, the oil business was already functioning. At Mohenjo Daro, among other things, was a reservoir or bath—eight by twenty-three by thirty-nine feet—with a bitumen layer an inch thick, and three large supply-and-drainage canals waterproofed the same way. The Mohenjo Daro'ans also painted some of their wooden buildings with asphalt.

Just where the inhabitants of Mohenjo Daro and Harappa got their bitumen we do not know, but many oil seeps and asphalt occurrences are reported in the Indus Valley. At Khatan, in Baluchistan, oil springs yield a thick, tarry oil. Similar springs are found in Rawalpindi district and elsewhere in the Punjab, and springs yielding a lighter oil are reported in the Sulaiman Hills. All of these are on or near tributaries of the Indus.

Most of the drainage basin of the Indus was separated politically

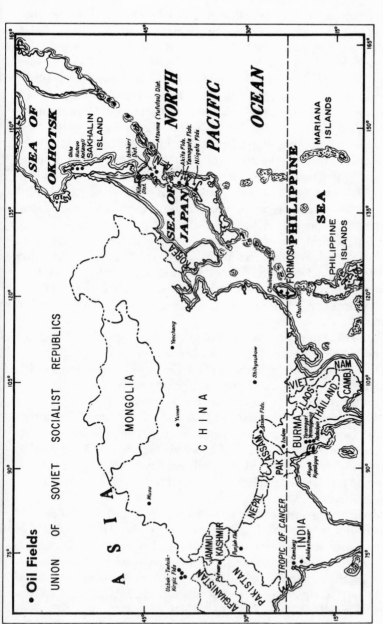

India, Far East and S.E. Asia

from old India on August 14, 1947, when Britain recognized Pakistan as an independent nation. Pakistan also includes a northeast zone around the lower part of the Ganges River east of Calcutta, sandwiched between a piece of India next to Burma and the main Indian subcontinent.

With the partition of Pakistan from India went some of the oldest oil fields of the Far East. Exploration begun in the 1880's finally resulted in a little production at Khaur and Khatan and Dhulian in the Rawalpindi district of the Punjab before the inception of the new state. Since then a booming oil and gas industry has developed in Pakistan mainly through the efforts of eight British and American companies. Production in 1960 exceeded 2 million barrels from the Joya Mair, Balkassar, Dhulian, Khaur, and Karsal fields in the Punjab. Natural gas has outstripped oil for the lead in the country's economy, providing over 70 per cent of the industrial fuel requirements of the nation. The Sui gas field, one of the largest in the world, is said to have more than $5\frac{1}{2}$ trillion cubic feet of gas reserves. This field and other smaller ones nearby deliver gas to Multan, Karachi, Rohri, Kahipur, Hyderabad, Nawabahah, and Dabeju, where gasoline plants, carbon-black plants and petrochemical industries are springing up like mushrooms.

East Pakistan already boasts two gas fields—Chhatak and Sylhet—connected to cement and fertilizer plants in the area.

Oil exploration began in the 1800's in the former British India region but the only semblance of success appeared in Assam. Here in the far northeast corner of present-day India, between Tibet and Burma, the Digboi field was discovered in 1890 and still produces 3,000 barrels a day. Later on, Assam contributed some 300 million barrels of reserves through the discovery of the Nahorkatiya and Moran fields. These fields will deliver 55,000 barrels per day just as soon as the new 720-mile pipeline is completed to carry oil to the refineries at Gauhati and Barauni.

Moderate successes were also chalked up in the early days of exploration in the Rawalpindi district of the Punjab, but this area has now taken a back seat to new discoveries at Cambay in western India. The new western area fields are shut-in awaiting an outlet, but the area gives great promise for the future of India's oil economy. Details of some very recent discoveries at Rudrasagar and Ankelsvar have not been released.

Other production of the British Indian countries comes from Burma, which, in 1937, was formally separated from India to become a separate unit of the British Commonwealth.

If you travel up the Irrawaddy River from Rangoon to Mandalay, while you listen to the paddles "chunkin," you will pass the principal oil fields of Burma: Pyaye, Yenanma, Minbu, Yenangyaung, Chauk, and Yenangyat-Lanywa-Singu. Below Mandalay you pass the mouth of the Chindwin River; two hundred seventy miles up the Chindwin is another field at Indaw. The output of these fields, combined with that from Assam, gave Burma and India about 8 million barrels in 1960— roughly the production of Alabama. More would have been produced if the markets would have taken it. The new pipeline will change this, for India at least.

The Yenangyaung field is no youngster. The name is said to mean "town through which flows a river of earth oil." How far back the oil from the Yenangyaung springs became an article of commerce we do not know, but in his *Chemistry*, written in 1724, Boerhaave says it was so scarce that it was "kept by the Princes of Asia for their own use." By 1800, the field was the world's principal source of oil.

Major Symes, in *Embassy to the Court of Ava in 1795*, gives a picture of the production methods of those days: "We found the aperture about four feet square, and the sides lined, so far as we could see down, with timber. The oil is drawn up in an iron pot fastened to a rope passed over a wooden cylinder, which revolves on an axis supported by two upright posts. When the pot is filled, two men take hold of the rope by the end and run down a declivity which is cut in the ground, to a distance equivalent to the depth of the well. Thus, when they reach the end of the track, the pot is raised to its proper elevation; the contents, water and oil together, are then discharged into a cistern, and the water is afterward drawn through a hole in the bottom."

Production by more modern methods began some seventy-five years ago and the field is now operated electrically in the most up-to-date fashion. Despite the field's age, extensions and deeper horizons, recently discovered, have expanded the original reserves. Some of the other Burmese fields have developed similar extensions, and the end of Burmese development is by no means in sight.

The oil is processed at a number of refineries, the principal ones being at or near Rangoon and Chauk, and the products, except those consumed locally, are marketed throughout the Far East and as far west as England.

Indonesia

Indonesia is ninth among the oil-producing countries of the world. The Moluccas, redolent with memories of the spice trade, Java and Sumatra, famous for coffee and tobacco and rubber, and Borneo with its head-hunters produced 150 million barrels of oil in 1960, which was within a few thousand barrels of the production of Wyoming.

The recorded history of these islands, which are strung like beads along the equator, goes back less than a thousand years, and there is little to indicate the extent to which, before the advent of the white man, the Polynesian and Malaysian inhabitants noticed or used the numerous gas seeps, oil springs, and asphalt deposits scattered through the archipelago. On the Mota Mutika River on the island of Timor are both an oil seep and a burning gas-vent. From earliest recorded time the natives have regarded the fire as sacred and have used the oil in lamps. It is reasonable to suppose that similar occurrences on the other islands received similar attention.

Not ancient but modern history, however, gives the East Indies their petroliferous fame. In them the world's second largest oil organization had its beginnings.

In 1883 an enterprising Hollander named A. J. Zijlken obtained an oil concession in the Langkat district, near Pangkalan Brandon, in northwest Sumatra. By 1885 he had a small well. By 1890 he had enough production to call for transportation and a refinery. In Holland that year, J. B. A. Kessler had organized and managed to finance a company called the "Koninglifke Maatschappij tot Exploitatie von petroleum-bronnen in Nederlandsch Indie" (Royal Dutch Company for the Exploitation of Petroleum Wells in the Netherlands Indies). The new company began by taking over the Zijlken concessions. In 1896, Kessler employed as head of his sales department a young man named Hendrik Deterding, who had come out from Holland to the Indies to seek his fortune as a bank clerk, and who was working as factor at a trading post at Penang on the Malay Peninsula. When Kessler died in 1900, young Deterding, who had already shown extraordinary ability, succeeded him as president. By 1905 the company had developed a large production in Sumatra, Java, and Borneo, had adequate refining facilities, and had its eyes on attractive markets, but it lacked one very necessary thing: it needed ships.

Meanwhile an old English shipping and mercantile concern, the Shell Transport and Trading Company, had in 1898 gone into the oil business in Borneo. The firm had its beginning in a curiosity shop in Houndsditch, London, and took its name from the shells in which its founder, Sir Marcus Samuel, used to trade. It had production in the Balikpapan and Sanga Sanga or Kutei fields, as well as a refinery at Balikpapan. The Royal Dutch Company was in these fields also. The two companies fought each other and both fought Standard Oil, which had established itself in the markets of the Orient. Shell production was overshadowed by that of the Royal Dutch Company, but Shell was strong in the one spot where Royal Dutch was weak: it had ships.

The financial genius of Deterding, who eventually became Sir Henri Deterding, brought about the affiliation of the two companies. With its headquarters in London and its operating office in The Hague, the Shell-Royal Dutch group did not long confine itself to the East Indies. Today there are few important producing countries and even fewer important markets in which its subsidiaries and affiliates are not active. It has not hesitated to compete with American companies on their home grounds.

Indonesia has many producing fields and much promising territory still untested. The oil comes largely from three main districts: Java, where two fields produce over 2¼ million barrels a year; Kalimantan, in Borneo, where seven fields with 50 to 60 years' history make 7¾ million barrels a year; and South Sumatra with 17 fields putting 33 million barrels into the pipelines.

Only four major producers in Indonesia—Caltex Pacific Oil Company, Royal Dutch Shell, Standard-Vacuum Petroleum Mij, and Permina—carry away the honors for the 150 million barrels produced in 1960. Refineries serve these fields and the products are marketed throughout the Orient and Europe.

Much remains to be explored for oil possibilities in Java, Sumatra, and Borneo; the islands, such as New Guinea, which are more remote from the main centers of European settlement have hardly been touched. In Dutch New Guinea intensive aerial mapping and geologic work have found three fields: Klamono, Wasian, and Mogoi. Together they produce 4,200 barrels a day.

World War II operations in the southwest Pacific made headlines for many Indonesian oil fields and refineries, thus dramatically pointing out the strategic importance of oil to the conduct of war. Refineries at places like Balikpapan, Soerabaja, Bendoei, Palembang, Pladjoe, Tjepoe, and Kapoehn were either totally destroyed, severely damaged, or sharply curtailed in production during the battle for the East Indies. In spite of the widespread demolition, Japan was able to produce 3¼ million barrels from the Java fields and about 10 million from the Sumatra fields for their war machine. Control of oil was indeed a major factor in the battle for the Pacific.

Indonesia will doubtless continue for many years to be an important factor in the world's oil production.

British Borneo

Great Britain owns a slice of Borneo, and the British have not been backward in the search for oil. In 1866, a well off the northwest coast on the island of Labuan was drilled to the great depth of nineteen and a half feet. It produced a small quantity of oil for many years.

The real production, however, comes from Sarawak, a crown colony, and Brunei, a protected sultanate, both on Borneo itself. The production in 1960 was 35 million barrels, about the same as Utah, making British Borneo the nineteenth country in oil production.

Labuan had the earliest discovery but Sarawak holds the distinction of having the oldest continuously-producing field. Discovered in 1911, Miri field has produced over 76 million barrels of crude oil and is still going at 1,100 barrels a day. Two fields in Brunei—the older (1929), Seria field, and younger (1955), Jerudong field—produce over 90,000 barrels every day. Two-thirds of the production is consumed "at home" and the other third is shipped out by tanker.

Australasia

Australia is a country as big as the United States, with fewer people than lived in the United States when the British burned the White House during the War of 1812. Except for a fringe several hundred miles wide around the western, southern, and eastern coasts, it has a climate like that of the Mohave Desert and is about as populous. Throughout most areas in the interior, rock outcrops are few and geologic work is uncertain.

No vigorous and alert people like the Australians, however, could be deterred from the search for oil by such obstacles, and millions of dollars have been spent in drilling in several of the six states of the Commonwealth. Some of the early-day exploration was done with

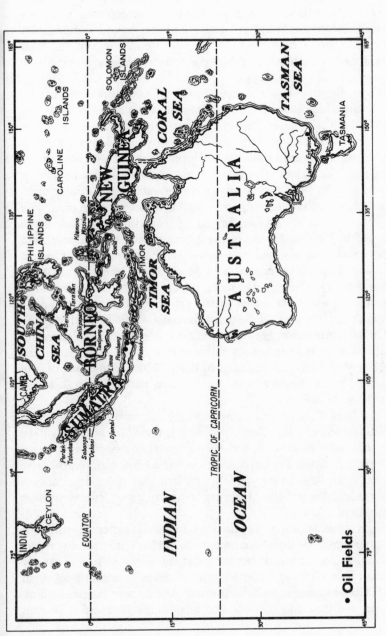

S.E. Asia and Australia

• Oil Fields

little or no geologic guidance and the net results were dry holes and teasers—wells with showings of oil but no commercial production. A few wells, located on competent geologic advice, have yielded no better results. However, these dry holes no more condemn Australia than those drilled in Nevada before oil was discovered elsewhere would have condemned the United States.

Within the last decade, interest in the oil possibilities of Australia has mounted almost to fever pitch. The Petroleum Search Subsidy Act of 1957–1958 encouraged many foreign operators to delve into almost every Australian sedimentary basin. On November 4, 1959, the Act was revised to include subsidies on geophysical surveys, drilling for stratigraphic data, and off-structure tests. The results have been most encouraging. In 1959, Australians anxiously watched the drilling of 59 wells but to no avail; only noncommercial shows were recorded. The year ended with no commercial oil on the continent. New hope bloomed in 1960 with a few of the 17 exploration tests finding good shows of oil. One gas field was discovered and confirmed in Queensland and another confirmed in Victoria. Two wells found small amounts of gas in New South Wales; it looks like the boom is on in the newest continent to be added to the list of gas-producing countries. By the time this book is off the press, Australia should be producing oil.

Although Australian geologic conditions are less favorable, on the whole, than those of many other countries, it is reasonable to expect that many oil fields will be found in this vast area. A fair part of it is underlain by sedimentary rocks, and in many parts of it oil showings have been found.

New Zealand is another country where the story is not yet told. About two-fifths of the Commonwealth is underlain by sedimentary rocks that may contain oil, and only a moderate amount of exploration has been done. The only results to date are four wells on the west side of North Island, each yielding four barrels or less a day, but geologic work and some wildcatting are being carried on with reasonable prospects of success.

New Guinea is an island in the East Indies. Its eastern, British half is administered by the Government of Australia, the southern part (Papua) as an Australian territory, and the northern part (formerly German New Guinea) under a United Nations Trusteeship set up in 1949. Head-hunting cannibals still roam the interior jungles of New Guinea, and crocodiles, malaria, and blood-suckers are said to be even more serious deterrents to exploration, but oil geologists are hard to discourage. A number of promising structures have been found, but remote locations and drilling difficulties have prevented their successful development.

The exploration story reads like that of early-day Australia—a few good shows of oil and gas, but none commercial.

Philippine Islands

The Philippines lie to the north of the East Indies and are topographically and geologically a part of the same archipelago. The President of the Republic can stand on his southernmost island and see the Borneo domain of Her Majesty, Queen Elizabeth.

In *Ores and Industry in the Far East* W. B. Heroy says, "On all of the ten larger islands of the Philippine Group, except on Palawan, there are definite indications of the presence of oil. Luzon, Leyte, Cebu, Mindanao, and Panay have definite oil seepages and gas seepages, and oil shales occur in the other larger islands as well as in many of the smaller islands. . . . Geologic mapping indicates that the oil-producing areas have been folded into anticlines and thus are structurally favorable for oil accumulation."

The earliest efforts to find oil near these seeps began in 1896; since then 107 wells have been drilled. One early test on the Bondoc Peninsula of Luzon gave promise for a while, but not until 1959 was the first commercial discovery made near Toledo in west-central Cebu. A 1959 gas discovery stimulated exploration in the Cagayan Valley, and now there are 26 companies hunting oil and gas in 30,000 square miles of concessions. The Philippines use about 50,000 barrels of oil a day—all of it shipped in, so a few good fields are needed.

Formosa (Taiwan)

Many favorable aspects of the geology of Formosa suggest that it has the possibilities of being an important oil-producing area. The whole of the island is covered with sedimentary rocks which may be sources of and reservoirs for oil. Exploration, dating almost from antiquity, has produced a little oil from small wells. In 1905, drilling activity picked up in 54 districts; the result was a few small and short-lived producers. The determination behind the search for oil is shown by the fact that, in 1937, the Japanese drilled a wildcat well to 11,502 feet—the deepest well in the world outside the United States. Drilling activity has continued since the Nationalist Chinese regained control of the island in 1945. Fields have been deepened to newly productive reservoirs and expanded in area, but the discovery rate does not quite keep up with the depletion of the existing wells. In 1960, 24 wells in six fields were producing oil and gas, but production totaled only 39 barrels a day. The search goes on with some success. New gas reserves have been found and the search is now becoming feverish.

Communist China

No doubt you have seen statements of Russia's objectives in China, and prominent among them references to China's "rich oil resources." It would appear, to judge from editorial comment about China, India, Iran, Ethiopia, and Bulgaria, that any country subjected to conquest is *ipso facto* rich in oil. The fact is that China had no known oil resources prior to World War II worth the dispatch of a regiment, much less an army.

This is not to say that China has no oil wells. In this, as in so many other things, the Chinese may well claim to have anticipated the achievements of Occidental civilization. They may have been drilling oil wells a couple of thousand years before Colonel Drake drilled his famous well in Pennsylvania in 1859. The origin of the business, too, seems to offer an interesting parallel to that in the United States, though in this we must stray somewhat into the realm of conjecture.

In Szechuan are salt beds; centuries before the Christian era, the Chinese were drilling wells to obtain salt. Using a primitive method of raising a bit and letting it fall by means of a "spring-pole" (a similar method was used for drilling water wells in Europe and America many years ago) they drilled wells to the salt beds, poured water down the holes, let it saturate itself with salt, pumped the brine back to the surface, evaporated the water, and had the salt. In some wells they encountered gas, and discovered, perhaps by accident, that the gas would burn. In time they learned to conduct the gas through bamboo pipes for heating and illuminating and to evaporate the brine.

In some of the wells the recovered brine was polluted with a greasy substance, which was later also found to burn and to help cure skin diseases and rheumatism. Eventually—how long ago no one knows—the Chinese began to burn it in closed lamps. The first lamps of China were not introduced by John D. Rockefeller and Sir Henri Deterding, though until these giants began their struggle for the Chinese market the number of lamps was small, for the simple reason that there was little oil to supply them. Despite the fact that Chinese ingenuity succeeded in drilling wells as much as four thousand feet deep with the crudest of equipment, all the wells of Szechuan never produced more than a few barrels, or at most a few hundred barrels a day.

In northwest China are oil seeps and the Chinese may have recovered and used some oil from them. Many years ago a few shallow wells were drilled with crude equipment, and before World World II some wells had been drilled with Chinese and American capital.

Since the take-over by the Communists, the reports of the Chinese Republic have been quite uncertain and sometimes conflicting. It has been reported that Japanese technicians drilled 214 wells and developed three areas during the war. These are about equally spaced across northern China. Starting with the Wasu field in the northwest, the others are: the Yumen field in the central region, and the Yenchang field on the Yellow River. Wartime production was probably not over a few thousand barrels a day from all fields.

The Communist influence and industrial control of the country have cloaked the Red China production reports in vague and misleading terms; however, some sources say that the country produced 5½ million metric tons of crude oil in 1960—or about 110,000 barrels of oil a day.

Except perhaps for western Mongolia and Sinkiang, about which little is known and which may have possibilities, the most promising territory in China is still in the northwest and central part where the old development took place. Most of the rest of China is underlain by igneous or metamorphic rocks, the types in which commercial oil is hardly ever found.

In Manchukuo, China acquired lots of room in which to drill but not a great deal in the way of promise. The western part is made up of thoroughly metamorphosed sedimentary rocks. These are overlain in part by younger, wind and stream sediments, typically devoid of organic matter, hence not likely to have oil accumulations.

Eastern Manchukuo is underlain by igneous and metamorphic rocks in which are basins of younger sedimentary rocks containing coal and oil shale. No oil seeps are reported from these basins, and, though important oil fields may be developed in some of them, the odds seem to be against it. Some oil is produced by mining oil shales and distilling their kerogen into oil—a process that costs more money and requires more labor and greater investment than producing oil from wells.

Red China drew no petroleum prize in Korea (Chosen). The peninsula is made up almost wholly of igneous and metamorphic rocks, and the few sedimentary areas have been explored without success.

Japan

Japan is doubtless more eager than any other nation to develop a large supply of oil. With her densely populated country rapidly being industrialized, she needs oil desperately. The need has been recognized for many years and few countries have been more thoroughly explored than Japan.

In the province of Echigo, across the island of Honshu from Tokyo, is a place called Kusodzu. The name is said to be a corruption of the words *Kusai midzu*, "stinking water," which is early Japanese for petroleum. The term for natural gas is *kazakusokzu*, which means "stinking water wind." At Kusodzu are both oil seeps and gas vents. Japanese history reports that "burning water" was found in Echigo nearly thirteen hundred years ago. In 1613, a man named Magara obtained oil in Echigo and attempted to refine it to obtain illuminating oil. What success he had is not recorded.

The Kusodzu seeps are in a belt of favorable sedimentary strata on the western coasts of Honshu and Hokkaido islands, the two largest in the archipelago. Seeps seep at other places here and there in this belt, and about 31 oil fields have been developed. The wells average less than two barrels a day, which means that Japan's production, despite intensive search and development, is distinctly in the stripper class. The 5,100 wells produce only 7,900 barrels a day, but this is improving each year.

Sakhalin

Sakhalin Island, close to the Siberian coast, and once again all-Russian territory, is the northernmost main island of the Japanese Archipelago. At the close of the Russo-Japanese War the south half was ceded to Japan, but it proved to have practically no oil possibilities. Then, after World War II, it was surrendered to Soviet Russia, and since 1951 all production figures have been shrouded in Russian censorship. The north half, which contains large lakes of oil and many oil and gas seepages, has most favorable geologic conditions. In 1938 some 3.9 million barrels were produced—about 50 per cent more than the combined production of Japan and Taiwan.

Known production from this small patch of ground in the west Pacific totaled over 101 million barrels before its identity was buried in questionable Communist statistics. The island has great potentialities and its development under Communist control would be interesting to watch if we could only see it.

USSR

The USSR has been more fortunate than Japan. Some of her oil fields in the Caucasus are among the greatest in the world. Areas of large promise, most of them explored only slightly, if at all, lie scattered across her vast expanse from the Crimea to Kamchatka. In 1960, with many possibilities still untouched, the USSR produced 2.96 million barrels a day: more than any other country except the United States. This was 13.8 per cent of the world's total: more than twice as much as Texas, and as much as California, Kansas, Louisiana, New Mexico, and Wyoming produced together.

The Caucasus Mountains, from which the white race is named, run from the Black Sea to the Caspian, and separate the rest of "Russia" from the Soviet states, Georgia, Azerbaijan, and Armenia. Beyond these lie Iran, Iraq, and Turkey. At the eastern end of the Caucasus, in Azerbaijan, the Apsheron Peninsula projects into the Caspian, and on the Peninsula is the city of Baku. On the Peninsula, and on nearby islands in the Caspian, are large oil springs and asphalt deposits which have been known and used since ancient times, and numerous gas seeps that have played a major part in one of the great religions of the ancient world.

The religion of the Fire Worshippers was already old when Zoroaster reformed and reconstituted it somewhere between 1000 and 600 B.C. It was the religion of Persia in her days of glory and it has left a deep impress on Judaism, Christianity, and Mohammedanism. The Magi of Scriptural days were Zoroastrian priests and astrologers. The religion's elaborate ritual centered about the "eternal fire" that burned perpetually in or adjacent to the temple, and the principal temple was near Baku. One of the early Iranian myth-hymns, long antedating Zoroaster's reformation, refers to an eternal fire that needs no feeding, on the shore of the Caspian Sea.

In A.D. 636, before the Saracens swept the Persians from their mastery of the Apsheron, the Baku neighborhood was dotted with Zoro-astrian temples, and thousands of pilgrims visited the sacred region every year. The Moslem invaders extinguished Zoroastrianism in Persia, but a few devoted pilgrims continued to come from India until recent years. Hanway wrote in 1754, "What the Guebers or Fire-Worshippers call the Everlasting Fire is a phenomenon of a very extraordinary nature. . . . There are several ancient temples built with stone, supposed to have been all dedicated to fire. Amongst others is a little temple at which the Indians now worship. . . . A little way from the temple is a cleft of rock, in which there is a horizontal gap, two feet from the ground, nearly six feet long, and about three feet broad, out of which issues a constant flame, in color and gentleness not unlike a lamp that burns with spirits, only more pure. . . . They do not perceive that the flame makes any impression on the rock. This also the Indians worship, and say it cannot be resisted, but if extinguished will rise in another place."

Several writers earlier than Hanway refer to the Baku oil. Marco Polo says, "To the north of Armenia lies Georgiana, near the confines of which there is a foundation of oil which discharges so great a quantity as to furnish loading for many camels. The use made of it is not for food, but as an unguent for the cure of cutaneous distempers in men and cattle, as well as other complaints; and it also is good for burning. In the neighboring country no other (oil) is used in their lamps, and people come from distant parts to procure it."[1]

People were still coming from distant parts to procure it when Peter the Great conquered the Khanate of Baku in 1723, and Peter sent a master of refining to take possession of the supply and see to its transportation up the Volga. In more recent times, Joseph Stalin worked in the Baku oil fields, became there a revolutionist, and rose thence to the dictatorship of the USSR.

The glory of Baku's deposits has not departed. Some of the world's largest wells, with initial productions up to 120,000 barrels a day, have been drilled there. Azerbaijan produced nearly 80 per cent of

---

[1] Another translation reads: ". . . inasmuch as a hundred ship-loads might be taken from it at one time." It says nothing about camel-loads; perhaps a ship and a camel look alike in medieval Latin. In this translation "the men" and "the other complaints" are absent, the "cattle" become camels, and "cutaneous distempers" become simple mange. The way of the translator must be hard!

Soviet oil in 1938; it produced 340,000 barrels a day in 1959. One would think that such fields as Bibi-Eibat, Balakhany, and Sarakhany, not to mention myriads of smaller fields, would exhaust the possibilities of the Apsheron Peninsula. But new fields, extensions of old fields, and new producing sands are still being discovered, and there are said to be many structures still untested. Inland from Baku, along the south side of the Caucasus clear to the Black Sea, are other promising but untested areas. During 1948, new discoveries balanced the decline in older wells, so that the 1948 production was up 5 per cent over 1946, the last year for which good figures are available.

Across the Caucasus from Baku, on the shores of the Caspian, are the Daghestan fields of Isberbash, Achi-Su, Kaya Kent, and Berkei. Inland along the north flank of the Caucasus is the Grozny district, which contains two oil fields called Old and New Grozny, and a number of younger fields. New fields are still being found and old fields are still being extended in both districts, though production fell off materially between 1939 and 1944.

On the north side of the Black Sea is the Maikop district; still farther west, on the peninsula that stretches out toward the Crimea, is the Kuban district. Kuban began producing early in the history of the modern oil industry; Maikop's first production was in 1891. Despite such venerable age the production of both districts apparently has decreased notably only in the last few years.

Farther north, between the Ural and Volga rivers, is a great geologic basin which has scarcely started to be explored. Its possibilities, however, are already indicated by such fields as Chusovski Gorodki, Krasnokamsk, and Polazninski in the north, Ishimbaevo and a cluster of others in the east, Syzrian and Yablonvo in the west, and Tuimaza and Buguruslan in the interior. The basin appears to occupy some 300,000 square miles. Its production for 1944 was nearly 8 million barrels, the last figure available.

Farther west, in the Ukraine, a new salt dome basin was discovered at Romi. Twelve new salt dome fields in 1958–1959 may give the great plains area of the southern USSR important production.

In the northern USSR, not far from Archangel, unimportant production has been obtained in an area containing outcrops of asphaltic sand. Little exploration has been done, and whether the area has any real possibilities remains to be seen.

Northwest of the Caspian Sea, in Kazakh—the land of the Ural Cossacks—is the Emba salt basin which first produced in 1911. Soviet geologists are optimistic about the basin's possibilities as they claim to have mapped four hundred salt domes and anticlines of which fifteen are fairly important fields. However, non-Soviet geologists do not consider the area too promising.

Away to the north of Emba, on the eastern side of the Ural Mountains, lies a larger salt basin, said also to contain salt domes and anticlines. It may or may not have been tested and may or may not be productive.

Southeast of Emba, in Turkmen, the USSR has a group of fields in production on the eastern shore of the Caspian, nearly opposite the Apsheron Peninsula and close to the border of Iran. Farther east, in the Bukhara Fergana region, 22 per cent of the USSR's production has been obtained on the banks of the Oxus, now the Amu Darya, where the Macedonian Phalanx once battled the Parthian hordes. Emba, Turkmen, Bukhara, and Fergana are in the vast lands of Central Asia, on the borders of Iran, Afghanistan, India, and China, from which the Mongols under Genghis Khan once conquered the most advanced culture of their day. This Central Asiatic region has great potential.

In the remainder of Soviet-Asia, an area twice that of the United States, virtually no exploration has been done. Surface evidences are reported as follows: in southern Siberia at Minusinsk on the headwaters of the Yenisei and Kuznetsk near the head of the Tom, and near Lake Baikal; in eastern Siberia on the Aldan River and between the Lena and the Vilyui and on the Amur where it turns north from the Manchurian border; on the Bay of Khatanga, an arm of the Arctic Ocean; and on the Kamchatka Peninsula, jutting into the Pacific above Japan. Tests have been drilled in the Lake Baikal, Aldan River, and Khatanga areas; beyond this the whole vast area is yet untouched.

Such is the USSR, the world's second largest oil producer, with new oil being found in her oldest producing areas, with new fields being found in new areas, and with most of her eight million square miles still practically unexplored. Her exploration is being guided by modern geologic methods, and development is proceeding at fantastic rates.

Recent reports suggest that to prospect the country systematically, the Soviets plan to drill 984 million feet of deep holes in the next 20 years. They reported drilling 25 million feet of holes in 1960, the same as in the 1959 record year. Production has mounted rapidly from 388,-000 barrels a day in 1945 to 2.96 million barrels a day in 1960 when the USSR exported 502,000 barrels a day. She is fast becoming an oil exporting nation. Her drive for political and economic domination of the world overlooks a serious consideration: the USSR system of rating management in terms of barrels a day, in disregard of the principles of conservation, is forcing the waste of reserves that comes from producing wells too rapidly.

To compare the United States and the Soviet Union: the USSR is about where the USA was in 1929, but is using the geology, geophysics, drilling technology, logging, and completion tools of 1960. Most of these are borrowed from the United States.

Is the Soviet Union likely to swamp the world with oil, some day soon? Probably not. If she succeeds in her program of industrialization, and if she builds the roads that an industrial organization requires, she will do well to develop her petroleum fast enough to supply her own needs—factories eat fuel oil and roads drink gasoline. Already she has to follow wasteful practices to supply her needs and 97 per cent of her satellite country requirements.

Poland

The sedimentary formations that are so productive in Rumania extend more than two hundred miles along the northern flank of the Carpathians across Galicia and contain a number of important fields. Here, as in Rumania, are many oil seeps and oil springs, and the production of oil from pits goes back beyond recorded history. Some of the old timbered shafts can still be found. A crude kerosene produced from Galician oil was used for lighting in Prague for a short time before 1818. Then in the early 1850's refining was again started, and its products began to compete with candles as far away as Vienna.

Active modern development began in 1874 and the country reached its productive peak in 1909—14.9 million barrels. Since that time the production has declined steadily, despite fairly active drilling, to just over 1 million barrels in 1960. Intensive development and exploration during 1938 to 1944 increased the production but did not discover a single new field.

New fields are hard to find. The latest discovery at Mielec produces 89 barrels each from 2 wells on a daily basis.

Practically all of Poland's pre-World War II oil fields are located southeast of Krakow between the Wisha and San rivers. Almost four-fifths of her production was grabbed by the Soviets when eastern Poland was annexed at the end of the war, leaving Poland virtually destitute of a domestic supply. Falling from third place among oil-producing nations in 1909, she is now compelled to buy 90 per cent of her needs from the Soviets. From something like 3,550 wells, this Communist satellite produced only 4,000 barrels a day in 1960, just over a barrel a day per well.

Outside of Galicia, in the plains of northern and western Poland, no commercial oil has been found but four salt domes are known. In one of these domes, oil showings have been found. Geological and geophysical work continues and may lead to commercial discoveries.

The Netherlands

Holland waited long and patiently to produce her own oil, while depending upon imports to satisfy her thirsty industries. Not until 1943 did she work up to her first 1,000 barrels of crude for the year. The first discovery at Schoonebeek, in eastern Holland near Coevorden, has now expanded to a 284-well field. Schoonebeek actually drapes over the German border where it is known as Emlichheim. For many years the Schoonebeek field was the Netherlands' sole producer; then came a flurry of 13 discoveries in western Holland, all centered around Rotterdam. Industrial Holland took a "shot in the arm," and began building more refineries, new plastics plants, a network of pipelines and a big 24-inch crude transmission line from Rotterdam to the Rhine-Rhur refineries. The combination of a good seaport, a large crude-oil line direct to refineries, and abundant domestic production will be hard to beat in building Rotterdam into a major world oil center.

The Dutch consume about 164,000 barrels a day, the 1960 rate, which has increased 10 to 15 per cent each year since the western fields were discovered. Schoonebeek still produces about seven million barrels a year, almost twice the total of all the smaller western fields. Holland's yearly total is twelve million barrels but the twelve million barrels is only a fifth of the local demand. Many new fields in both western and eastern Holland will probably help to meet the energy demand which is now supplied by oil. In the meantime, exploration for more oil is going on at a frenzied pace with most encouraging results.

Great Britain

Britannia rules the waves with oil-burning ships, but the fuel they use is not produced in the British Isles. In the past fifty years a considerable number of deep tests have been drilled, and, though quite a few have had showings of oil and gas, the only commercial discoveries prior to 1940 were small wells. Near Hardstoft, in the south of England, a small producer was drilled in 1919, and a ten-barrel well was discovered in Scotland in 1938. A seven-barrel well was completed at Formby in Lancashire, and a 240-barrel well found oil at Eakring in Nottinghamshire—both drilled in 1939. The 1941 exploration in the land of Robin Hood brought in a 420-barrel well. Since then new discoveries have barely kept pace with the depletion of the older fields. Some 260 wells in all of England produced 1,740 barrels a day in 1960 —a small part of the 895,000 barrels a day required by cars and trucks and factories and "The Queen's Navee." England now stands in second place in world consumption, just one jump behind the United States.

Encouraging signs for more production are being found near Kimmeridge, in Dorset, where a 1959 discovery produced 18,500 barrels in the first year. The oil fever is running high on the belief that the Paris basin reservoirs extend across the English Channel into Dorset. Offshore exploration is witnessing a boom in the hope of finding submarine oil fields. It appears probable that the British Isles will find more and more of their own oil.

England continues to be dependent on oil from Kuwait, Iraq, Iran, Venezuela, Bahrain, Qatar, Colombia, and the East Indies, in all of which her nationals have large interests.

To all this there is one minor addition. In Scotland a small amount of oil has been produced by the distillation of oil shale.

Oil was distilled from coals and shales in Great Britain as far back as 1694. In 1850, when the diligence of the whaling fleets had depleted the supply of whales so that sperm oil for lamps and paraffin for candles were becoming scarce and expensive, one James Young obtained a patent on a process for distilling oil from coal. He had the benefit of experimental work and commercial production from oil shales in France during the preceding twenty years, and of personal familiarity with whale-oil refining, and he started an industry that within two years had one hundred thirty plants in Great Britain and sixty-four in the United States.

The completion in 1859 of the Drake well in Pennsylvania and the flood of cheaper oil that followed it almost ruined the shale-oil industry, but it managed to pay its way in Scotland until 1923. Since then it has subsisted with government subventions.

The production of oil from shale and coal is interesting as an indication of what may be done when, in some distant day or under special circumstances, oil from wells becomes scarce or unduly expensive, but it makes no important contribution to Great Britain's present supply.

France

France has moderate oil production and prospects of a material increase. In 1938 she produced 516,000 barrels, but concerted efforts brought this to 2 million barrels in 1951 and 14 million in 1960. Of this, nearly 90 per cent came from the Aquitaine basin, the remainder from the Paris basin and the Rhine graben. Of the production from the depleted Pechelbronn field, which was discovered in 1948, about 55 per cent came from wells and 45 per cent from shafts into which the oil drains through drifts that are run in the producing sands. Formerly wells drilled into the sands produced until they would yield no more; the oil now flowing is what the wells left in the sand.

Exploration finally turned France from a "have-not" nation in the 1940's to one of the fastest-growing producers in the world, with petroleum discoveries in the Paris and Aquitaine basins. Southern France is the leading French gas-producer and one of the world's largest. Some 300 million cubic feet of gas per day now pour from the Aquitaine basin into ever-expanding French industries and new petrochemical plants. More is yet to come.

Germany

As early as 1436 oil from a spring near Tegernsee in Bavaria attained such medicinal fame that it was known afar as "St. Quirinus's oil." However, Germany has not continued to earn fame as an oil producer.

The east-west partition of Germany at the end of World War II gained very little in the way of oil reserves for Soviet Russia. East German oil fields apparently accounted for a pitiful 1,700 barrels of crude in 1960—not much of a plum for the Communist machine. Soviet and German geologists and seismologists are reported to have found a field at Halberstadt which produces from exactly one mile in depth. Another recent discovery, at Ludwigslust, is supposed to be the best well so far in East Germany. The details are not known.

West Germany has fared considerably better, particularly since the end of World War II. The near-total destruction of Germany's producing and refining capacity combined with the over-production demanded for the Nazi war machine left a permanent mark on some of West Germany's older fields. The large prewar Nienhagen field, for example, which produced over two and a half million barrels a year under wartime acceleration, gave up only 940,580 barrels in 1947.

Intensive exploration and drilling campaigns have brought West Germany to eighteenth place on the list of oil-producing nations. Her output of nearly 39,000,000 barrels in 1960 was still only enough to satisfy about a fifth of her own demand. She marked up a 23 per cent increase in production over that of 1959, but her consumption rose 25.6 per cent. Four-fifths of her crude oil requirements are shipped from the Middle East, Venezuela, USSR, Colombia, Brazil, Algeria, and the Far East.

Most indigenous German oil comes from the area between the Wesser and Elbe rivers in the Hanover Plain.

The Hamburg-Hanover Plain is underlain by a large number of salt domes, which for many years have been the world's chief source of potash, besides producing a large amount of salt. Small quantities of oil have been encountered in many salt and potash wells. As in early China, the oil polluted the brine and was therefore cased off, if possible, or else the well was abandoned. Germany's present production is mainly the result of successful attempts in recent years to develop oil on and around these domes.

It sometimes takes many wells to find the major pool in a salt-dome field, and as a result, new types of fields are also discovered. The Knesebeck field in the Broistedt-Gifhern Plain is one of these. Oil was discovered there in a series of fault blocks between salt domes, where the structure looks like several displaced slices in a loaf of bread. This new field produced a third of a million barrels in 1960 while it was still being developed.

A pinch-out from sand into shale is responsible for the oil accumulation in the Bramhar-Wettrup field, one of the newest West German discoveries. It produced over a half million barrels from 40 wells in 1960, nearly tripling the 1959 rate.

Deeper drilling and more geophysical studies will probably continue to boost West Germany's stature in the world of oil. Where will the new oil come from? Probably from three favorable areas. One is the Northwest German basin, the second is the fault graben of the Upper Rhine Valley, and the third is the sub-alpine or mountain-flank area.

Czechoslovakia

Czechoslovakia was not blessed with a great deal in the way of oil resources, although further exploration of her territory may result in increased production. The only commercial oil to date is from small fields in Moravia and Slovakia, in the northern Vienna basin. In 1960, these fields produced about 800,000 barrels, a little more than the production of the Rangely field of Colorado.

Czechoslovakia joined the oil-producing nations in 1913 with the discovery of the Gbley field in Western Slovakia (Moravia). Oil in this and the nearby Hodonin field comes from 550 to 800 feet; the fields were still producing in 1960, some 47 years after discovery. Eastern Slovakia has contributed the Mikalovce and Semplin oil fields where the greatest exploration efforts are concentrated today.

An interesting development in Czechoslovakia is an idea for mining oil sand and extracting the oil. The deposit lies two hundred and sixty feet below the surface. This was mentioned in Chapter 9 on "Producing the Oil—The Mechanics of It."

Hungary

In 1937 Hungary became an oil producer for the first time. A promising field was discovered at Budafa-Puszta in southern Hungary not far east of the Danube and about thirty miles west of Lake Balaton. The first well produced 53 barrels of high-grade oil a day, and a little gas. The second well came in for about 400 barrels, and by the end of 1938 there were eight wells in the field with an aggregate daily capacity of 1,400 barrels. The field is on a large anticline found by torsion balance; geophysical work may lead to other important fields there.

Another field was found in 1937 at Bukkszek in the north. The wells were abandoned before 1947, but the finding of even this small field on the edge of the Hungarian plain has led to a third of all 1960 drillings being concentrated in that great area of wheat fields and vineyards.

All the fields in Hungary produced 8¾ million barrels in 1960—over 24,000 barrels a day.

Rumania

In about A.D. 106 when the Romans conquered Rumania, a territory that still bears their name, they must have noticed many of the oil springs that dot the country. Probably they gathered and used the oil, for they were accustomed to its use. What happened during the Middle Ages we do not know, but in the second half of the sixteenth century oil was being produced from pits in the Prahova district of Walachia, and as early as 1650 oil shafts, dug by hand, were producing in the Bacau district of Moldavia. In 1750, according to a writer of that day, oil was being used to light courtyards, treat diseases of cattle, and grease the axles of carts. The first wells were drilled in 1860 in the Bacau district, and in 1863 in the Prahova district.

During World Wars I and II, the Rumanian fields were among the principal objectives of the Allies, but the wells had been systematically and scientifically ruined by the time they reached them. After the two wars, the fields were redrilled and recompleted, and today Rumania ranks as the eleventh producing country in the world. Production in 1960 was 91.5 million barrels, about a tenth that of Texas.

The producing formations are the same as those of the Russian Caucasus. The main producing fields lie in a belt extending from the famous Iron Gates of the Danube eastward around the outer edge of the Transylvanian Alps and the Carpathian Mountains, across Walachia, thence northward and westward across Moldavia and Bukovina to the Polish border. In Transylvania, enclosed within the half-circle of the Alps and Carpathians, are a number of important gas fields referred to as the Targul Mures fields.

In the outer belt are some thirty-six producing fields, mainly in the Ploesti or Prahova district of Walachia and the Bacau district of Moldavia, the two districts in which oil was first produced from pits and in which the first drilling was done.

The production in 1960 was nearly 7 per cent above that of 1959, after 2.4 million feet of drilling. Rumania has probably seen her best days as an oil producer because each year sees a lower and lower rate of gain in spite of increased drilling. The geology of the country seems favorable for finding of new fields, however, and the decline in production may be due to the socialistic control of the country and a former government attitude that did not foster systematic exploration. A more favorable attitude is now apparent with the establishment of a geological institute for research and mapping.

The Balkan States

Ancient oil sources included the oil springs on the island of Zante (now called Zakynthos) off the west coast of the Peloponnesus in southern Greece. According to Herodotus a myrtle branch would be lowered by a pole into a spring and drawn up dripping with oil. The oil was accumulated in a cistern and then transferred into jars. Oil seepages are reported at various places on the Greek mainland; the one most mentioned was near Delphi where gas seeps may have something to do with the famous oracle. However, the oil springs of Zante remain Greece's sole contribution to commercial production.

Exploration has gained momentum following passage of a 1959 oil exploration law that offers a 50-50 split of profits with private investors. The new law appears to be working, if we can judge by the flurry of concessions granted in 1959 and 1960.

Oil springs and asphalt deposits in Albania were described by no less a scientist than Aristotle and by such early historians as Strabo and Pliny. Burning gas seeps near the temple of Apollo, not far from Apollonia, were famous in the days of Greece's glory. In Albania, as in Zante, many attempts have been made to find commercial production by drilling; the Italians were particularly active prior to World War II.

One oil field at Kucove produced all of Albania's 489,000 barrels in 1938. Since the Communist coup in December, 1945, and the ensuing domination by Russia in June, 1948, production reports have become more valuable for what they conceal than what they disclose. "Plan Fulfillment" statements may be more propaganda than fact, but we are led to believe that Albania had recovered her prewar status as an oil-producing nation by 1950. From that time until 1960, production has risen erratically from 8,953 barrels a day in 1957 to 7,370 barrels in 1958, 9,588 in 1959, and probably 14,500 in 1960. The reports credit the increase to new fields in the Fieri and Valona areas and to other Russian drilling in the Patos and Mariz areas.

Oil-saturated outcrops are reported in European Turkey near the Sea of Marmora and from a few points in Yugoslavia; oil has been distilled commercially from Yugoslavian oil shales.

Yugoslavia joined the oil-producing countries in 1939, with 21 barrels of oil a day. Development during the German occupation brought the 1944 daily rate to 2,000 barrels from the Peklenco-Selencia regions in the central part of the republic. Production declined rapidly after the wartime exploration frenzy, then climbed gradually to pass 3,000 barrels a day in late 1952. Later discoveries at Struzec-Osekovo, Ferdinandovac, Elemir, and at Novska and Kikinda—mostly north of the Sava River near Zagreb—tripled production in only four years. Yugoslavia produced 18,836 barrels daily in 1960. Geologists are watching the area north of the Sava and Danube rivers as well as the Adriatic coast for the next big development.

The Balkan situation sums up to this: most of the region is underlain by igneous and metamorphic rocks, with minor areas of sediments. Some of the sediments are petroliferous and give rise to oil springs and asphalt deposits and a few gas vents, but the sediments, in the main, are so folded, faulted, and shot-through with igneous intrusions that the favorable areas are small.

Italy—Sicily—Sardinia

Italy is a country with much oil history and little oil production.

The Romans were enthusiastic users of oil and its products; they or their Greek predecessors found a supply in a number of oil springs in Sicily and on the mainland. The most famous, though perhaps not the largest, of the Sicilian springs was near Agrigento, and "Oil of Agrigentum" seems to have been something of a trade name for all liquid oils. It is mentioned by Strabo in his *Geography* and by Dioscorides in his *Botany* and *Materia Medica*. Pliny, in his *Natural History*, written about A.D. 60, discusses it at some length.

Oil was used in lamps. The poorer people used it, or asphalt, as a protective coating for their furniture and other wooden articles, after the fashion of royal and priestly Egyptians. It was highly regarded as a medicine for boils, ringworm, gout, epilepsy, blindness, toothache, and colic. The Roman ladies, no less interested in beauty culture than their sisters of today, used it to color and beautify their eyebrows and also used it as smelling salts. They or their menfolk seem even to have bathed in it, for at the Bath of Nero, near the ancient Roman watering place of Baiae, was a Petroleum Oil Bath.

Nor were the Romans alone in their interest in oil products. The Carthaginians did a thriving business collecting asphalt on the Lipari Islands off the north shore of Sicily.

Skip a thousand years, and the Lombards of northern Italy were no longer satisfied with the oil they could skim from the hundred-or-so oil springs in the valley of the Po. In 1640, they dug a well twelve miles from Modena and skimmed oil from water that collected in it. In succeeding years other wells forty to sixty feet deep were dug, and the Province of Emiliana in the Po Valley has produced small quantities of oil ever since. The first drilled well was completed in the 1870's. The pre-World War I production peak was reached in 1908 when Emiliana produced about 50,000 barrels during the year. A second record was set in 1933 when all Italy produced 208,000 barrels, but that mark was not matched again until 1953. Production began to rise steadily in 1955 as the result of new discoveries. Eight new fields—Ragusa, Gela, Fernandia, and Vittoria in Sicily, and Tramutola, Vallecupa, Cigno, and Polesine on the mainland—have largely replaced the older fields. The Po Valley is still an important prospecting and producing area, particularly for natural gas.

Italy can now boast of 2,000 barrels of oil a day and Sicily of 33,000 barrels. Altogether Italy can put out 614 million cubic feet of gas a day, mostly from 47 fields in the Po Valley.

It goes without saying that in recent years strenuous and effective efforts have been made to find more.

Morocco

The monarchy of Morocco, not much larger than California, lies on the northwest tip of Africa adjacent to the fabulous new oil country, Algeria. Morocco has not fared quite so well as a producer despite her dignified position as the second earliest (1918) producer on the continent. During her first thirty years as a producing country, geologists thought the rocks near the Atlas Mountains were too intensely folded and faulted to favor the existence of good oil reservoirs. Five fields were discovered in the Rharb basin during that period, but most of the wells produced only a few barrels a day. One well, discovered in 1947, did make 113 barrels. The geologists appear to be right so far. The old fields have now been abandoned and thirteen new ones have taken their place, mostly farther west in the basin, away from the mountains toward Casablanca. The thirteen fields barely make 2,000 barrels a day, but intensive exploration still goes on in several concession areas. The "Road to Morocco" oil and gas has not been bright, but around any curve may be something new and startling.

Libya

Libya is another Cinderella in the African oil picture. Adjectives like prodigious, astronomical, and exciting are common in the 1960 Libyan oil-reporters' commentaries. Even though Libya has not shipped any oil out during the two years of drilling since the first discovery in January, 1958, she lists seventeen producible fields. Each year has seen a redoubled effort over the previous year. In 1960, 167 wells were drilled which totalled 923,604 feet. If the Arabs could cut it all into two-foot post holes they could set 622 miles of barbed-wire fence stretching from Bengasi to the shores of Tripoli or from Chicago to Wichita. Libyan oil men estimated that the desert oil fields had 115,171 barrels a day of shut-in capacity in 1960. No wonder that three large pipelines were under construction between the Zelten, Dahra, and Mabruk fields and the seaports on the Gulf of Sirte.

One of Libya's greatest problems with her new-found oil is that of being a newly independent nation[2] with no established marketing ties to the European Common Market. Once again, the principle of supply and demand will probably iron out this problem so that Libya can start tankers steaming out of the Gulf of Sirte to world markets.

Egypt

Egypt has been an oil consumer nearly as long as Mesopotamia, but ancient Egypt was an importer, not a producer. Because of the expense of importing and because the Egyptians built more with stone than with brick, they made only minor use of asphalt mortar.

Asphalt was used for waterproofing basketware as in Mesopotamia. Moses' mother "took for him an ark of bulrushes, and daubed it with slime and with pitch, and put the child therein; and she laid it in the flags by the river's brink" (Exodus 2: 3).

One oil use long claimed by Egypt has partly disintegrated under chemical analysis. The black, asphalt-like substance with which bodies were mummified has been assumed to be bitumen for several thousand years. Diodorus, who wrote during the time of Julius Caesar, mentions the Egyptian use of asphalt for embalming, and hundreds of writers to

---

[2] Libya was approved as a sovereign state by the United Nations in 1949. National independence became effective on January 2, 1952, with King Idris I as the ruling head of the hereditary monarchy.

the present day have made similar statements. The word mummy is derived from an old Syrian-Arabic word, "mummia," meaning bitumen or asphalt. A few hundred years ago, powdered mummy was widely prescribed by Egyptian apothecaries for wounds and bruises, and mummies were also used for fuel. In time the supply ran low and was augmented by wrapping newly-deceased corpses in bandages, dipping them in bitumen, and exposing them to the sun. The neo-mummies thus produced seem to have been accepted by the apothecaries, the bruised and wounded, and the fuel-gatherers without doubt as to their authenticity. However, the asphalt-like substance of several hundred ancient mummies turns out, on analysis, to be resin of various types. Not until Ptolemaic time, apparently, was bitumen extensively used for mummification and then perhaps only for lower-class mummies. Certain words of doubtful meaning suggest that small quantities of bitumen may have been used in more ancient mummies, and additional analyses may show that Diodorus and his successors were not wholly wrong after all.

Whatever may or may not have been the use of bitumen for mummification, it had a wide variety of other uses. As in Mesopotamia, it was used to waterproof cisterns, basements, bathrooms, granaries, and coffins. It was used to damp-proof dwellings, palaces, temples, and tombs against the inundations of the Nile. Artists used it to color metals and to make imitation precious stones. It was applied as a preservative coating to valuable wooden articles, such as mummy cases, jewel cases, and statues. Among the finest articles found in the tomb of Tut-ankh-Amen were two statues of King Tut himself, superbly carved from wood, coated with bitumen, and covered with beaten gold.

So far as the ancient accounts show, the asphalt used in ancient Egypt was imported from the Dead Sea of Palestine. The Egyptians must not have been production-minded, for at Jebel Zeit, near the mouth of the Gulf of Suez, are asphalt deposits and seepages of oil and gas that must have been known from earliest times. The Egyptians could have worked these, but the Romans apparently were the first to develop them—only on a limited scale, however. The name Jebel Zeit means "Oil Mountain," and the Roman geographers called it Mons Petrolius. The oil and gas showings led to the present Ras Gharib, Hurghada, and former Gemsah fields in the vicinity along 90 miles of Red Sea coast.

Oil seepages and asphalt deposits are also known across the Gulf of

Suez on the Sinai Peninsula. This is the land where Moses tended the flocks of his father-in-law, Jethro, and "the angel of the Lord appeared unto him in a flame of fire out of the midst of a bush: and he looked, and behold, the bush burned with fire, and the bush was not consumed" (Exodus 3: 2). This reference may indicate the existence of a natural-gas seep that apparently burned intermittently and may thereby have been the phenomenon of inspiration for the communicants of the Old Testament.

Not until 1885 was serious thought given to Egypt as a possible oil producer, and not until 1910 did the country produce commercially. The Hurghada field, first explored in 1903, has been the major producer, but in June, 1938, a new field was discovered at Ras Gharib. By the end of the year Ras Gharib had five completed wells and was outproducing Hurghada. The Gemsah field, where the first showings were found about 1869, was abandoned in 1929. The crude oil is shipped to Suez for refining, and the products are marketed mainly around the eastern Mediterranean.

Across the Red Sea from Ras Gharib at Abu Durba, on the western shore of the Sinai Peninsula, small wells were completed in 1923. The field was shut down for several years because of its picayune production. It was reopened in 1938 but was never an important producer. Revised governmental regulations spurred oil interests in 1937, and by the end of 1938 some 846 exploration licenses were on file. The Sudr field was discovered in 1946 and was followed by the Ras Matarma, Asl, and Wadi Firan fields in 1948, all on the west coast of the Sinai Peninsula. Exploration moved southward along both coasts of the Gulf of Suez through 1955, locating new fields such as Rudeis and Belayim in the east side, and Ras Bakr and Karim on the west.

In 1960, Egypt produced 24.5 million barrels, a little more than Nebraska. Liberalization of the law for obtaining oil rights is resulting in active exploration of the country by the geologists of several leading American, Dutch, Italian, and French companies, and is making the Land of the Pharaohs an important world producer.

Algeria

At Ain Zeft oil has been produced from springs and pits since the days of Aristotle. Perhaps Aristotle fathered the Algerian oil business when he wrote about one of the seepages which proved to be the starting point for a search that led to one of the major oil discoveries of the world. Some 2,275 years later, farsighted twentieth-century geologists took up the lead of Aristotle's Lyceum lectures. They mapped and predicted great oil potential in the marine embayment on the North Africa coast. Their prophecies were well founded. Algeria and Libya alone have enjoyed the investment of 2,500 million dollars in the five years from 1955 to 1960. The expenditure has not all been wasted on dry holes.

In 1949, there were only two producing countries in Africa—Egypt and Morocco. In 1960, 185 wells were drilled in Algeria alone; 140 were completed as producers; the total became 322 active wells. Algeria chalked up oil runs of 60 million barrels from the Edjele area through the new 100-million-dollar, 24-inch pipeline to the Port of Skirra in Tunisia. In 1960, more pipelines were being built, more wells being drilled, more dock facilities springing up on the Mediterranean coast, more concessions being granted, and more geological and geophysical crews were plodding the sand dunes.

The crowning glory of the year was the 52.4 million barrels produced from the gigantic Hassi Messaoud field, 350 miles inland from the sea. The 160,000 barrels a day from this field are moved through a 410-mile, 24-inch pipeline to Bougie on the Mediterranean coast. Here some of it is loaded into tankers and moved to an Algiers refinery; the rest goes to France where it helps to quench the thirst of the bustling mass of little cars. It looks as though the 15,000-barrel daily production rates of some of Algeria's wells will go a long way toward making France an oil-sufficient nation. Her oil-products output is already ahead of domestic consumption, but the law of supply and demand will probably even this out in due time. In 1960, the French government was toying with the idea of reducing the 74 per cent tax on gasoline, which should make quite a difference to all the Renault and Citroen drivers in Paris and Marseilles.

Algerian oil is changing the whole way of life in the French Sahara, as well as back home in Armentières.

Nigeria

Thirty-four million Nigerian people gained their independence from Great Britain on October 1, 1960, just four years after Shell Oil Company and BP Petroleum Development Company found oil in the Niger delta. Shell's convictions, 20 years of exploration, and 200 million dollars finally bore fruit with the discovery of the Oloibiri field in Nigeria. This good fortune was quickly followed by four more bonanzas in the Port Harcourt region and one in the Western Region Province at Ughelli. Oil men generally divide the Niger delta into the Eastern and Western regions relative to the course of the river. Operating conditions in both segments of the delta are like those of our own Gulf Coast and Louisiana bayou areas. Expensive barge-mounted drilling rigs and offshore drilling platforms are responsible for Nigeria's startling rise into the ranks of important oil-producing countries. In 1960, the jungle swamps were producing an average of 17,500 barrels a day, with double this amount recorded for certain days near the end of that year. There is no telling where Nigeria will stand in the ranks of the oil nations by the time you read this. She certainly is a contender for a prominent spot among the coastal states of equatorial Africa.

Gabon

The Republic of Gabon in French Equatorial Africa forms the "Adam's apple" on the east coast line of the neck of the continent. Most of its 103,000 square miles covers a plateau of igneous rocks lying at 2,000 to 3,000 feet above sea level. The country is about the size of Colorado and has an estimated population about equal to that of Wyoming.

A "whisker" growing out into the Gulf of Guinea from this lump on the throat of Africa is named Cape Lopez. It resembles Cape Hatteras and Cape Cod, but differs from these in being the site of seven oil fields. The Cape Lopez and the Ogooue embayment fields were producing 16,600 barrels of oil a day during 1960. This put Gabon in fourth place as a producer on the African continent. She produces nearly 6 million barrels a year of 13°–33° API oil from depths of 1,700 feet to 5,700 feet. Like most other countries with prolific coastal fields, she is now working on extensive offshore exploration in the hope of extending the chain of onshore fields into the Gulf. The narrow but productive coastal plain, the Cape, and the delta of the Ogooue River have thus far given up fifteen landlocked oil fields to seismic exploration. Seismic equipment is now going to sea, hunting for more "on trend" fields.

The Republic of the Congo

LEOPOLDVILLE

The Republic of the Congo is our newest promising oil country, not because of the date of the first oil discovery but because of the 1960 date of its independence from Belgium. Although no oil fields had been found there by 1960, the Lake Kivu gas field is estimated to contain two trillion cubic feet of natural gas. As yet, there is no market and therefore no production, but where there is so much gas there is probably oil.

BRAZZAVILLE

This part of French Equatorial Africa gained independence along with Gabon, her oil-producing neighbor to the north. As a matter of fact, their common border looks like the outline of a jig-saw puzzle piece, and so we should expect the Congo to be oil country. Seismic exploration along 90 miles of coast from Gabon to Cabinda has been quite successful.

Discovered in 1959, the Pointe Indienne field near the delta of the Kouilou River produced almost 400,000 barrels in its first year. As in many other areas of the world, oil accumulation is directly related to thick salt beds. As much as 1,300 feet of salt have been found to overlie the 200-foot Chela sand. Production depths are about 2,900 feet. This success will probably stimulate an active geophysical survey program with the seismograph, the airborne magnetometer, and the gravimeter.

Angola (Portuguese West Africa)

A narrow strip of African Coastal plain south of the Congo River had developed into a compact little oil province. The Cuanza basin of Angola has been enjoying increasing success each year since 1955 and now it has five oil fields capable of producing half a million barrels a year. The fields have names with a typical Portuguese ring to them— Benfica, Luanda, Cacuaco, Calinda, and Bom Jesus. Most of the 1,300 barrels produced each day go to the coast by pipeline, where the oil is refined in a 4,000-barrel refinery along with imported crudes.

Angola also boasts an *asphalt rock* operation on the northeast edge of the Cuanza basin at Libongos and Caxito. Additional localities are awaiting development at Calucala and Quilungo near the town of Zenza do Itombe on the eastern edge of the coastal plain. It looks like little Angola will cut a wider and wider swath in the oil statistics columns as the years go by.

Adieu to the Old World

That ends our rambling journey through the Old World. We have not visited Switzerland, Finland, Scandinavia, Thailand, or Indo-

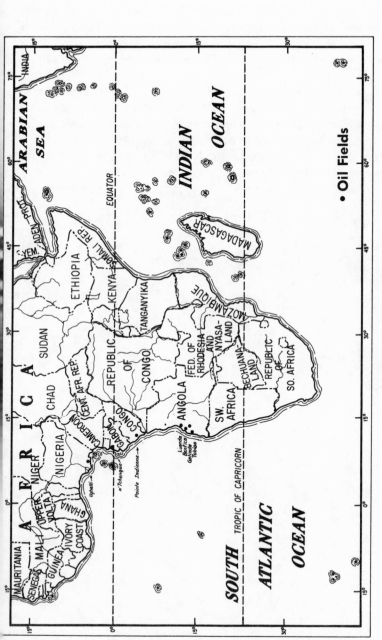

Sub-Saharan Africa

China, because none of them seems to have much chance of developing important production, though parts of Denmark, Timor, Yemen, Africa, and the Alpine foreland of Switzerland may have possibilities. Most countries of the world now have exploration parties working— so time will tell.

We have not mentioned the small shale-oil industry kept alive by government subsidy in France, or the somewhat larger shale-oil production built up behind the iron curtain in Estonia, because neither has any important bearing on the world's oil supply.

Now, if you like, we sail for the New World to visit the countries from which most of the world's oil comes.

CHAPTER 17

Oil in the New World

The United States

THE United States has been, and is, the world's leading oil-producing country. In 1960 it produced over two and a half billion barrels—twice as much as the USSR, its nearest competitor. By the end of 1960 its total production was over 65,000 million barrels, more than a third of the world total. Two of its states—Texas and California—yielded more oil in 1960 than the USSR.

Moreover, although the oil business is six thousand years old, the modern industry originated in the United States, and American methods, drillers, and equipment have played a large part in the development in the rest of the world. Accordingly, to get a balanced view of the world's oil economy, we must give extended consideration to the United States.

Most of the country's oil fields lie in a half-dozen or so great producing areas generally referred to as districts (formerly called fields). The Appalachian district includes the individual fields of New York, Pennsylvania, West Virginia, eastern Ohio, eastern Kentucky, and Tennessee. The Lima-Indiana district is in northwestern Ohio and eastern Indiana; Illinois, northwestern Kentucky, and western Indiana are in the Illinois basin district. The fields in the Lima-Indiana and Illinois basin districts and those of Michigan are sometimes grouped as the Central Fields, and sometimes everything east of the Mississippi is given the single designation, Eastern Fields.

The Mid-Continent district includes Kansas, Oklahoma, Arkansas, northern Louisiana, and northern Texas. Southern Louisiana and the salt-dome fields of southern Texas are in the Gulf Coast district; other fields in southern Texas are in the South Texas district.

The West Texas or Permian Basin district includes the fields of eastern New Mexico and western Texas, and excludes the Texas Panhandle, which is in the Mid-Continent district. Colorado, Wyoming, Montana, northwestern New Mexico, and Utah are in the Rocky Mountain district; this district also includes the fields in the Dakotas and western Nebraska. California is the California district. The boundaries of some districts are indefinite, and those who use the district names do not always agree on the boundaries; but in a general way the classification is definite enough to be convenient.

Now for a bit of history and a tour of the Western Hemisphere.

The Needs of the Industrial Revolution

The Industrial Revolution started a few years before 1800. In a hundred years, by the substitution of machine power for muscle power, it changed men's industrial habits more than they had been changed since the days of Ur and Mohenjo Daro. The shift from cottage handicraft to factory manufacture called for more and better artificial illumination, and, until the invention of the electric light, the oil lamp and the candle were the only sources. Machinery, the agent of the Revolution, required lubrication; without adequate and abundant lubricants the Revolution could neither function nor progress. The Revolution had brought an urgent need for something that only oil could supply.

During its early decades, the new mechanized industry got along with castor oil and other vegetable oils, but its requirements soon outgrew the amount that vegetables could provide at reasonable cost. Then came sperm oil and the great days of the whaling fleets, but by the 1840's whales had become scarce and the price of sperm oil soared.

Before the crisis became acute, French scientists had begun to advocate the production of oil from oil shale. By 1845 they had developed a small commercial production. In 1847, James Young of England extracted paraffin from the oil of a seep in Derbyshire, and by 1850, he had perfected his process for extracting oil from coal and oil shale. Boghead coal, cannel coal, and the oil shales of Scotland, Pennsylvania, West Virginia, and Kentucky were found to be good sources, and in a few years there were more than a hundred thirty plants in Great Britain and sixty-four in the United States. Even the proud whaling port of New Bedford, Massachusetts, had a plant operating on a Scotch boghead.

Meanwhile, an American here and there was wondering if oil could not be produced more cheaply from wells than through distillation from shale and coal. From time to time, wells drilled for salt had encountered oil and had been abandoned because the oil ruined the brine. Why not drill wells for oil, instead of for brine? That oil flowed from certain springs had been known for generations; would it not flow more copiously from wells that tapped the hidden sources?

Prehistoric and Historic Uses

Back of these thoughts lay several centuries of American oil history. At least a thousand years ago the Indians of California used oil and asphalt for caulking their board boats, waterproofing baskets, covering wrappings, cementing beads to background material, making decorative designs, and even covering the eyes of the dead. The Iroquois, Algonquins, and particularly the Senecas, gathered oil from the springs of what is now southern New York and northern Pennsylvania, and esteemed it highly for the treatment of rheumatism, bruises, toothaches, coughs, and burns. When the white men came they acquired the Indian's enthusiasm for this sovereign remedy and considerably extended the list of human and equine ailments for which it was used. One early writer says of a spring on Oil Creek, near where the Drake well was drilled, "The troops sent to guard the western posts halted at this spring, collected some of the oil, and bathed their joints with it. This gave them great relief from the rheumatism with which they were afflicted. The water, of which the troops drank freely, acted as a gentle purge." Another writer, a doctor, says, "it seems to be peculiarly adapted to the flesh of horses and cures many of their ailments with wonderful certainty and celerity."

As the white men extended their exploration they found many other oil sources in the western Alleghenies and the Ohio Valley, but the most famous springs continued to be those in the Seneca country —particularly those on Oil Creek, Pennsylvania, and near Cuba, New York. The Senecas had long used the oil as an article of trade with other tribes, the term "Seneca Oil" coming to be applied to all oil obtained from the earth. The time came when Seneca oil, under various trade names, was sold at most drugstores in the United States and at some in Europe, priced at forty cents and upward a vial.

As had happened in so many older countries, men sought early to augment the supply of oil by digging for it. Among the springs on

Oil Creek are many pits, dug either by the Indians or by the early French explorers. Many of the later pits were dug for salt rather than oil, which brings us to another chapter of the story.

Salt and Oil

Men have always wanted salt, especially meat-eating men who were without the benefits of refrigeration. Their livestock wanted it also. It was natural as settlers pushed westward into and beyond the Alleghenies that they should gather salt from the salt springs or "salt licks" which were plentiful in so many places. When the deposits around the springs were exhausted, settlers began evaporating the brine from the springs. When this furnished insufficient supplies they began to dig pits. The first pits were merely enlargements of the springs, but as time went on the settlers dug wells away from the springs.

Digging and cribbing wells by hand is a laborious and uncertain business. In 1806 two brothers, David and Joseph Ruffner, started something revolutionary. They wanted to sink a well at the Great Buffalo Lick on the banks of the Kanawha above Charleston, West Virginia. They had sixteen or seventeen feet of ooze and gravel to penetrate before they could reach bedrock. They cut a length from a hollow sycamore tree with an inside diameter of four feet, and stood it upright on the ground. Then a man inside dug the ooze and gravel, which was hoisted to the top in half a whiskey barrel, letting the sycamore casing follow down. When the well reached bedrock and did not yield satisfactory brine they decided to go deeper. Taking their cue from the hand drill used to make holes for blasting, they attached a two and a half inch steel chisel bit to a long iron drill, and raised and dropped it by means of a rope and a spring pole. By 1808 their hole was forty feet deep in the bedrock and had the quality of brine they wanted.

Then, in order to get the brine to the surface without dilution, the Ruffners took two long strips of wood, whittled them into half tubes, fitted them together into a tube, and bound them together with a wrapping of twine. This they placed in the hole and thus cased off the "top water." Without benefit of the Chinese, of whose drilling achievements they probably never had heard, the Ruffners had invented cable-tool drilling, casing, and tubing, and had started the drilling of wells in the Western Hemisphere.

The new method was adopted with a speed that showed how much it was needed. Wooden casing and tubing quickly gave place to tin, then to copper, and finally to iron and steel. Drilling tools were improved. Six years after the Ruffners' achievement, wells had been drilled below 475 feet. By 1859, hundreds of salt wells had been drilled.

In many localities the salt men had to contend with an ubiquitous nuisance. Oil polluted the salt springs, seeped into their hand-dug wells, and was encountered in even larger quantities in their drilled wells. When they could, they cased it off; when they couldn't, they, perforce, abandoned the well. Oil-tainted salt had no value.

Some of the early salt wells encountered much oil. On Little Rennox Creek, near Burkesville, Kentucky, a well drilled in 1829 is reported to have been good for 1,000 barrels a day, to have flowed by heads[1] to a height of twenty-five or thirty feet, and to have covered the whole surface of the Cumberland River for many miles. A contemporary record says: "About two miles below the point on which it touched the river it was set on fire by a boy, and the effect was grand beyond description."

The Dawn of Recognition

Twenty years later even the salt men would probably have forgotten their pursuit of brine and have tried to save such a well as that on Little Rennox Creek for an oil well; but, in 1829, whales were still plentiful and the need for another source of oil was not yet acute.

Already, however, the value of petroleum oil and gas for illuminating purposes was beginning to be recognized. Houses in Fredonia, New York, were lighted with gas in 1821. In 1826, Dr. S. P. Hildreth of Marietta, Ohio, wrote, "Petroleum affords considerable profit and is beginning to be in demand for lamps and work shops and manufactories. It affords a clear, brisk light when burnt this way and will be a valuable article for lighting the street lamps in the future cities of Ohio." In 1845, oil from a salt well at Tarentum, Pennsylvania, was mixed with sperm oil and used successfully as a lubricant in a Pittsburgh factory. A few years later Tarentum oil was refined in Pittsburgh and sold for use in lamps.

By this time the demand for oil had become extreme. Shale-oil and

---

[1] A well that flows intermittently, subsiding until another "head" of gas has accumulated, and then flows again, is said to flow by heads.

coal-oil plants were numerous and some of them were large. The Lucesco plant near Pittsburgh could turn out more than a thousand barrels of oil a day, selling at thirty to forty dollars a barrel.

The Birth of the Modern Industry

With oil at such prices, and with the experiences of the salt industry before them, it is not surprising that men began to think of drilling wells for oil. In 1857, a group of New Englanders, encouraged by Professor Benjamin Silliman, Jr., of New Haven, formed the Seneca Oil Company to drill among the oil springs on Watson's Flat beside Oil Creek. In 1858, Colonel E. L. Drake assembled the equipment and drove a pipe down to bedrock. In 1859, drilling began; and when work stopped on Saturday night, August 28, the well was 69½ feet deep. On Sunday morning the hole was nearly full of oil, and the modern oil industry had been born.

The Drake well was not the first well to produce oil; oil had been produced from hand-dug wells for several thousand years. It was not the first drilled well to produce oil, for many wells, drilled for salt in both the United States and China, had produced oil. It was not even the first well drilled for the specific purpose of obtaining oil, for the Chinese probably did this before the time of Christ. What then was unique about the well drilled on Watson's Flat by E. L. Drake, erstwhile railroad conductor, "Colonel" by courtesy, and field superintendent for the Seneca Oil Company? Simply this: it was the first well drilled for the specific purpose of obtaining oil in a prolifically oil-bearing region at a time when the lamps and machinery of a rapidly industrializing world were in need of a cheap source of illuminants, fuels, and lubricants.

The Early Days

Those were uproarious days up and down Oil Creek and the Allegheny Valley. Wells were drilled as fast as money and equipment could be found. Men poor today were rich tomorrow—or were poorer than before. "Struck oil" became an Americanism for sudden riches. In 1859 the production was two thousand barrels, and in 1862, 3.06 million barrels. From nothing to three million barrels in less than four years! No wonder the price dropped from twenty dollars a barrel in

September, 1859, to ten cents in December, 1861. There had been an urgent need for a new source of oil. The new industry was not only meeting the need; it was drowning it.

Striking oil was not the only way to make a fortune; speculating in oil was almost as good. In Oil City, and elsewhere, were exchanges on which oil was bought and sold amid scenes that would make the Chicago wheat pit seem tame by comparison. The price rose and fell like an elevator. A new well would send it down; a dry hole might send it up. Important wells were watched by scouts who rode hell-for-leather to carry the news to their principals. Back-door tips and whispered rumors swept the market up and down.

Something had to be done with so much crude oil, and primitive refineries sprang up like mushrooms. Only yesterday the chief source of kerosene had been coal and oil shale, and so they called their kerosene "coal oil"—the name still sticks. Before long, when tank cars began to replace barrels, refineries were built in every city within reach of the oil fields. They were scarcely less crude than their country cousins, and none of them—city or country—turned out a uniform product. One batch of kerosene might be 10 per cent gas-oil, the next might be 10 per cent gasoline. When a man lighted a lamp he never know which it would do—smoke or explode.

Into this disorderly scene there presently came an orderly young man named Rockefeller. His discerning gray eyes saw three things: that the business needed a more stable price structure, that it needed uniform standards for its products, and that these things could be controlled through control of refining and transportation, leaving to the producers the hazards of exploration and discovery. Because its products were to be made to uniform standards he called his company the Standard Oil Company. Some of his methods seem ruthless and unfair when judged by the business standards of today; whether they violated the business ethics of his time is a question we need not try to answer. It does not affect the one important fact that stands out so clearly: more than any other man, John D. Rockefeller transformed a gambling game into an orderly business.

Westward Ho!

While all this was happening the tumultuous new business had burst out of Pennsylvania and started across the country. Men remem-

North America

• Oil Fields

bered where oil had polluted the salt wells, and 1860, the year following the Drake discovery, saw production in West Virginia, southeastern Ohio, and eastern Kentucky. In the same year a show of oil was encountered in a shallow well near Paola, Kansas, so close on the heels of the Drake well that a monument marks the spot and patriotic Kansans claim it as the original oil well.

In 1862 a small well was brought in near Florence, Colorado. In 1867, minor production was obtained in the Hilliard field in southwest Wyoming, near an oil spring where the Mormons and the "Forty-Niners" had greased their covered wagons. In 1873, oil was found on the Derby-Dallas anticline in central Wyoming. By 1875, oil was produced from two small fields in the Los Angeles basin of California. A significant point is that in sixteen years the oil business had crossed the continent and reached the Pacific.

Since then, the theme of the American oil business has been the discovery, in rapidly increasing number, of one field after another, from Pennsylvania to the Coast, and the correspondingly rapid growth in the market for the output. Most of the fields have been small, but the tale is highlighted by the discoveries of spectacular fields, each in its turn dominating the industry and its economics.

The Eastern Dominance

For more than twenty-five years after Drake's discovery Pennsylvania overshadowed the rest of the country and the production elsewhere was unimportant. Then, in 1886, oil was found in northwest Ohio and in Illinois. The new Ohio field was soon extended westward into eastern Indiana, becoming the Lima-Indiana field. The Illinois discovery opened an area that, after many ups and downs, is today another stable producer. By 1887 the discovery at Florence, Colorado, had developed into a commercial producer.

The significant event in 1887, though it looked trivial at the time, was the bringing in of small production near Nacogdoches, Texas—the puny birth of the present giant of the industry.

Kansas' first well came in at Paola in 1860. It fell short of being commercial, but, in 1889, more oil, negligible in amount as yet, was found in the shallow *shoe-string* fields of the southeast; in a few years the exploration had extended across into Indian Territory, presently Oklahoma. It was 1897 before Kansas, and 1900 before Oklahoma, broke into the recorded production column.

In 1895, Ohio passed Pennsylvania as the leading producer, but two discoveries farther west foreshadowed the eventual displacement of both Pennsylvania and Ohio. The first real field in Texas was discovered near Corsicana in 1895, and 1896 saw the discovery of Coalinga, the first of the great fields in the San Joaquin Valley of California.

The Leadership Goes West

The year 1901 brought one of the major events of oil history; this was the discovery at Spindletop, Texas, of the most prolific single field that had yet been opened—the first field in the Gulf Coast district, and the first field on a salt dome. Within a year, Texas had passed Pennsylvania and was crowding Ohio in production, and the Spindletop discovery had led to Louisiana's first field at Jennings.

It was not Texas but California that took the leadership away from Ohio. Coalinga led to McKittrick, Kern River, Maricopa, Midway, and Sunset—all spectacular fields—and from 1903 through 1906 California was the leading producer. Then she was displaced by a newcomer.

While attention was focused on the salt domes of the Gulf Coast and the spectacular fields of the San Joaquin Valley, production had been working steadily southward in Oklahoma. Most of it was small but profitable. Then, in 1905, a wildcatter found the famous Glenn Pool. By 1907 Oklahoma was the leader, and for twenty-one years the leadership oscillated between Oklahoma and California. In 1928, Texas first captured it, and Texas still has it. Louisiana finally bested California for second place in 1959.

With Glenn Pool, in 1905, came also the discovery of the Caddo field, in northwest Louisiana. Four years later, oil at Salt Creek focused attention on the Mountain States, established Wyoming as an important producer, and led to the Teapot Dome controversy many years later.

In 1912, attention swung sharply back to Oklahoma with the discovery of the Cushing field, which was big enough to wreck the market just as the flush production of Spindletop and the San Joaquin fields had previously done. The flood of oil from Cushing and the discoveries to which it led had not abated when Augusta and Eldorado, the first of the big Granite Ridge fields of Kansas, were discovered. That was in 1916, and no one knows what might have happened to the crude market if, in 1917, the United States had not entered World War I.

The Amazing Decade

The year 1920 saw the discovery of Cat Creek, the first field wholly within Montana; 1922 brought the discovery of Hogback, the first field of consequence in New Mexico; and 1923 brought the discovery of Wellington and restored Colorado prominence. These Rocky Mountain fields could not compete in importance with discoveries elsewhere, however. Arkansas became an important producing state in 1921 with the discovery of Eldorado (no relation to the Eldorado in Kansas), followed by the discovery of Smackover the next year. The years 1921 and 1923 brought the first commercial discoveries in the Texas Panhandle, in south Texas, and in the great Permian Basin of West Texas and New Mexico. Amarillo opened the Panhandle, Laredo opened south Texas, and Big Lake opened the Permian Basin. Within a few years each of the three districts was producing more oil than any eastern state.

Nor was this all. In the years 1920 to 1922, the Los Angeles basin became a major producing district, with such fields as Signal Hill (Long Beach), Santa Fe Springs, and Huntington Beach. The years 1920 to 1925 brought a new outbreak of salt-dome fields as a result of the application of geophysical exploration to Gulf Coast geology. In 1920, the discovery of Mexia rejuvenated the Balcones Fault area of Texas, site of the original Corsicana field. Michigan got its first production of consequence near Saginaw in 1925, and followed it two years later with a field at Muskegon, on the opposite side of the state.

In 1926 Oklahoma took the spotlight with the discovery of Seminole, which proved to be a whole group of fields so prolific that they led to the first officially-enforced proration. In 1929, that state came up with Oklahoma City, her biggest field so far.

In 1930 East Texas came in—the biggest field yet.

The decade from 1921 to 1930 was the Golden Age of oil discovery and the "Headache Age" of the oil producers. So many new fields were found so fast that, although consumption was accelerating rapidly, production potential ran miles ahead of it. The industry would think it had regained its balance after a discovery in California or Oklahoma, only to be bowled over by a new discovery in Texas or Kansas.

The New Stability

Only the growth and application of proration kept the producing end of the business from ruin and prevented enormous waste of oil and gas. By the middle thirties, it appeared that proration had the situation well in hand. Despite the fact that numerous and important discoveries continued to be made, a reasonable balance between supply and demand had been achieved. Into this picture came Illinois with a succession of new fields in a region that had long ceased to be important—and with no conservation law. At about the same time, Michigan, similarly strategically located for the great Chicago market, uncovered a number of new fields. Even this added production was fairly well absorbed, however, thanks to further restrictions in the Mid-Continent and Gulf Coast states.

By the end of the amazing decade, with the increased discovery rate and the adoption of conservation policies, the oil business seemed to be in a new economic phase. For the first time in its exciting history, the industry enjoyed an adequate, fairly long-term supply of raw material—known to exist, but left underground to be drawn on as needed. This is far different from the former feast-and-famine situation when reserves were produced as rapidly as possible after their discovery, when excess supplies were either wasted or expensively stored in aboveground tankage, and when few manufacturers could be sure of an adequate and sustained supply of raw material.

The present situation requires the producing end of the business to carry a much larger investment per barrel of daily production, but the investment per barrel of ultimate production will probably prove to be less. It requires a larger investment in underground reserves, but a reduced investment in aboveground storage.

It is tough on the producer who wants to get his money out of his wells as fast as possible, but, by assuring a dependable supply of crude, it enables the refiner and the integrated producer to accept a smaller return on their investment. It has taken the industry out of the riches-or-ruin class and into the stability enjoyed by other industries with dependable rather than fluctuating supplies of raw materials. Because a stabilized business can operate on a smaller margin of profit than an unstable one, the new state of affairs is all to the good for the consumer.

A sharp increase in consumption combined with a failure to keep discoveries abreast of consumption, or an increase in tariff barriers on foreign oil, might upset all this and throw the industry once more into a scramble for crude. This would cause consequent high prices for refined products. Barring such a development, however, the outlook for the industry and its price schedules is relative stability.

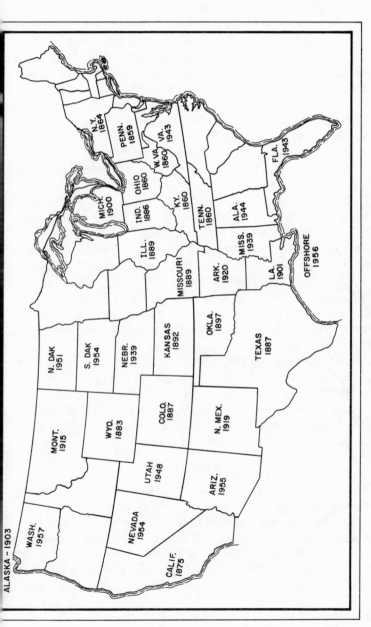

ALASKA – 1903

WASH.	1957
MONT.	1915
N. DAK	1951
N.Y.	1864
PENN.	1859
MICH.	1900
WYO.	1883
S. DAK	1954
OHIO	1860
W. VA.	1860
VA.	1943
NEVADA	1954
UTAH	1948
COLO.	1887
NEBR.	1939
ILL.	1889
IND.	1886
KY.	1860
CALIF.	1875
ARIZ.	1955
N. MEX.	1919
KANSAS	1892
MISSOURI	1889
TENN.	1860
OKLA.	1897
ARK.	1920
MISS.	1939
ALA.	1944
FLA.	1943
TEXAS	1887
LA.	1901
OFFSHORE	1956

First Commercial Oil Discovery Dates by State

Pennsylvania

The early followers of Colonel Drake soon learned some unexpected things about oil wells. The wells that produced more than 3 million barrels in 1862 did not produce 3 million barrels in 1863. By 1864, despite the bringing in of a multitude of new wells, the production of the state had dropped to 2.1 million barrels. Men hadn't known that the production of an oil well declines rapidly; now they knew it. It was sixty years before they recognized the fact that the total production of a well will be greater if the oil is produced slowly, than if the well is allowed to produce to capacity in its early stages. In the meantime, the only way to compensate for the decline was to find more fields and drill more wells.

The Pennsylvanians did it; and for thirty-four years after Colonel Drake struck oil, theirs was the leading producer state of the Union. In 1891 it hit its high of 31.4 million barrels. Then the inevitable decline set in, and by 1912 the production was down to 7.8 million barrels. It stayed there until 1924, after which it began to rise again, due chiefly to the rejuvenation of the Bradford field by water-flooding. By 1937 it was 19.1 million barrels. This second production peak, somewhat greater than the combined output of India, Burma, and British Borneo, put Pennsylvania in seventh place among the oil-producing states for that year.

It should be noted that not all Pennsylvania Grade oil comes from Pennsylvania. Practically all of the oil from Pennsylvania, New York, West Virginia, and southeast Ohio is of the same general character and is called Pennsylvania Grade crude. Because of the high quality of the motor oils and kerosene made from it, it commands a 50 per cent higher price than any other American crude.

New York

Oil men, through some queer quirk of herd psychology, have been great respecters of political boundaries. When oil is found in a state, they are likely to search that state with enthusiasm and thoroughness, ignoring areas of equal promise just over the state line. Perhaps this explains why no oil was produced in New York until five years after it was discovered in northern Pennsylvania. In 1864, the state line was crossed and New York broke into the producing column with a total production of 109 barrels. The oil-bearing area does not extend far into New York, however, and the state has never been a rival of its neighbor to the south. It produced its all-time peak, 6.69 million barrels, in 1882. After a fluctuating decline to 809,000 barrels in 1918, the output climbed again, due mainly to water-flooding. In 1938, New York fields yielded 5.05 million barrels, ranking New York the thirteenth producing state that year, with about 600,000 barrels more than Germany. New York's production has dwindled since to something over a million barrels in 1960.

An interesting development during the 1930's was the discovery of natural gas in southern New York and northern Pennsylvania, within reach of such great markets as New York, Pittsburgh, and Philadelphia. The production potentials have proved disappointing over the years, but recent discoveries are keeping the region active on the exploration lists.

West Virginia

West Virginia had oil wells before 1861, but the Civil War almost ruined the properties and recovery was slow. Production dropped from 120,000 barrels in 1876 to 90,000 barrels in 1884. Then development began in earnest, and in 1899 the state produced 13.9 million barrels—more oil than Pennsylvania. In 1900 it yielded its most—16.2 million barrels. By 1938 production had declined to 3.7 million barrels—somewhat less than that of Poland—thus putting West Virginia in fifteenth place among the producing states. By 1960 it was nearly at the bottom of the list, with only 2.3 million barrels.

Ohio

Oil in salt wells had been a nuisance in Ohio also, and when the Drake well inaugurated oil drilling in Pennsylvania, Ohio was not far behind. Within a year the Muskingum Valley in the southeast yielded oil—from the same geologic basin as the fields of Pennsylvania and West Virginia. Production was small until about 1885 when it jumped to 662,000 barrels. In 1886 Ohio produced 1.78 million barrels, most of it from the southeast, and a new field was discovered at Findlay, in the northwest. It was followed by other fields to the south, and west toward Lima, and the discoveries finally extended into Indiana and became known collectively as the Lima-Indiana field. These fields brought oil men a new problem. The oil was high in sulphur—from which Pennsylvania Grade oils are notably free—and some time elapsed before a process was devised to remove it. Then development went ahead rapidly and augmented the supply being obtained in the southeast.

From 1895 through 1902, Ohio was the leading oil producer. Then California passed it, and it never regained first place. Its production reached 23.9 million barrels in 1896 and declined to 5 million, somewhat less than that of Bulgaria and Albania, in 1960.

Michigan

Most of Michigan is covered by a mantle of sand, gravel, clay, and boulders. This is *glacial drift* brought down from the north by the continental ice sheet that once covered so much of Canada and the northern United States. The drift covers the rocks that might contain oil, so that no one can tell from surface evidences whether or not the geology is favorable to oil accumulation.

On the well-known theory that what can't be seen can't exist, most oil men and many geologists contended for years that no oil would be found in Michigan. Even the showings found in salt wells, and the few barrels of production from 1900 onward, did not shake them. Then, in 1925, a field was discovered at Saginaw, in the east-central part of the state, followed in 1927 by a field at Muskegon, in the western Lower Peninsula. In 1925, production was 4,000 barrels. In 1938, the state's production was 19.2 million barrels from fields scattered across the central Lower Penninsula from Lake Michigan to Lake Huron. It placed Michigan eighth among the producing states, with production nearly twice that of Bahrain and Egypt combined.

Nor is this the end of Michigan's story. In 1938, nineteen additional fields were discovered, the most striking, though perhaps not the most important, being southwest of Grand Rapids where wells good for 10 to 5,000 barrels a day were found at depths of one to two thousand feet. Michigan production rose again to a near-peak in 1942, then declined steadily from a half to three-fourths million barrels a year until 1959. Then drilling activity picked up, several discoveries added to the glamor of a recovery year, and Michigan seemed to be on the comeback trail with 15.6 million barrels in 1960. The ten-year cycle seemed about to repeat itself for the second time. Deeper sands of the same ages as some prolific Oklahoma producers are largely responsible.

Indiana

We have already seen that Indiana became a producer in 1886 by the extension of the Lima-Indiana field across the state line from Ohio. In 1889 more oil was found in the southwest, close to the Illinois line, and the state produced 33,000 barrels that year from the two areas. By 1904 production had climbed to 11.3 million barrels, but then came a long, fluctuating decline, and, in 1938, Indiana was eighteenth among the producing states with 995,000 barrels. This was more than nine times as much as Italy produced.

By 1950, Indiana had taken a new lease on life. In ten years she had risen from less than a million to more than ten million barrels yearly. Her production has hovered around eleven million barrels a year since.

Illinois

Oil was discovered in southwest Illinois in 1886. The state's production reached 1,460 barrels in 1889 but declined to 200 barrels in 1902. For two years the state recorded no production, and then new fields were discovered in the southeast counties along the Wabash River. Later, less productive fields were found in the southwest within fifty miles of St. Louis. In 1908, production reached 33.7 million barrels; then the decline set in. In 1936, production was down to 4½ million barrels.

Next came one of the revivals that make the oil business interesting. Improved geology and geophysics, with some help from Lady Luck, began to find new fields. In 1937, the new fields produced 2.34 million barrels. In 1938, Illinois produced 18.9 million barrels, and in two more years production peaked at 147.6 million barrels.

Production had increased 33-fold in only four years; in consequence, the Illinois basin, which includes part of western Indiana and northwestern Kentucky, received more geologic attention and more new drilling than any like area in the United States. Illinois producers, between the Mid-Continent fields and their main markets, pay about twenty-five cents a barrel less for pipeline transportation. No wonder the producers of Kansas and Texas worry about Illinois competition.

Illinois sustains her oil yield at around 80 million barrels, about half that of her 1940 peak.

Kentucky and Tennessee

Tennessee complained of oil interference with its salt industry in 1825, and Kentucky had a reported thousand-barrel well as far back as 1829—abandoned because its owners wanted salt, not oil. More than fifty years went by before Kentucky had as good a well again.

In oil statistics tables, the entries for states that produce only a little are in columns labeled "Unclassified" or "Others." Although drilling, begun in Kentucky immediately after the Drake discovery, resulted in a little oil, Kentucky and Tennessee failed to get out of the "Others" column until 1883, when they produced 4,755 barrels (nearly all of it from Kentucky). Thereafter their yearly production varied from 300 to 9,000 barrels until 1899, when new fields in east Kentucky raised production. Some 1.2 million barrels were produced in 1905. Tennessee's part was too small to be segregated and, after 1907, it ceased until 1916. Meanwhile Kentucky's production had declined to 437,000 barrels in 1915.

In 1916 Tennessee established itself in its own column with 677 barrels, which it boosted to 60,000 by 1927, and which thereafter dwindled until 1935 when the state passed back into the "Others" column. It has been in and out ever since.

In 1916, Kentucky started a new series of discoveries, mostly in the east. By 1919, production reached 9.28 million barrels. That was the top year until 1944 when Kentucky broke the record, advanced into the 10-million-barrel-a-year class, matched Canada, and produced about half as much as Rumania.

The fields in the Appalachian foothills no longer produce most of the oil. Sixty-five of the state's one hundred twenty counties now produce. Two-thirds of the oil comes from the blue-grass country in the west, in the region where the Rennox Creek salt well found its oil in 1829. Many fields are subject to secondary recovery: without it, Kentucky's production in recent years would have been considerably less.

Kentucky's future, however, is not wholly dependent on her older fields. Many of the major oil companies are looking seriously to the east where they hold large acreage blocks. New reservoirs are becoming apparent in the older rocks of Cambrian and Ordovician age, and stratigraphic traps in the Silurian and Devonian look mighty interesting around some of the known anticlines. In 1960, Kentucky's production was climbing.

Kansas

One would never accuse Kansas of being either mountainous or secretive, but beneath her prairies she conceals a buried but real mountain range, known as the Nemaha ridge, made up of sedimentary strata arching over a granite core. Eventually erosion wore the sediments down and exposed the granite. Then the whole area sank beneath the sea, and younger sediments buried the granite ridge and the truncated edges of the sediments along its flanks. Now the ridge lies snugly concealed across the state from Winfield on the south to Seneca on the north. A series of low anticlines and domes is its only surface expression.

The fields on or associated with the Nemaha ridge are spectacular; they include the famous Eldorado, Augusta, Florence, and Peabody fields. Their drilling and the outburst of wildcatting that followed in 1916 to 1918 disclosed the presence of the buried, secret mountain range, and jumped the state's production from less than 3 million barrels in 1915 to more than 45 million in 1918. Thereafter, production declined to 28.3 million barrels in 1923, and would have continued to shrink if the western third of the state had not been found productive.

The Nemaha ridge divides the oil fields of Kansas into three general classes; most of the fields in all three classes are in the southern two-thirds of the state. To the east are many small fields with oil in sandstone-lens traps. In the anticlines along the ridge and its subsidiaries are other fields with oil in anticlinal crests or in truncated reservoirs on the flanks. West of the ridge, scattered across the state almost to Colorado, the fields are mainly on anticlines, most of them roughly parallel to the ridge.

The early development was east of the ridge. These old fields are known as the *shallow fields* and most of the wells are now strippers, though new wells and new fields are brought in from time to time. The sandstone lenses were probably laid down as sandbars near the shores of ancient shallow seas. Because of their sinuous character they are known as *shoe-string sands*. When the area was submerged the sandbars were buried by muds and oozes which in time became shales and limestones. Several lenses close to the ridge are large; they include such fields as Rainbow Bend and Oxford.

It was inevitable that attempts should be made to find production west of the Nemaha ridge; the early efforts were none too successful, but here and there a new field was discovered close to the ridge at first, then farther and farther west. Each new field stimulated the search for others, and the campaign went on at an accelerated rate though slowed down temporarily by the depression. More than forty fields were found in 1937, 142 in 1954, and 142 was about the average each year from 1954 until 1960.

The state produced more oil almost every year, rising from 36.5 million barrels in 1917 to 124 million in 1956. The potential has been much greater, for Kansas practices proration. But even with proration Kansas was in fifth place among the states in 1938, producing 1.46 million more barrels than Europe, not including Russia. She was still in fifth place during her biggest year, 1956, but relinquished that spot to Wyoming in 1960. Wyoming's 135 million barrels were not pro-rated, but Kansas' prorated 113 million barrels still held her in sixth place.

Oklahoma

In the minds of most people Oklahoma and oil are synonymous, and well they may be. In 1938 the state, though surpassed by both Texas and California, produced 175 million barrels of oil—about nine per cent of the world's total and greater than the combined production of Iran, Iraq, Saudi Arabia, the Dutch East Indies, Japan, and Sakhalin.

Then, just to show how the scales can tip, in 1960 Oklahoma's 192 million barrels were only half the production of Iran and slightly more than half that of Iraq; about a third of Saudi Arabia's, a little more than the Dutch East Indies; and sixty-four times greater than those of Japan and Sakhalin.

Northeast Oklahoma has the extension of the shoe-string sand area of Kansas, but the Oklahoma sandstone lenses are more numerous and are generally larger and more productive. By 1891, a little drilling activity had crept down across the Kansas border, but Oklahoma's production by 1897 had only risen to 625 barrels. Then for two years the state had no recorded yield. In 1900 drilling began in earnest, and production began to climb. First Nowata and then Bartlesville became towns whose names were familiar throughout the oil industry. By 1904, with the discovery of new fields farther south, production had jumped to 1.37 million barrels.

In 1905 the famous Glenn Pool was discovered, producing from the biggest lens of them all. By 1907 the field had about 4,000 wells producing 120,000 barrels a day. In 1907 and 1908 Oklahoma was the leading oil state, and her production for 1908 was 45.8 million barrels. The crossroads village, Tulsa, became the oil capital of the world.

Productive lenses were found west and south of Nowata and Bartlesville, and the discoveries crept across the Osage country until in 1920 a large productive lens was found at Burbank near the western border of Osage County. Osage County is owned by the Osage Indians, and royalties, rentals, and lease bonuses—distributed among them share and share alike—made them the richest people per capita in the world. Cultivated and cultured Osages took their wealth quietly like other sober folk; other Osages still in the blanket stage bought Cadillacs and Pierce-Arrows and abandoned them beside the road if a tire blew out. One Indian, shopping for the grandest car obtainable, drove proudly away at the wheel of a hearse. Those were lush days for the tradesmen of Pawuska and Hominy.

Meanwhile, anticlinal fields were being discovered; before long they had become the leading producers. First came a number of small fields scattered through the central part of the state. In 1912 came Cushing —Oklahoma's greatest single field except for Oklahoma City—and Cushing crude became the typical oil on which Mid-Continent oil prices were based. In 1913, away to the south across the Arbuckle Mountains, came Healdton, thus making Ardmore an oil center.

So much oil was developed in Oklahoma and Kansas that the crude market crumbled. By 1916 the posted price of Cushing crude was down to fifty-five cents, with much of it actually bringing as little as twenty-three cents. Oil men and investment bankers were sure, as they had been during the San Joaquin Valley boom of 1908–1912 and the Spindletop boom of 1901–1902, that demand would never overtake supply and that the oil business was permanently ruined.

Within a year or two men were talking "shortage," and oil went to three and a half dollars a barrel.

One after another new fields were found, and year by year (with a recession from 1918 to 1920) production mounted. From 1920 to 1922 and again in 1927 Oklahoma was once more the leading oil state. In 1926 oil was discovered in the Seminole district in a group of fields rather than a single field, and 1927 production rose to 278 million barrels. This threatened another market collapse, and proration, already governing Salt Creek in Wyoming, was introduced at Seminole, at first on a voluntary basis and later under state control. It was soon extended to the other flush fields of the state. Since 1927, Oklahoma has not produced at full capacity. Despite the 1928 discovery of her most prolific field, Oklahoma City, her production has never even equaled that of 1927. In 1943, under wartime conditions, it was down to 123 million barrels. By 1956 it was back to 216 million barrels, though the market forced conservation authorities to prorate it down again.

Some people, especially Texas oil men, will tell you that all the important Oklahoma fields have been found, and that nothing but a gradual decline lies ahead. Year by year Oklahoma oil men refute them by finding new pools—some among old fields, some southeast and west of the older fields. In 1938, 18 fields were discovered; in 1953, 110; in 1960, 51 new pools were added to the list. How wrong can prophets be?

Arkansas

Arkansas was still a stepchild to the oil business long after its sisters on the west and south were in the big producer class. Not until 1920 did anyone have the hardihood to find oil in the state. However, in 1921 a real field was developed at Eldorado, and Arkansas produced 10½ million barrels. The next year a wildcatter discovered a field with the charming name Smackover. It turned out to be one of the major fields of oil history, and Arkansas was on the oil map in a big way. In 1925, production was 77.4 million barrels.

Even fields like Eldorado and Smackover do not maintain their initial production for long, especially when they are produced to capacity in their early days. By 1936, despite new discoveries in the meantime, production was down to 10½ million barrels. In 1937 the Rodessa field, sprawled across the Texas line, was discovered; and, in 1938, production was 18.1 million, making Arkansas tenth among producing states. Twenty years later she was still holding doggedly onto tenth place with 28.7 million barrels. By 1958 Arkansas had produced 1,032 million barrels since the Eldorado discovery, which figures out to 1.71 per cent of United States production to that date. Still producing 28.7 million barrels in 1960, she was finally bested by Colorado, Utah and Montana with over 30 million barrels each.

All of the fields found so far are in the west. Northwestern Arkansas is occupied by the Ozarks—not promising oil territory. Just south of the Ozarks is the Arkoma basin; it yields mostly gas from Pennsylvanian-age sandstones. South of the Arkoma basin lie the Ouachita Mountains, which geologists discount as significant oil country. South again from the Ouachitas lie the smaller uplifts and basins of the upper Gulf Coastal province where we find the Eldorado, Smackover, and Rodessa field discoveries. In southwest Arkansas, each year has seen the recording of a half dozen or more new fields such as Sandy Bird, Lewisville-Old Town, Genoa, and Champagnolle. The formations that produce in southwest Arkansas thin to the east and some even pinchout, and so the possibilities of the remainder of the state seem to be none too promising to many geologists. Seismic surveys and wildcat drilling may still find oil where the Arkoma basin plunges under the Mississippi embayment of the Gulf Coastal plain in eastern Arkansas.

Arkansas has been one of the states that thought proration unnecessary, but the rapid development of Rodessa and the consequent dissipation of gas energy made her think again. The state adopted a strong conservation law, and, after the usual course through the courts, conservation became effective. The steady, year-after-year regularity of production is a tribute to good practice.

Louisiana

Louisiana is in the Gulf Coast geologic province. In the north Louisiana shares the upper Gulf Coast with Arkansas; the oil is found on anticlinal structures of the usual type. In the south, largely populated by the "Cajun" descendants of Evangeline's fellow exiles from Acadia, the oil comes mainly from salt domes.

Production started in southern Louisiana with the discovery of the Jennings field in 1901; other southern discoveries followed. In 1905 drilling found oil on the shores of Caddo Lake, in the north, and soon a good part of the lake was spotted with derricks. Louisiana's production was 9.1 million barrels in 1906, but the new field could not compensate long for the decline of the southern fields. Production was only 3.06 million barrels in 1909. Since that time new fields in both north and south have given the state an exciting history. In 1936 and 1937 a number of important salt-dome fields were developed along the Gulf Coast. The 1938 production was 95.2 million barrels—20 million barrels greater than that of Iran, Bahrain, India, and Burma—ranking Louisiana the fourth oil state that year.

In 1958, producing more than three times that volume of oil, Louisiana remained in fourth place but was 11.2 million barrels a year behind the annual production of the same countries. In one year she had taken over second position on the state list but lacked 723 million barrels to keep up with developments in the same countries. 1960 was a banner year, but the foreign crude output far outstripped Louisiana's.

Louisiana's oil development has a considerable aquatic element, with Caddo an underwater field in the north, and the bays, swamps, and bayous of the Gulf Coast, and many wells are drilled from barges or man-made islands. The oil men make their rounds in boats or helicopters, and geologic and geophysical crews use amphibious *marsh buggies* and surplus Navy landing craft.

Mississippi

Some years ago a moderate-sized gas field was developed near Jackson, the wells showing a little oil. However, numerous other Mississippi wells were dry or had only small oil or gas shows. The first oil well in the state was completed in September, 1939, near Yazoo. Now, after twenty years, the old field map of southern Mississippi has so many black spot symbols it is difficult to count them all. Mammoth oil fields such as Gwinville, Cranfield, Jackson, and Baxterville are so large that their solid black-patch shapes on the map look like someone tipped over the ink bottle.

Production in Mississippi has grown from 107,000 barrels in the discovery year to over 52 million barrels in little more than twenty years. As a relative newcomer to the list of oil states, she has done well to gain ninth position in the 1960 statistical tables.

Although the most feverish exploration activity has been concentrated in southern Mississippi for many years, it now looks as though the northeastern quarter of the state will get a fair run for the money. The Black Warrior basin, straddling the Alabama-Mississippi line in the north, has been drilled with success, discovering important fields such as Siloam, Muldon, and Amory. Across the state line in Alabama are the Hamilton and Fayette fields. The entire basin will probably become more active as new techniques of drilling with gas and of reservoir fracturing are applied.

Texas

It is not easy to present all of Texas under a single heading. The state is as big as Austria, Czechoslovakia, and Germany; from Texarkana to El Paso or from Dalhart to Brownsville is about as far as from

Berlin to Rome. In this great area are about 190,000 wells, including some of the world's largest. In 1962 Texas produced 932 million barrels—36 per cent of the United States production, about three times that of Louisiana, the strongest competing state, and equal that of Russia, the strongest competing country. Small wonder that Dallas and Houston think Tulsa should surrender her claim to "The oil capital of the world."

Production began with a few puny wells near Nacogdoches in 1887 and received an increment from like sources in the San Antonio district in 1889, but the best the proud state of Texas could do from 1887 to 1895 was forty-five to sixty barrels of recorded production a year—not much more than enough to oil the guns of the Texas Rangers. In 1895, a small field was discovered at Corsicana, in east-central Texas, and production climbed to 1,450 barrels for the following year.

The discovery at Corsicana and its duplication in 1900, at Powell, eight miles east, were important because they eventually directed attention to a geologic feature that has been responsible for many million barrels of production. Trending northeast-southwest through this part of Texas is a belt of folding and faulting known as the Balcones Fault zone. In the belt are anticlines, domes, and places where the beds are arched up against fault planes. On these structures, in the earlier years, were found such fields as Mexia and Luling.

Captain Anthony F. Lucas, a retired Austrian naval officer, a graduate of Gratz Polytechnic School and the Imperial Austrian Naval School, had come to the United States in the late 1880's, had been for a time in the lumber business in Michigan, and had practiced mining engineering in Colorado and Louisiana. In 1896 he drilled an artesian well for Joseph Jefferson on Jefferson Island, Louisiana, where the famous actor spent his winters. Captain Lucas' well encountered a thick bed of rock salt, and thereafter the Captain was busy drilling salt wells at various points in Louisiana. In some wells he found showings of oil and sulphur; consequently he developed the theory that commercial accumulations of oil and sulphur are associated with the salt domes of the Gulf Coast.

Meanwhile Savage Brothers had found oil showings in shallow wells on the Sour Lake and Saratoga salt domes in Texas and had made an unsuccessful attempt to test the Spindletop dome near Beaumont, Texas—a low swell, ten to twelve feet above the surrounding plain. Captain Lucas decided to make his supreme effort at

Spindletop, and in late 1899 or early 1900 he started a well. He put every cent he had into the venture. When the going got tough he and his patrician wife lived in a shack furnished with cracker-box chairs and a rough board table. The well got a showing of oil at 575 feet in March, 1900; then the casing parted and the hole was lost. Refusing to be whipped, the Captain obtained some financial support and started another well in October, 1900. He had been using the rotary method in Louisiana salt wells, and a little rotary drilling had been done in Corsicana. Its use at Spindletop overcame the quicksand that had stopped Savage Brothers, and on January 10, 1901, the Lucas well came in for an estimated 75,000 to 100,000 barrels. At that, the Captain almost missed the field; he later stated that if the well had been drilled sixty-five feet farther northwest it would have been dry. By being sixty-five feet on the right side of the oil-water contact he had discovered Spindletop, one of the greatest producers in American oil history. Of perhaps equal importance, he had discovered the first commercial salt-dome field and had drilled the first important oil well ever drilled with rotary tools.

Like many another salt-dome field, Spindletop held a great deal of oil beneath a very small area. The field was leased in tracts about large enough for a drilling rig, and nearly every tract was occupied. There was no room for boilers, most of which stood in a ring around the edge of the field with steam lines running from them to the wells. A man could walk across most of the field without being off a derrick floor.

Sour Lake and Saratoga were drilled promptly and became important fields. By 1902 the state's production had jumped to 18 million barrels—and the price of oil was down to five cents a barrel. The Baston field was discovered in 1903 and the Humble field in 1904. In 1905 the state's production hit 28 million barrels, but the crest soon passed, and by 1910 production had dropped to 8.9 million barrels.

In the meantime, exploration was going on in other parts of the state. In 1902 oil was discovered at Brownwood in west-central Texas and in Jack and Montague counties in north Texas. In 1904 the Petrolia field, also in north Texas, which had been producing small quantities for several years, began to ship oil. These districts away from the coast were comparatively quiescent, however, until Electra was discovered in 1911, Burkburnett in 1912, and Strawn in 1915. These touched off an active development campaign, culminating in the discovery of Ranger in 1917 and of Desdemona and Eastland in 1918. In

1920 the Balcones Fault zone came back to life with the discovery of Mexia. In 1919 the state's production rose to 79.4 million barrels, and it has increased erratically every year since. In 1921–1923 three major districts were opened. Oil was discovered in the Panhandle northeast of Amarillo, near Laredo in south Texas, and at Big Lake in the Permian basin of West Texas.

The Panhandle, like Kansas, has a buried mountain range. This westward continuation of the Arbuckle-Wichita mountain uplift is called the Amarillo Mountains. On the side of the buried granite core, in granite debris (washed off the range when it towered above the surrounding country) and in sedimentary beds tilted up against its flanks are a number of prolific oil fields and large gas fields which serve gas to such widely separated cities as Denver, Omaha, Minneapolis, Chicago, Indianapolis, and Detroit.

South Texas also has developed a whole string of fields, from Mc-Mullen and Live Oak counties on the north, to the Mexican border at the Rio Grande.

The Permian basin of West Texas and eastern New Mexico is a major and extremely productive geologic feature. An early well here came in with 184,920 barrels a day initial productive capacity. The basin is as large as an eastern state, and its possibilities are still being explored.

It must not be thought that the Gulf Coast was being neglected while north, south, and west Texas were receiving so much attention. New salt-dome fields were being brought in throughout the years, and deeper productive zones were discovered in some of the old fields. Spindletop had a second productive period that rivaled the first. Geophysical methods, first applied in the early 1920's, greatly accelerated the discovery rate. More salt domes were found in two or three years by the torsion balance and seismograph than had been found in the preceding ten. The Gulf Coast kept pace with the rest of Texas.

Texas was doing well enough when, in 1930, over toward the Louisiana border, Dad Joiner discovered East Texas, the biggest field of them all. He was drilling blind, without benefit of geology, and his well was an edge well. He had already lost two holes, and each time he had moved a location east to start his new hole. Had he lost his third hole and moved yet another location east he would have missed the field. It would not have been a disaster, however, for only a few miles farther north another well, located on geologic advice, was already drilling; in a few weeks it confirmed the discovery.

East Texas is not an anticlinal field; it is a huge sand lens. When the Sabine uplift in western Louisiana and eastern Texas was being eroded, and the sand from its erosion was being laid down in the sea to the west and north, a great sandbar formed along the beach. As the sea encroached on the uplift, both the bar and the sand extending west, and finally the uplift itself were buried beneath several thousand feet of mud. By the time the area emerged, the mud was shale, and the sand, with the sandbar at its eastern edge, was a poorly consolidated sandstone. Oil and gas migrated up the sandstone from the basin to the west and accumulated in the sandstone lens that the sandbar had become. In the sandbar is the thirty-mile-long East Texas field which once supported nearly twenty-five thousand wells. This single field produced 3,352 million barrels in the twenty-eight years ending in 1958—more than Europe (excepting Russia) has produced yet.

East Texas was discovered in the worst of the depression. The posted price dropped to ten cents, and isolated batches brought as low as a cent a barrel. Texas joined the ranks of the states enforcing proration. Despite litigation, evasion, and the running of much *hot oil* across the state line in the boom days, the East Texas field has never been allowed to produce more than a small percentage of its capacity. Nevertheless, its peak is past.

East Texas is not the end of the Texas story. A field is discovered somewhere in the state every few days; 113 were discovered in 1938, 306 in 1958, and 276 in 1960. Curiously enough, the average daily yield of the Texas wells is lower than we might think—21.4 barrels in 1930, 14.5 barrels in 1940, 18.3 barrels in 1950, and 12.9 barrels in 1960.

Texas' annual rate is really no measure of the potential; remember that the East Texas field has been prorated since its infancy. From this grew the state-wide proration plan administered by the Texas Railroad Commission. The Commission divides the state into eleven petroleum districts and keeps detailed records on every field and every well—a staggering amount of data for the 189,511 oil wells and the 23,875 gas wells (1960).

Engineers digest all these data to determine what is happening to the reservoirs as they are depleted and the depths and costs of the wells. Then they consider the factors of supply and demand, export trends, import volume, refinery and pipeline capacities, and future demands, to finally come up with monthly proration figures. Over the years, the *Texas allowable* has let the wells produce only a third to a quarter of the time each month. In 1959, for instance, the average was 10¼ days a month. In 1960 it was 8.67 days.

New Mexico

New Mexico is a relative newcomer to the oil states. A few barrels of oil were produced from shallow wells here and there beginning in 1919, but this merely put New Mexico in the "Others" column on most lists. Not until 1924 did it break into its own producing column. Michigan entered a year later, and Mississippi, Missouri, and Nebraska joined the ranks in 1939.

The first commercial field was discovered in 1922 on the Hogback structure in the San Juan basin in the northwest. By 1924 the Rattlesnake and Table Mesa fields had been discovered nearby. The combined production was enough to warrant a pipeline, and marketing began. These fields are in the Navajo country, where red mesas turn purple after sunset and Shiprock bears down with all sails set. The first fields were small; other early small ones in the San Juan basin are Hospah, Aztec, Bloomfield, and Kutz Canyon. All produced more headaches than oil in their infancy.

These fields of the Navajo country, profitable though a few of them are, would never have brought New Mexico a reputation as an oil producer. That honor was left to the southeast corner, New Mexico's part of the Permian basin, where Billy the Kid used to ride, and here again the early years were not spectacular. Late in 1923 oil was found east of the Pecos, not far from Artesia. The wells were not large but they held up; the area is big and altogether the field has proved to be a good work horse. Farther east, the Maljamar field, discovered in 1926, added some to the state's yield and reserves, though not a great deal. As an oil state, New Mexico was getting mostly nowhere.

However, close by in Texas, Hendricks field was discovered the same year. Hendricks field is one of the great fields of the West Texas Permian basin. It is scarcely more than a stone's throw south of the New Mexico border and on a structural feature that obviously extends into New Mexico, yet for nearly three years the state line stopped the oil men cold. It was 1929 before southeast New Mexico began to burgeon with Jal, Eunice, and Hobbs, and the thirty years since have added prolific producers. Names like Anderson, Caprock, Dayton, and Loco

Hills were making the headlines in the oil news columns of 1940. New Mexico chalked up 39.1 million barrels for the year, still mostly from the Permian basin.

In 1950 lists of New Mexico oil fields it is hard to find a name not in Lea county or Eddy county. They produced most of New Mexico's 47.7 million barrels that year.

It was not until 1954–1955 that headlines began to feature the opposite corner of the state. Then McKinley, San Juan, Sandoval, and Rio Arriba counties became familiar names in the industry. *Gallup sandstone* became a magic word-combination, almost synonymous with bonanza. In fact, many of the Gallup fields were found by drilling to the Gallup between shallower-reservoir gas wells. Accordingly most of the new oil field names include producing-formation names to distinguish them from the overlying gas fields. The Bisti Gallup, Horseshoe Canyon Gallup, Verde Gallup, and Hospah Gallup fields gave up 11 million barrels in 1960.

The new oil province helped bring New Mexico's 1960 production to 108.4 million barrels, about twice that of 1951. Proration affects the figures; imagine what the figure would be if all the wells were opened up.

New Mexico is not nearly so well known for oil as for the San Juan basin gas, the subject of quite a story. It includes the discovery of helium in a step-out well from the old Rattlesnake field and the helium-extraction plant the United States Bureau of Mines built in the long shadow of Shiprock. The helium didn't amount to much, but the fields now support mammoth pipelines across the desert to Los Angeles and a longer one to the Pacific Northwest. New Mexico truly feeds the furnaces of the West.

Colorado

In Colorado the Rocky Mountains are highest and widest; twenty-four mountains in the state are higher than Pike's Peak. The mountains are mainly composed of igneous and metamorphic rocks in which no informed person looks for oil. The east half of the state, however, is a great plain underlain by sedimentary rocks, and the western third is largely a rolling, semi-mountainous region also underlain by sedimentaries. Infolded among the igneous and metamorphic rocks, moreover, are intramontane sedimentary basins that have oil and gas possibilities and several oil fields. Eastern Colorado is a continuation of western Kansas; the Western Slope, the intramontane basins, and the foothills belt along the east front of the Rockies are more closely akin to the productive areas of Wyoming.

Colorado is one of the oldest oil-producing states. The Florence field, discovered in 1862, in an embayment in the east front of the Rockies, is still producing. It is not an orthodox field; the oil is not on the crest of an anticline, nor is it in a stratigraphic trap like the East Texas sandbar or the shoestring sands of Kansas and Oklahoma. On the contrary, the oil is in a syncline, in sandy shales that have been shattered and made porous by faulting.

For many years Florence was Colorado's only producing field. Then, in January, 1901, a small field was developed at Boulder. It didn't amount to much, although it still produces a few barrels a week. Another small field, producing from sandy lenses and fractures on a pronounced dome, was developed on the Western Slope near the Utah border at Rangely in 1902, and it, too, is still producing.

Other years went by with no new discoveries. Wyoming became a great producing state and was actively explored, but the state-line taboo was working. Anticlines near Wellington and Fort Collins were discovered in 1917 and thoroughly mapped in 1918. They were within forty miles of the Wyoming line and were underlain by some of the same sands that produced in Wyoming, but not until 1923 could anyone be persuaded to drill an adequate test. The first set of leases, held by one of the great world-wide oil companies, was allowed to lapse. The men who re-leased the area were turned down by practically every Rocky Mountain company before they were finally able to persuade Union Oil Company of California to drill. Union got a good field on the Wellington dome in 1923 and another on the Fort Collins dome in 1924.

The Wellington discovery touched off drilling on the Western Slope—testing geologists had recommended for three or four years. The Moffat and Tow Creek fields were discovered in 1924 and the nearby Iles field in 1925. By 1927, Colorado's production, which had been 86,000 barrels in 1923, had risen to 2.83 million barrels.

Other fields have been found on the Western Slope and in the intramontane basins. The Wilson Creek field, discovered on the Western Slope in 1938, produced 7.36 million barrels from the Morrison formation in the next ten years. The Morrison is the renowned dinosaur-fossil formation of the West. Three years after the discovery well was drilled, the operators found oil in the next deeper formation, the Entrada sand. The Wilson Creek Entrada reservoir produced 9.9 million barrels by 1951. The Entrada is flippantly called the "Slick Rock" where it crops-out in Colorado and Utah because of its smooth, rounded outcrops.

Rangely anticline—a bald, clearly defined structure—was an early producer, as we have mentioned. In 1933 the California Company drilled a significant, deeper test. This well flowed 300 barrels of oil a day from the Weber formation. After a few months of production the well was shut-in.

Not until 1945 did wartime demand, particularly on the Western Slope, produce the Rangely boom. By 1949, 480 Weber oil wells catapulted Rangely into Colorado's largest field. Like the national debt, production figures mounted year after year from 15,050 barrels in 1943 to 1.37 million barrels in 1945, to ten times this in 1947, and finally to 22.6 million barrels in 1953.

Three geologic phenomena make the Western Slope of special interest: One is the widespread but fickle nature of the Weber formation as a producer. More Rangelys are sought eagerly in trap after trap, year after year, and the Weber produces here and there in Colorado, but nothing has been found to compare with the bonanza in Raven Park, as Rangely was known to the old-time ranchers.

The second item of note is the Paradox Valley salt dome, one of the largest salt domes known. Although several wells have been drilled on it, including one that penetrated more than five thousand feet of salt, the dome is still a non-productive enigma. The region contains other monstrous salt domes, not so obvious at the surface.

The finding of oil and gas in nonmarine beds in the Piceance Creek, White River, Hiawatha, and Powder Wash fields is the third phenomenon. Many geologists have believed that oil can be formed only in marine beds, not in beds laid down in shallow, inland seas and lakes. The Powder Wash and Hiawatha discoveries negated this theory and made attractive the great basin of nonmarine sediments in Colorado, Wyoming, and Utah.

Eastern Colorado is about as flat as Kansas. In its kinship to Kansas, it even has its own north-south buried granite range, the Las Animas or Sierra Grande Arch. Several dozen wells drilled in this plains area before 1950 found only one oil field, the small Greasewood field northwest of Fort Morgan. Eastern Colorado, however, is as big as the whole state of Pennsylvania, and the dry holes in it did not discourage the oil men. Nevertheless, the state line between Colorado and Kansas was a serious psychological barrier until March, 1950, when oil was discovered in Logan County. Colorado's second big oil boom in five years was under way, and the Denver-Julesburg basin was a place to reckon with in every oil man's plans. The *D-J Basin* soon came up with its whopper, the Adena Feld in Morgan County. Adena has been good for 41 million barrels to date.

Now, 3,607 wells on both sides of the Continental Divide contribute their share in keeping Colorado in tenth place among the oil states.

Wyoming

In 1836, the tide of emigration began its westward sweep to save Oregon for the United States. In 1847 Brigham Young led the Mormons to the shores of the Great Salt Lake, and in 1849 the gold-seekers rushed to California. Most of these pioneers crossed what is now Wyoming by the Oregon Trail, later known as the Mormon Trail, and still later as the California Trail. Many of them greased their wagons at an oil spring near Poison Spider Creek, west of where Casper now stands. Many of the Mormons and gold-rushers also used the Carter Oil Spring in the Hilliard-Spring Valley country in extreme southwest Wyoming, beyond the point where the Oregon Trail branched to the northwest.

In 1867 a little oil was obtained from wells at Hilliard and, in 1873, small quantities were discovered in what later became the Dallas-Derby field in the Wind River basin. By 1894 Wyoming was in the producing column with 2,369 barrels of oil, and production gradually increased year by year. In 1900 commercial oil was found in Spring Valley, though it never became an important field.

A prospector named Cy Iba, who had followed the gold rush into the Black Hills of South Dakota during the eighties, struck out westward across the Powder River basin intending to prospect in the Big Horn Mountains. Camping one night on a nearly dry stream called Salt Creek, he saw oil seeping from the bank. When he found other oil evidences along the creek and its tributary gullies, he forgot about possible gold in the Big Horns and plastered the surrounding country with oil claims.

In those days Bradford, Pennsylvania, was the oil man's rendezvous, and in time Iba's find interested a Bradford man named Shannon. In

1890 Shannon brought in the discovery well of the Shannon pool. American oil companies were not interested in oil in so remote a region, but eventually Belgian, French, and Dutch capital was obtained and the pool developed. Its wells were small but shallow. A small refinery was built at Casper, the nearest town, and the oil was hauled fifty-five miles overland in wagons.

The enterprise was losing money, and in 1906 the Dutch company in control employed the eminent Italian geologist, Dr. Caesar Porro, to advise whether to drill more wells or abandon the project. Dr. Porro promptly reported that the Shannon pool was on a slight terrace on the north edge of a pronounced dome, and that the place to drill was a few miles south along the road to Casper. The Dutch company started a well, and a little later Fitzhugh and Henshaw of California started a well in the same locality. By 1909 both wells had been completed as good producers, and the Salt Creek field—one of the major fields of oil history—had been discovered. For about fifteen years the owners of the Shannon pool had been hauling their oil across it without suspecting its presence.

Salt Creek brought oil fever to Wyoming, and many other good fields were discovered, though none as good as Salt Creek. The state's production, only 9,339 barrels in 1907, was 44.8 million barrels in 1923. After that it declined steadily to 11.2 million in 1933. In 1934, the discovery of new fields and of new producing sands in old fields started the curve upward again, so that by 1960 Wyoming's production was 136 million barrels, and Wyoming was fifth among oil states. Wyoming oil men had drilled 19,710 wells to do it.

The Lance Creek field in the east was responsible for most of the 1930 to 1940 production increase. Lance Creek is a good example of what can happen in the oil business. It was discovered in 1918 and oil men, thinking they had another Salt Creek, acted on their belief and drilled and equipped accordingly. It was a sad story. Lance Creek's area was smaller than expected, and the flashy production declined rapidly. For many years Lance Creek produced more red ink than oil.

Finally, in 1930, a well was drilled to a deeper group of sands (Sundance), and the dreams of 1918–1919 came true. In 1937, because a still deeper and apparently more productive producing horizon was found, Lance Creek started crowding Salt Creek for first place among Rocky Mountain producing fields.

Another Wyoming field has had a great deal of public attention, although it is not otherwise important—Teapot Dome—so named because the south end of the Salt Creek anticline is crossed by Teapot Creek. In the belief that the structure was a true dome, separated by a saddle from the Salt Creek closure, the Navy set the area aside as a Petroleum Reserve. In time, drilling on the Salt Creek field extended to the Naval Reserve boundary and showed that part of the Reserve lay on the Salt Creek structure. We won't rehash the controversy, nor try to decide whether the Salt Creek wells were draining the Reserve's gas pressure or whether the Reserve should or should not have been leased. But one fact, known to every oil man, seems to surprise everyone else; instead of being the $100 million oil field of newspaper headlines, the Teapot Reserve was practically a dud. Despite the drilling of more than eighty wells, production was only a few thousand barrels a day.

Wyoming's oil story is by no means told. Around the edges of her great sedimentary basins many anticlinal structures are so easily recognized that even the sheepherders used to report them in the halcyon days of 1910 to 1925. Those days are gone. But within the basins, obscured by younger rocks, are other structures that time and patience, geological studies and good geophysical work have discovered.

All of Wyoming's major basins are now productive. Some produce mostly gas, others oil. The once remote and nearly trackless Green River basin in the southwest became known first for its gas. In recent years the emphasis has switched, and now oil men are pushing for more big oil fields like Desert Springs, Patrick Draw, and Table Rock. After its discovery in 1959, Patrick Draw became one of the state's largest producing fields in just over a year.

We mentioned the *sheepherder anticlines* of the halcyon days and we spoke of buried structures, but we should not neglect the stratigraphic traps—oil caught by sand pinch-outs. New discoveries are being marked up every year, and most of them in the last decade or two have been made in traps formed by permeability differences within the reservoirs. Meticulous geologic work with well logs of all types has helped to find a whole string of new fields in the eastern Powder River basin. Donkey Creek and Coyote Creek and Flat Top are just three of the many fields found in the Powder River basin recently by detailed subsurface geologic work. There seems to be no end to the oil possibilities of Wyoming as long as geologists keep on their toes and come up with fresh ideas.

Montana

One of the good Wyoming fields found during the post-Salt Creek boom was Elk Basin, discovered in 1915. It lies athwart the Montana line; production sneaked across into Montana before the state-line taboo wakened. In 1916 the state's daily 45,000 barrels came entirely from Elk Basin.

You might suppose that Elk Basin would stop the state-line ghost, but it didn't. In 1920, when a well was proposed on the Cat Creek anticline in central Montana, a well-known geologist is reported to have bet that he could drink all the oil the state would ever produce, outside of Elk Basin. Cat Creek made a good field; so did Cut Bank, Sumatra, Kevin-Sunburst, and Pine, Bigwall, Poplar, and a host of others right up to 1960.

Montana has produced 371 million barrels since the Cat Creek discovery. When you subtract the 51 million barrels from Elk Basin, the remainder is a copious draught, even for a geologist.

Montana's second biggest field is a 1922 discovery, the Kevin-Sunburst field near the Canadian border. The structure is a broad, low arch, and productivity is controlled mainly by variations in porosity in the reservoir beds. More than one producer in the field has been ringed with dry holes, and more than one dry hole has been surrounded by producers. Despite the erratic pattern, however, the field vindicates itself by producing both oil and gas from five different formations. Who knows, maybe geologists will find several more productive horizons before the first-discovered reservoir is fully depleted.

A more recent field, Cut Bank, lies farther west and equally close to Canada. Cut Bank is a stratigraphic trap field with the oil in the up-dip edges of three lenticular sandstones and a limestone. Discovered in September, 1926, Cut Bank has added a ninety-one-million-barrel gulp to that geologist's penalty and has accounted for a fourth of all the state's production. It is no longer the biggest yearly producer, however, because the Pine, Cabin Creek, and Poplar fields have taken over first, second, and third places. Each yields over a million barrels each year.

In 1960 Montana produced 30.2 million barrels—more than Austria and France, and more than all of the Far East—placing Montana twelfth among the oil-producing states.

Montana's geological conditions and her great sedimentary areas hold many surprises for those who think her oil fields have all been found.

California

California is the third largest state in size and oil output. Most of her oil comes from a comparatively small area west of the Sierra Nevada and San Bernardino Mountains and south of Monterey and Fresno—an area remarkable for its many fields and striking yields per acre.

Some California sand reservoirs are 1,500 feet thick, much thicker than those in most of the rest of the United States where typical producing sands are a few feet to a couple of hundred feet thick. California also seems to have been blessed with richer source beds. These two factors give many California fields productions per acre not equaled elsewhere in the United States except in some of the Gulf Coast salt-dome fields.

Most fields lie in one of the two great geologic basins: (1) the San Joaquin Valley, between the Sierra Nevada and the Coast Range, the southern part of the Great Valley of California, and (2) the Los Angeles Basin, south and west of the Coast Range, the Sierra Madre, and the San Bernardino Mountains. Besides these, some important fields are scattered northwest from the Los Angeles Basin to the Ventura Basin, Santa Barbara, and still farther north through the Santa Maria Valley almost to Monterey Bay.

California's oil history predates the first white men and even the first Indians. Within the extensive city limits of Los Angeles is a group of large oil springs known as *La Brea*, from which water and sticky, asphaltic oil spread over several acres.

For thousands of years, birds and animals have come to these springs to drink and have become entangled and finally engulfed by the asphalt, which has almost perfectly preserved their skeletons. The first overland expedition to Monterey Bay in 1769 made note of the springs. Within recent years, thousands of bones have been exhumed from the springs, including those of the saber-toothed tiger and other animals that became extinct, so far as we know, before the first human beings migrated to America.

La Brea is only one of several prominent oil seepages. There are others in the Los Angeles Basin, the San Joaquin Valley, and along the coast, and in the northern part of the Great Valley. No other state has so many surface evidences of oil.

Naturally the California Indians used oil and asphalt, and the white men took an early interest in oil possibilities. By 1875, shallow wells had been drilled in Pico Canyon and the ex-mission locality, both in Los Angeles County, and in 1876 the state produced 12,000 barrels. In 1880, oil was found in the Brea-Olinda field east of Los Angeles. It is still a good field; in 1960 it produced 5.9 million barrels—as much as Japan and Pakistan. The Moody Gulch field just south of San Jose was found in 1880 and produced about 85,000 barrels before it was abandoned.

In 1886, a little oil was found at Half Moon Bay, not far south of San Francisco, but, like the few other northern discoveries made so far, it amounted to little. In 1892 the Summerland field was discovered near Santa Barbara. It is now depleted, but its offshore extension, first drilled in 1958, is still going strong.

In 1896, with the discovery of oil at Coalinga, there began a rapid development of the great San Joaquin Valley fields, which for many years overshadowed the rest of the state. The McKittrick field came in 1898—an interesting exploration venture downdip from an old seep. McKittrick piques geologists because oil is trapped by tar that seals the pores in a sandstone reservoir—a rare sort of stratigraphic trap. The Kern River field north of Bakersfield was discovered in 1900, and the Maricopa, Midway, and Sunset fields were found and extended within the following few years. By 1908 the San Joaquin Valley fields had flooded their market, and for a time the price of oil was down to ten cents a barrel. By 1914, California's production was up 99.8 million barrels.

In the early 1920's the Los Angeles Basin took the play away from the San Joaquin Valley. Highly productive fields were developed at Huntington Beach, Long Beach (Signal Hill), and Santa Fe Springs; thus in 1923 the state's production hit the new peak, 263 million barrels. In 1928 the play returned to the Valley with the discovery of the great Kettleman Hills field on an anticline that had been drilled (but not deep enough) many years before. In 1929 the state produced its pre-World War II peak of 292½ million barrels.

After 1929 many fields and many deeper producing horizons were found, but voluntary proration—California has no proration law— tended to hold back production. Production dipped to 172 million barrels in 1933, then climbed steadily for twenty years to its all-time peak of 365 million barrels in 1953. It fell off gradually to 304 million barrels in 1960—a figure third among the states' and a little less than that of Iraq.

Fields will doubtless continue to be found for many years in the San Joaquin Valley, the Los Angeles Basin, and along the coast. Four oil and thirteen gas fields were found in 1960. Exploration has pushed northward in the Great Red Bluff Valley and has developed important gas fields as far north as Red Bluff, 170 miles north of San Francisco. The northern part of the Great Valley, drained by the Sacramento River, is more difficult to explore and somewhat less promising for oil than the southern part that has yielded such spectacular results, but recent gas discoveries are encouraging oil hunters. If the Sacramento Valley should develop major production, Texas might have to look to her laurels. Even without it, California is and will continue to be a great oil state.

Alabama

Alabama is a "war baby" in the oil-state family. Her first commercial oil came from the Gilbertown field in February, 1944. One might say that this was also the discovery of Mississippi's Longsdale field just across the state line. The first year's 43,000 barrels probably helped power the 835 L.C.T. landing craft which hit the Normandy beach on June 6, 1944—D-Day. One year later, the new Falls City pool and the Gilbertown pool had helped boost Alabama's war contribution to 181,000 barrels.

After the war came the big Pollard field in 1952 and the huge Citronelle field in 1955; by 1960, ten fields had produced 36.8 million barrels. Alabama has added about a million barrels of production to her tally sheet each year since 1955—a respectable record for any infant.

Florida

Florida, like Alabama, was a staunch supporter of the war effort. She came through gracefully, in time of need, with her first oil field in 1943. The Sunniland field between the Everglades and Lake Okeechobee has produced 6½ million barrels from Lower Cretaceous lime-stones below 11,535 feet. This is almost a thousand feet more than two miles deep—a deep field anywhere.

Eleven years of exploration in the swamps and everglades passed by before the Forty-Mile Bend field was found in Dade County, forty miles west of Miami. This oil, too, is deep—11,322 feet from a Lower Cretaceous reef. Two wells have produced 32,888 barrels in the seven years since the big day of discovery, and so we may be sure that the geologists and engineers are trying to figure out what to do next with Forty-Mile Bend. There may be many more pools hidden below the Mangrove swamps of Florida, but it is a tough place to explore.

Nebraska

With the discovery of oil in the southeast near Falls City, Nebraska became a producing state in 1939. The discovery well gave up only 2,000 barrels the first year, but in 1940 Nebraska chalked up 276,000 barrels of crude oil and added two more fields in Richardson County. For a few years it looked as if this county had a corner on Nebraska's oil. Development in southeast Nebraska, in the Forest City basin, was not extensive and the state had a fruitless ten-year search for more oil. Then, in 1950, rigs drilling in the opposite corner found the Huntsman, McLernon, and Dorman fields in Cheyenne County, kicking off an oil boom in seven counties of the Nebraska panhandle. The state-line psychosis had been broken again because just two months had passed since the discovery of oil across the line in Logan County, Colorado.

After ten years of aggressive exploration Nebraska looks back on a phenomenal rise to fourteenth place among the oil producers. In 1941, two years after the first discovery in Richardson County, the state had 70 producing oil wells, and in 1950, 140 wells.

Development in the Nebraska panhandle part of the Denver-Julesburg basin boomed along after 1950, doubling the 472 holes drilled in 1954 with 918 holes in 1956. Then, typically, the boom passed and the Denver-Julesburg basin slumped toward a normal development. But then, again typical of the fascinating oil business, activity leaped cross-country, striking Red Willow County. McCook became another overnight boom town. Pennsylvanian limestones, lapping on the west flank of the long-known Cambridge arch, contained oil below an un-conformity. Finding where the Lansing and Kansas City formations rose to the unconformity had been a problem to oil geologists and geophysicists for many years. Now they knew, and the boom was on. Drilling rigs moved in in droves, oil fields spread out into Hitchcock and Dundy counties, and the well-completion rates rose to the all-time high, 951, in 1959. At the end of 1960, 1,642 producers were pouring 66,502 barrels a day of green oil into the pipelines and refineries. Nebraska produced 24.5 million barrels in 1960, and her cumulative total jumped to 141 million barrels. The Cambrian Reagan sand, also lapping up on the arch, looks like a good bet, and so do several other western Nebraska formations. Nebraska bears watching and you can bet the geologists are watching it.

Nevada

Nevada nosed out South Dakota by four months in completing a producing well. Shell Oil Company announced the discovery of Nevada's first oil in Nye County on February 14, 1954, but did not start producing until June, 1954. (The Buffalo field in Harding County, South Dakota, started producing in May, 1954.)

For many decades, geologists, geophysicists, promoters and "water well drillers-turned-expert-oil-finders" have been searching the flat-bottomed desert valleys of Nevada's basin and range country, but they were not the first.

Exploration dates back to the white man's first notes on the country, such as those of Jedediah Strong Smith of the Lewis and Clark Expedition in 1826 and 1827. The first geologic reports on Nevada's later-to-be oil country were written in 1859 by Henry Engleman, geologist and engineer with Captain J. H. Simpson's party of the United States Army Corps of Topographical Engineers. More geology was written on the great basin of Nevada in the forty years between Simpson's reports and those of Josiah Edward Spurr in 1899 for the United States Geological Survey Bulletin No. 208 than in the next forty years.

Earnest efforts to find oil in central Nevada began in September, 1920, on the Illipah Anticline in White Pine County. Four holes were drilled in a span of seven years; each one reported oil shows, but none produced a gallon of oil.

The gambling spirit for which Nevada is famous seemed to subside for almost twenty-three years. Then the epidemic broke out again in 1950 with Standard Oil Company of California, Continental Oil Company, Gulf Oil Company, Richfield Oil Company, and finally Shell Oil Company getting into the Nevada "game." Seven did not come eleven in this game, but "unlucky thirteen" did. Shell's No. 1 Eagle Springs Well in Nye County was number thirteen in the play for a winner. The well went into the production tables in June, 1954, making 275 barrels a day from rocks that are not supposed to produce oil. The productive formation is tuff, a volcanic ash. Shell drilled another; it produced from a fresh-water limestone, not supposed to be favorable oil rock either. Shell drilled nine holes to develop the field, but only one was successful; this produced from a marine limestone almost 200 feet deeper than the discovery. Shell finally cashed in its chips in 1956 to rest on its winnings in the Eagle Springs field.

The three-well Eagle Springs field, which is still Nevada's only producer, has made a bulge in the economy. Oil lease fees paid to the state treasurer's office since 1948 total 2.9 million dollars—six times more than the value of the 1959 mineral production. Very few mining areas can equal this dollar value, and so Nevada is looking ahead hopefully. Nevada is a mining state and a gambler's paradise, but a few more oil gamblers with a little luck could make her an oil-conscious state.

North Dakota

North Dakota is another newcomer to the oil states. When we look at the depth of the discovery well we can understand why this is so. But first, let's look at the geology and see why oil men thought it a good place to explore.

This great wheat state lies athwart a broad shelf extending from the Rockies on the west to the pre-Cambrian shield in Minnesota and Ontario. In the midst of this shelf a basin has been sinking and filling with sediments almost continuously since pre-Cambrian times. The deepest part of the basin is under the town, Williston, from which Dr. W. Taylor Thom, Jr., coined the name, Williston Basin, in 1923. The southwestern boundary is the Cedar Creek anticline, an oil producer. Out in the middle of the depression, on the Nesson anticline, the first commercial well came in on April 4, 1951. Amerada Petroleum Company drilled the test to 11,955 feet, reporting good recoveries of oil during drillstem tests in three producible zones. This famous Clarence Iverson well No. 1 opened the Beaver Lodge field, and was completed for twenty-five barrels an hour at 8,520 feet. Now it is easier to see why cable-tool rigs, and even the early rotary rigs, had not found this North Dakota oil.

Only the latest seismic and drilling techniques and equipment stand much of a chance to explore successively deeper formations as the years go by; at Beaver Lodge technology finally found oil in 1951. By 1960 the deepest hole in North Dakota was 15,135 feet deep, although deeper ones were then possible.

Since her spectacular but belated beginning, North Dakota has added a hundred oil fields spread over thirteen counties and can boast of 22 million barrels of oil produced in 1960. She is rapidly catching up with Nebraska and challenging all comers for a high-ranking place on the list of producing states.

South Dakota

South Dakota, the Mt. Rushmore state, is truly an infant in the oil sorority, and must still prove herself worthy of recognition. She probably will. The first commercial oil was found in 1954 by extension of the North Dakota Williston Basin play; about 19,000 square miles of the Basin is in South Dakota. Oil boom contagion spread north, from the Denver-Julesburg Basin too, and South Dakota was caught in a squeeze of discovery excitement from two directions. The Buffalo field in Harding County was discovered in May, 1954, at 8,587 feet on the south end of the Cedar Creek anticline. Fifteen wells in the Buffalo field produced 810,024 barrels of oil to 1960.

About a year and a half went by before Custer County brought the play down the west end of the state, off the south tip of the Black Hills. The two-well Barker Dome field was found at 1,390 feet. Compare its depth with that of Buffalo or the 11,000-foot Beaver Lodge field in North Dakota. Here, too, a structure makes the difference. This one, named the Hartville Uplift, is a wrinkle that branches from the Rocky Mountain front range at Wheatland, Wyoming, to the Black Hills and brings old, deep formations, higher and shallower than they are in the adjoining basins. Barker dome oil comes from the north (Powder River basin) side of the uplift; Barker is the outpost of a string of fields in Wyoming.

South Dakota's score was about 750 barrels a day in 1960 and it had run up a total of 643,000 barrels by the year's end. The score could change abruptly with new exploration.

Utah

Utah oil history reads like a prospector's dream in wide-screen technicolor. "Seek and ye shall find" has been the watchword since the Mormon pioneers trudged across the mountains to the dry valleys of Utah. The Mormons were spreading out to settle the beehive state about the time the Drake well in Pennsylvania came in, during 1859. Water to sustain life and to irrigate the fertile land was their first need, and they brought drilling skills with them to the desert lands. The first true oil prospect was drilled in Grand County in 1891, forty-four years after the first settlers moved in. It was dry. Sporadic drilling has been going on ever since, finding a little oil in the Virgin field in 1907, and at Mexican Hat in 1908. Extensive exploration has been carried on in widely separated parts of the state. Let's take a look.

In northeastern Utah is the major part of a great basin of nonmarine sediments that also includes a large area in southwestern Wyoming and northwestern Colorado. This is the region of the great deposits of Green River formation oil shale, in which the oil-forming material is still mainly in the *kerogen* stage, requiring heat or pressure to convert it into oil. In the Utah portion of this Uintah-Piceance basin are more oil seepages than in any other area of equal size in the United States outside California. In places at least there must have been enough folding to generate the heat and pressure necessary to convert part of the kerogen into petroleum. In the White River, Hiawatha, Piceance Creek, and Powder Wash fields of Colorado, the oil-shale beds contain gas. On the eastern rim of the basin, also in Colorado, gas and oil accumulations have been found in nonmarine beds older than the oil-shale.

Utah's first good field was found in the Uintah-Piceance basin at Ashley Valley near Vernal, in 1948. Ashley Valley has produced 10.9 million barrels from three different reservoirs between 1,430 and 4,400 feet, but not from the oil-shales.

Since 1948, deeper drilling has found sixteen oil pools in the shale beds, and today 54 oil fields in the Utah part of the basin produce from many different formations, some as deep as 11,100 feet.

In east-central Utah, along the Colorado River, are the western extensions of the Colorado salt domes. For forty years people have drilled in and around them, and for forty years the tests have found shows of oil and gas but no commercial production. Drilling in the area is intense in these early sixties, however, and the newly discovered field at Lisbon is probably only the first of several.

Farther west, gas has been found on anticlines at Last Chance, Clear Creek, Joe's Valley, and on others. Harley dome just west of the Colorado line in Grand County and the Woodside field in northeastern Emery County contain gas that is part helium. These have been set aside from commercial exploitation as federal helium reserves.

Southern Utah has been a fine example of the state line prejudice. Although the Mexican Hat and Virgin fields have produced a little oil since 1907, it was once popular to debunk the idea and castigate the character and wisdom of anyone exploring for oil in Utah's southeastern deserts. Montana once heard the same kind of talk about her oil potential. Brave souls with courage in their convictions and ears closed to the unneeded remarks of their friends kept on working. Nature and circumstances supplied plenty of discouragement; the area had no roads, no markets, no water, no pipelines, and no housing because there were no towns. As a matter of fact, there was no oil there, the "smart" ones said, on the day the Desert Creek No. 2 well was spudded in November, 1954. The shoe was on the other foot when Shell Oil Company brought in 1,128 barrels a day, flowing from a mile deep. The Paradox Basin boom was under way. Drilling rigs sprang up across San Juan County like spring tulips. Roads were blasted through to new locations almost overnight. Water wells, field camps, supply houses, and oil fields appeared in the desert at every turn in the road. By 1960, the new Aneth field had produced 29 million barrels, the McElmo Creek field, 23 million barrels, and the Ratherford field, 20 million barrels—just to name the three largest fields. Now the oil flows to Los Angeles refineries through a pipeline across Arizona and Nevada. Highways in and out of the oil fields have been paved, and the Navajo Indians, to whom the land belongs, are profiting by their good fortune. The tribe receives a royalty from every barrel produced and a fee for every oil lease sold, so the Navajo have new schools, missions, cars, and homes, and no one knows how many more horses and sheep.

Alaska

Political change, not exploration, added Alaska as the 33rd oil-producing state in 1958. The first field, a small one at Katalla on the south coast, has produced intermittently since 1903, but exploration largely ignored southern Alaska until the mid-fifties. One hundred forty-five wells have now been drilled in the forty-ninth state, most of them being concentrated in the three years since discovery of the Swanson River field in September, 1957.

Supply and transportation have always been problems in Alaska, and the consideration of what to do with the oil or gas that might be found has bothered and still bothers the oil operators. One good field like Swanson River can reverse pessimism overnight, and Swanson River did. The Kenai Peninsula, south of Anchorage, was an ideal place for the first good field. Cook Inlet affords easy access to shipping, and Anchorage, at the head of the inlet, is a ready, willing, and able city of 35,000 people, eager to serve and buy from a new industry. Trucks and ships hauled the first three years' crude oil to and from the Port of Seward. The Kenai Pipeline Company completed nineteen miles of eight-inch line and began oil deliveries in October, 1960. Thirty-one holes were drilled in 1960 for the all-time record. At the end of the year, Alaska could look back on 928,000 barrels produced from 26 wells with considerable pride and a keen sense of victory over the long-standing pessimism. A gas field on the Kenai and a nearly completed gas pipeline to Anchorage added materially to the boom spirit in southern Alaska in 1960.

The world's northernmost known oil seep is a large one east of Point Barrow. It seems that Alexander Malcolm Smith was the first white man to note the seeps near Cape Simpson in 1917. Four years later, Harry A. Campbell examined them in detail for private interests. The geological potential for oil appeared so great that President Harding issued an executive order in February, 1923, establishing Naval Petroleum Reserve Number 4 to preserve the valuable resources of the north country for any future national crisis. World War I had demonstrated that in time of war the Navy could create an astronomical demand for petroleum. Geological parties mapped the *north slope* until 1926, and then activity subsided until World War II revived a concern for strategic oil. In 1943, a Bureau of Mines party explored the area with the assistance of Simon Paneak, an Eskimo guide. They traveled by bush plane to several seeps never before seen by a white man. Three of these seeps are located on the Umiat anticline which was later to become the site of the discovery well in Naval Petroleum Reserve Number 4.

Umiat No. 1 was spudded in May, 1945, and drilling continued throughout the summer in sub-zero cold and frozen ground. Drilling was suspended in the fall of 1945, and when the crews returned to Umiat in May, 1946, they found thirty gallons of oil atop an ice plug between 775 and 920 feet. They deepened the well and it proved to be the discovery hole on the largest field in *Pet Four,* as the area is

now known. Umiat now has eleven wells ready to produce an estimated 70 million barrels of oil. Pet Four can boast of two more oil
fields, Simpson and Fish Creek, and six new gas fields. Navigation
to the reserve is open for only a few weeks each year, and pipelines
to Bering Strait or over the Brooks Range to the Yukon would have
to be laid on or in tundra and ice and over mountains. Getting Arctic
oil to points where it can be used presents something of a problem.

Parts of the great interior of Alaska are underlain by sedimentary
rocks with favorable oil possibilities, but very little drilling has been
done here yet. Oil men have plans, however, judging from the nearly 40
million acres of oil and gas leases they hold in Alaska.

Arizona

Arizona is the newest oil state, but opinions vary regarding the first
date of commercial production. Shell Oil Company's East Boundary
Butte wildcat in Apache County near the Utah border was completed
on December 13, 1954, as an oil well. Oil was not produced until 1955,
though some reports do not credit the well with commercial production until 1958. Northern Arizona's first test was drilled in 1921 in
Apache County, and many other tests in both the north and south
preceded the discovery. This long-sought success—the outcome of the
Paradox basin boom in southeast Utah—proved once again the fallacy
of state-line thinking.

Other oil and gas fields were developed in 1956 and given strange
names such as Bita Peak and Toh-Ah-Tin; Pinta in 1957, and Dry
Mesa in 1959. The names are mostly Navajo words because all of
Arizona's fields are on the Navajo reservation. Since all the petroleum
so far is under Indian land, the state has offered a big bonus to the
first operator who finds commercial oil in non-Indian ground. Drillers
are anxiously boring holes in both the southern tier of counties and
in the Black Mesa Basin south and east of the Grand Canyon. So far,
their quest has been encouraging but fruitless.

One of Arizona's finest resource contributions has been the discovery of the world's highest concentration of helium at Pinta Dome,
not far from Holbrook. This valuable gas is separated from other non-
flammable gases in a plant near the field, compressed, bottled, and
shipped to markets in the western states. From Johnny's balloon to the
complex nuclear energy industry, helium finds a hungry market, and
so Arizona is looking for more.

Virginia

Virginia is listed among the oil-producing states, but she barely holds a position there from year to year. Seven counties in the western tip of the state next to Kentucky, West Virginia, and Tennessee—from Cumberland Gap to Saltville—hold the honors for the state in oil circles. The Trail of the Lonesome Pine country gave up 2,000 barrels of oil in 1943, its first year of production. Oil flowed to market in greater and greater quantities until the peak year of 1947 when Virginia produced 61,000 barrels. Then the operators' dreams suffered a blow as the production rate dropped precipitously to about half in 1948. It fell off to about a tenth of the 1947 rate by 1954 and has held steady since. We can't say that the oil men haven't been trying to improve their record; they kept six or seven rigs running in 1960 and drilled 30,192 feet of hole. Some gas was found and some oil was added to the totals, but not enough to compensate for falling production rates in the oil fields. Patience and diligence probably will find more in years to come.

Other States

A few other states yield a little oil, several more have possibilities, and the balance will probably never be oleaginous.

New England has few if any prospects. The Gulf Coast states east of Mississippi and the Atlantic Coast states south of New York have distinct possibilities. Parts of them are underlain within drilling depth by formations of the same ages as those that are productive in Arkansas, northern Louisiana, and Mississippi. Evidences of anticlinal folding have already been found here and there. The dry holes drilled so far have not furnished an adequate test of the region; it is reasonable to expect production some day in the Florida Panhandle, in Georgia, and along the Atlantic Coast as far north as Delaware.

Missouri produces a little oil from shoe-string sands along her western border, adjoining the fields of eastern Kansas; much of the rest of the state and all of Iowa are underlain by sedimentary beds of the same ages as some of those that produce in the Mid-Continent. However, many geologists believe these formations have been subject to enough heat and pressure to destroy any oil they may once have contained.

The great interior basins of Washington, Oregon, and southwest

Idaho are mainly underlain by thousands of feet of lavas, and by lake deposits unlikely to contain source material, and, though a little gas and oil have been found, it seems improbable that this area will become important.

In the coastal parts of Oregon and Washington are many oil springs and a few gas seepages. Although the rocks here are of the same ages as producing horizons in California this region is more intensely folded and faulted. A number of tests, most of them in the Washington coastal region, culminated in the August, 1957, discovery by the Sunshine Mining Company of the Ocean City field in Grays Harbor County, Washington. In that year this field produced 5,000 barrels of oil. The subsequent years have been disappointing. In spite of continued exploratory drilling, Washington's 1960 production was only 898 barrels—all from Ocean City.

Several major companies have started offshore seismic work and submarine geology; the Pacific Northwest is an area to watch.

Canada

Canada trod on the heels of Pennsylvania in finding oil. Dr. Tweedel, a refiner from Pittsburgh, started drilling in New Brunswick in 1858 and came close to antedating Colonel Drake's discovery. Dr. Tweedel's project was not a commercial success, but in 1909 a new attempt developed a small oil and gas field at Stony Creek, not far from Moncton. By 1930 the field produced more than 5,000 barrels of oil a year, and careful development increased the rate to 14,148 barrels in 1960. This is small production, but it proves the presence of source beds and reservoir beds, and if these can be found under structural conditions more favorable than those at Stony Creek, more fields may result.

In some of the southwestern Ontario swamps, oil collected on the water and became mixed with the swamp soil to form what were called *gum beds*. In 1859 or 1860, enterprising individuals began to extract oil from the gum beds at a place called Oil Springs; by 1860 they had drilled a well which got such a flow of oil that a lively boom resulted. In 1865 a pumping well was drilled nearby at Petrolia, and the next year a flowing well started a boom that caused the virtual abandonment of Oil Springs. Now, a hundred years later, the well and the Petrolia field are still producing, as are a number of other small fields, almost as old, in the same region.

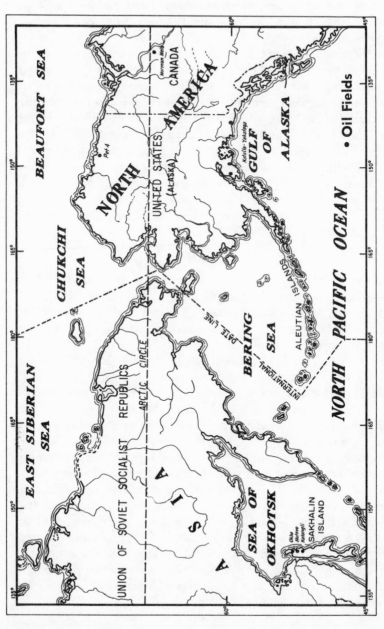

Siberia, Alaska and N.W. Canada

The seventy-five Ontario fields are not so productive nor so numerous as those of the Appalachian region in the United States, but they have produced a lot of oil. Starting with less than 12,000 barrels in 1862, their production climbed to 365,000 barrels in 1873, and, after many ups and downs, reached 913,000 barrels, the peak, in 1900. Since that time discoveries have failed to keep pace with declines. In 1958, Ontario produced 777,117 barrels.

With Ontario and New Brunswick making modest contributions, it remained for Alberta, in western Canada, to put Canada's annual production above the million-barrel mark.

In the foothill belt that separates the Rocky Mountains from the great prairies to the east is a long, narrow, sharply folded, heavily faulted anticline known as Turner Valley. Oil was discovered on it in 1913, and nearby Calgary proceeded to stage one of the wildest booms in oil history. Oil companies were formed by the score. Lawyers, doctors, clerks, and chambermaids neglected their businesses to gamble in oil stocks. Brokers' offices sprang up like weeds and took in money so fast that they stuffed it into waste baskets until they could find time to take it to the banks. Then came World War I, which absorbed all of Canada's attention. Many of the wells so wildly promoted were never drilled; many that were started failed to reach the producing sand; and those that did were not big producers. The first Calgary oil boom was over.

In 1924, a well drilled to a deeper horizon in the Mississippian-age Turner Valley lime found gas in large quantities under high pressure. Calgary staged a second boom, longer than the first and almost as wild. By the time the depression struck in 1929, Turner Valley was a tremendous gas field. The wells on the crest produced dry gas; the wells on the flanks produced gas so wet that considerable quantities of natural gasoline could be extracted from it. Unlike most natural gasolines which contain only the lighter gasoline fractions, some of the Turner Valley product contained the heavier fractions also, and the entire liquid output came to be called *naphtha*. The only way to get the naphtha was to produce the gas. The available markets could use only a small part of the gas; the rest was wasted into the air, burned in great flares that lighted the whole countryside. For more than ten years over 200 million cubic feet of gas—enough to have supplied the city of New York—were flared each day.

In 1936 a well was drilled farther down the west flank than its predecessors. It found oil, and a third boom took place, somewhat more restrained than the first and second, with drilling concentrated in the oil belt on the flank below the level of the gas. The field produced 2.3 million barrels in 1960. The gas wasted over the years would have been of great value in producing the oil; its loss has been felt severely and will be felt more as time goes on. The Alberta Government has forced a drastic reduction in the flaring; the gas and pressure saved will add years to the life of the field.

The excitement over Turner Valley led to many attempts to find oil in the rest of the province. First attention was naturally given to the Mississippian elsewhere in the foothills belt, and a multitude of other fields resulted.

Farther west in the foothills, so close to the mountains that the Turner Valley lime has been eroded away, is a series of anticlines in which the older Devonian beds are within reach of the drill. On one of these structures known as Moose Dome a well found a little Devonian oil, the forerunner of a host of Devonian fields.

East from the foothills the prairies extend nearly a thousand miles across Alberta, Saskatchewan, and much of Manitoba. The prospects in this immense territory are similar to those in the Dakotas and eastern Montana. Each province has become a major producer. In 1960, Saskatchewan listed 156 fields, many of them producing from several different reservoirs. Manitoba listed fifty-one fields in 1960, although some of them were abandoned by the end of the year.

British Columbia came into the act in 1943 with the Pouce Coupe field, and oil has since been found in 126 different reservoirs.

Canada has another distinctive field, albeit a small one. The Fort Norman field, on the Mackenzie River almost at the Arctic Circle, is Canada's northernmost producer. In 1960 it had 63 wells good for 468,545 barrels, a small refinery, and a market in the rapid development of the Far North.

In 1960 the Dominion's production, from New Brunswick through Ontario and British Columbia to Fort Norman, was 196 million barrels, enough for eighth place in world production.

Canada's greatest oil reserve, however, is not in any of these fields, present or prospective, but in the oil-sand of Northern Alberta, frequently referred to as the *tar sands*. According to the Dominion Mines Branch, the oil-sands contain at least 100,000 million barrels. The oil, apparently subjected to less heat and pressure than the oils produced from wells, is in the last stages of the transformation from organic material—through asphalt—into oil. It is neither fluid enough nor under enough pressure to make its way into wells, and the economical way to recover it is to mine the sand and then extract the oil. Once separated, the oil is highly sensitive to refining temperatures and can be converted into all of the conventional products. Out of the 100,000 million barrels in the deposit (a United States Bureau of Mines estimate places it at 250,000 million), perhaps 1 per cent can be easily mined; the remainder can only be won by shafting at considerable expense. The accessible 1,000 million can be considered a reserve of immediate importance, the rest as reserves against the day when oil prices are considerably higher than they are now.

Mexico

Along the Gulf Coast of Mexico, between the Gulf and the Sierra Madre Oriental, is a narrow strip of semitropical plain and foothill country which widens somewhat toward the south at the Isthmus of Tehuantepec, into the valleys of the Coatzacoalcos and San Juan rivers. In this belt, which is sprinkled with asphalt deposits and oil springs, some of the great oil fields of the world have been found.

Asphalt from these deposits must have been an article of commerce among the Indians long before the coming of Cortez, for the Aztecs, whose empire was inland across the mountains, used asphalt in their art and architecture; some of the splendid mosaics of the pre-Aztec civilization, often referred to as "Toltec," are set in asphalt.

In the early 1880's, a small shallow well was drilled at Cerro Viejo, west of Tuxpan, and in the early 1890's a small discovery at Ebano, forty miles west of Tampico, opened a field that is still producing. It was not until 1901, however, that Mexico's 10,345 barrels was enough to give her a place in the producing column. This amount, though scarcely enough to notice, attracted American and British experience and capital. Things began to happen.

The *Northern Fields* in the vicinity of Panuco were rapidly developed and yielded some big wells. In 1907, production passed the million-barrel mark, and in 1908 it was nearly four million barrels. In 1910 came the opening of a remarkable series of fields farther south, so spectacular that they are still called the *Golden Lane*, with some of the largest wells the world has seen. One of them, Potrero del Llano No. 4, had an estimated initial production of 100,000 barrels a day. By

1918, Mexico was the second-largest producer; in 1921 her production reached its peak, 193 million barrels—more than six times as much as any other country except the United States and nearly as much as Canada produced in 1960.

Mexican oil flooded the markets of the United States, at prices the domestic producers could not meet. Facing apparent ruin, they clamored for a tariff on imported oil. The idea of a tariff on a raw material that the United States had always exported was too novel to be adopted without extended argument, and before any action was taken, the Mexican oil menace began to wane.

Some of the big Mexican wells ceased flowing; some began, all at once, to produce water with the oil, and some went-to-water altogether. The troubled condition of the country, with successive revolutions bringing increasingly restrictive laws and regulations, was not conducive to new development, and production dropped year by year until, in 1932, it was down to 32.8 million barrels.

Exploration went on; a few fields were found, some among, and others to the south of, the old Northern Fields and the fields of the Golden Lane. In time, new fields on the Isthmus of Tehuantepec produced more than the Northern Fields. Most important, a field at Poza Rica, southwest of Tuxpan, is Mexico's leading producer. In 1937, the country's yield climbed to 46.9 million barrels—a little more than half that of Louisiana—thus placing Mexico seventh in the world. The chaos resulting from government expropriation of most oil properties dropped the 1942 rate to 34.8 million, a decrease of more than 28 per cent. Then, new discoveries and a somewhat rejuvenated industry boosted production each year, but Mexico has never returned to her 1921 peak. In 1960, she produced her largest output since 1925—108.8 million barrels—enough to give her tenth place as an oil country. The Golden Lane is still golden, but Mexico is now looking more seriously at the inland mountain areas and the already productive offshore waters.

The Caribbean Islands

Cuba is a teaser. There are oil seepages and asphalt deposits in almost every part of the island, but the net result of exploratory work to date is the 1956 maximum, 543,000 barrels, which collapsed to 115,000 barrels in 1960.

Curiously, part of Cuba's oil comes from highly metamorphosed rocks usually considered incapable of yielding oil.

Cuban production figures dramatically reflect the island's politics during the stormy periods of unrest in the 1950's. Immediately following General Batista's seizure of the government on March 10, 1952, oil production dropped to one-fourth the 1951 rate. It rose slowly under the dictatorship to the 1956 peak. The collapse followed the Castro coup in January, 1959. The "black years" began in October, 1959, when the Communist regime renounced all private concessions, took over the files and records, and confiscated the property of private industry. Is it any wonder that Communist Cuba looks like a powder-keg on our southern doorstep, just ninety miles from Florida?

The island of Haiti has oil seeps, and a few wells were drilled many years ago; the production was short-lived and the island is not a commercial producer.

Barbados likewise has oil seepages. In earlier centuries, Barbados tar was gathered from trenches and used in lamps and for medicinal purposes. Later, shallow hand-dug wells produced a little, but no commercial production in the usual sense has been found.

Trinidad may surprise you. In its 1,864 square miles are over twenty oil fields, several gas fields, a 114-acre lake of asphalt, two refineries with 285,000-barrels-a-day capacity, and ten operating oil companies. Asphalt from the natural lake has been an important export item for centuries; Trinidad has had an oil business quite a while.

The asphalt lake is one of the world's great oil seepages. Sir Walter Raleigh visited it in 1595 and wrote that he saw "that abundance of stone pitch, that all the ships in the world might be laden from thence; and we made trial of it in trimming our ships, and found it to be excellent good."

The lake is near the south shore. A stream of semi-liquid asphalt, fifteen to eighteen feet deep, flows slowly but steadily from it into the sea. At the surface the asphalt is firm enough to support teams and light trucks, and millions of tons of asphalt have been mined and shipped. Many an American and Canadian street is paved with Trinidad asphalt. The pits left by the removal of the asphalt soon fill with semi-solid material rising from below; it hardens on contact with the air, leaving the surface of the lake as it was before.

The island has many other oil springs and smaller asphalt deposits, and in 1858 a refinery was built at La Brea to distill *coal oil* and lubricants from the solid or semi-solid bitumen. The next year the Drake well started the Pennsylvania oil boom and American products put the refinery out of business.

Drilling for oil began in the 1900's, and in 1908 the island produced 169 barrels. The next year's figure was 57,000 barrels (and few years since then have failed to register substantial increases). Then 1911 brought the first flowing well, the establishment of the first refinery in half a century, and the shipment of the first products.

Excluding asphalt, little Trinidad has produced 720 million barrels of oil including the respectable 42 million barrels in 1960—good for seventeenth place in the world list.

One might think that Trinidad's 5,734 oil wells would be about as many as could be packed in, but 1960 found her drilling 321 holes. The completion of 277 of these as oil or gas wells shows an 89 per cent success ratio. That is hard to beat anywhere. It looks like Trinidad has an interesting future in exploration.

Venezuela

The coast of Venezuela is visible from Trinidad, and a few miles inland from the coast at the edge of the Orinoco Basin is the Bermudez pitch lake, larger than that of Trinidad though probably not as deep. In this vast basin of jungle and savanna, plagued with crocodiles and malaria mosquitoes, geologic work is slow and diffcult, but in spite of the trials and hazards, geologists began to find rewards in 1937 and 1938. In the latter year there were five fields ready to produce but very little market because there were no pipelines. Nothing succeeds like success, however, so the pipelines and fields blossomed in rapid succession like the buds on the bougainvillaea vines along the Orinoco. Today the fields of the Department of Anzoategui are too numerous to count and those of Nanagas are almost too big to see across. Their oil now flows through a multitude of pipelines to the refineries and docks at Puerto la Cruz on the Caribbean Sea. Oil writers and compilers of statistics find Venezuelan production so large and unwieldy that they divide the country into three areas. We have just looked at the eastern part.

Central Venezuela, principally the difficult jungle country south of Caracas, in the departments of Guarico, Aragua, and Miranda, was a late-comer to the producing rolls, but don't sell it short. The Las

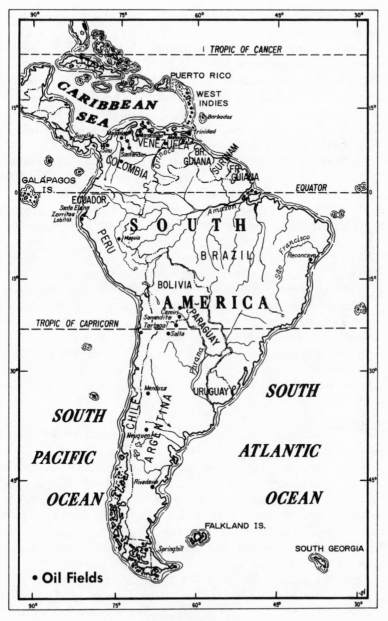

South America

Mercedes, Saban, and Tucupido areas are each in the multimillion-barrel-a-year class.

Most of Venezuela's oil comes from the Maracaibo Basin in the west across the Merida and Maritime Andes from the Orinoco Basin. There, beneath the waters of Lake Maracaibo, on its banks, and for short distances inland are the great oil fields which for some years have made Venezuela the world's third-largest producing country. Exploration in the basin and in the lake is still proceeding, and other fields are being found each year—many in deeper zones.

Venezuela's persistent problem is not how to find more oil but how to sell all it can produce. Lake Maracaibo is closer to New York by water than is Houston; consequently, Venezuelan oil could capture much of the eastern United States market were it not for United States import limitations and the low gravity of most western Venezuela crudes.

Venezuelan production, therefore, like that of Texas, Oklahoma, Kansas, and California, does not represent the country's productive capacity. Nevertheless, the 1960 total was 1,040 million barrels—about 22 million barrels less than that of the USSR and nearly 108 million barrels more than that of Texas.

Venezuela's first production was 120,000 barrels in 1917 and the annual yield has increased every year since. Whether or not the increase can be sustained depends as much or more on Venezuelan political stability as on Venezuelan geology and the world's oil thirst.

Colombia

In the upper Maracaibo Basin, in the state of Santander del Norte, lies the famous Barco concession. Here four oil fields—Petrolia, Rio do Oro, Carbonera, and Tibu—were discovered despite jungle, dysentery, hostile Indians, and title trouble. After a little more than ten years, the Barco fields had produced over 36.8 million barrels; this did not represent their full potential. Labor difficulties, then as now, have kept entire fields shut-in for months at a time. A pipeline north from the Barco fields to the port city, Corenas, has delivered 177 million barrels across the jungles to the sea.

Across the Sierra de Perija, in the Magdalena River basin, also in Santander, are other fields. La Cira has been a producer since 1926 and Infantas since 1915. The Tropical Oil Company has demonstrated what can be done in the tropics. At its camp at El Centro, equipped with modern residences, hospital, school, and complete sanitation, the standard of health is higher than in most United States cities. These properties are now operated under the name Ecopetrol.

From El Centro, a pipeline has been built through the foothills of the Sierra Cordillera and the jungles and swamps of the Magdalena valley to water transportation at Mamonal, nearly five hundred miles away, so that oil from La Cira and Infantas can be refined in Bayonne, St. Johns, or Montreal.

Colombia's first recorded production was 67,000 barrels in 1921. With only two of five developed fields open, she produced 22½ million barrels in 1938, thus ranking ninth on the year's list of producing countries. Thirty-six fields produce now, most of them strung like beads along the Magdalena River. Colombia's 1960 production was near the 56-million-barrel mark, and one more year would put her over a 1,000 million barrels cumulative. Although Colombia's all-time production is about equal to a year's output from Venezuela, she hangs doggedly onto fifteenth place among oil countries.

Ecuador

Oil seepages near Cape St. Elena on the west coast of Ecuador were known at least as early as 1700 when the Spaniards were recovering small quantities of oil from shallow pits, as the Indians may have done before them. Oil and water collected in many pits five to fifty feet deep. The Spaniards skimmed the oil, collected it in barrels, and rolled the barrels to a small refinery on the coast by burros or mules.

Despite her early start, Ecuador has not become a large producer. The sea coast along which most of the seepages occur is highly folded and faulted and the fields are small. Farther from the coast, bigger wells have been found recently, but they produce only a little because Ecuador's potential oil markets are already supplied with Peruvian crude.

Across the mountains, on the headwaters of the Amazon in the "Oriente" country that Peru also claims in part, many promising indications have been reported, but the country is difficult of access and hard to explore, and the cost of transporting oil to market would be high under present circumstances. The most reasonable pass across the mountains is 14,000 feet above sea level. The day will doubtless come when more of the region will be explored, and important fields may be found.

Ecuador's first recorded production was 57,000 barrels in 1917. This increased to 3.6 million barrels, the peak, in 1955. Since then new discoveries have not been able to keep abreast of the decline, and Ecuador supplied only two-thirds of her own 1960 requirements with 2.8 million barrels.

Peru

Peru probably had the first oil industry in the Western Hemisphere, at Punta Parina, near the village Negritos. Here at the western tip of South America is La Brea, an area of large oil springs. When Pizarro landed in 1527 the Inca-dominated Indians were collecting the oil in trenches, boiling it down to the consistency they desired, and using it to waterproof the earthen jars in which they transported and stored their fermented liquors. Here was an integrated industry, producing and refining oil and disposing of the products—and one of the rare situations where oil and alcohol have gone well together. How long the business had been in existence, no one knows.

The La Brea-Parinas field has become one of the two largest in Peru and one of the best-managed fields in the world. The discovery well was drilled in 1889, and in 1960 the field and its satellites produced 8.1 million barrels—about the output of another group of fields at Lobitos, a short distance north.

Peru's first commercial well was drilled at Zorritos, north of Lobitos, by a former Pennsylvania oil operator who became interested in a local seep in 1867, drilled, and before long had two shallow pumpers. In 1873 he deepened one of them, and got a flowing well. The Zorritos wells were small producers, and the field is now a stripper.

At the opposite end of the country, the Pirin field near Lake Titicaca produced from 1906 to 1910. In 1910, its yield was about 50,000 barrels. Pirin is no longer producing, and nearby exploration is less than intense.

Throughout the world we have noticed that many oil fields are found along coasts, many of them right on the beach. Peru is no exception. The coastal La Brea-Parinas-Lobitos-Zorritos fields have been extended more than a mile under the Pacific. In 1960, a drilling barge, dubbed the Rincon, completed fourteen holes in the ocean bottom and was thus responsible for Peru's first significant offshore production—155,889 barrels. The barge also set the world's first underwater *Christmas Tree* on the sea floor, 139 feet under water. The well controls are operated hydraulically from a control house onshore, 6,000 feet away.

The Oriente, the part of Peru that lies east of the Andes, in Amazon drainage, has important possibilities and two fields producing oil commercially. This "Montana" is accessible from the Atlantic via the Amazon and its navigable tributaries. Tankers and barges ply the rivers, hauling oil, drilling equipment, and other supplies on a six-to-eight-week trip each way. The Aguas Calientes field discovered in 1932, and the newer Maquia field have helped the Montana break the million-barrel-a-year mark.

In 1896, Peru produced 47,500 barrels. Production has increased rather steadily, with occasional recessions, until in 1960 it was 19.3 million barrels.

Bolivia

Bolivia is chiefly notable for the fact that her government has taken over the development and operation of her oil fields. Private companies, mainly American, had developed a few small fields, but these were remote from water transportation, production costs were high, and the companies were well supplied with cheaper crude from other countries. Whether for these reasons or because of the nature of the fields themselves, or both, the country's yield, which began with a few hundred barrels in 1929, was only 164,000 barrels in 1935. In that year Bolivia decided that, by taking over the industry, she would get vigorous development and badly needed revenue; and, in 1936, the Yacimientos Petroliferos Fiscales Bolivianes (Y.P.F.B.) assumed the development and exploitation of the oil resources. The new agency attacked its problems with enthusiasm, and despite recessions Bolivia enjoys a continual increase in producing rate. By 1943, the new oil regime had doubled the rate, and by 1949 doubled it again. In 1960 the country was producing 3.6 million barrels—more than twenty times as much oil as she could muster twenty-five years earlier. Under a new petroleum law, fifteen companies held concessions and were exploring, although the Y.P.F.B. controlled most of the 1960 oil runs. Most of the oil comes from the Camiri field in the Department of Santa Cruz.

Bolivia's fields are all east of the Andes in the basins of the Amazon and La Plata. This fact seems significant to the similar, trans-Andean parts of Peru, Ecuador, and Colombia.

Argentina

Oil springs and seepages have been known in northern and western Argentina since the first white settlements there; they were doubtless known to the Indians before the white men came. Drilling began in the 'eighties in the province of Mendoza, just east of the Andes, and resulted in a number of small wells. The oil was used partly for railway fuel and partly for light. In later years, a few shallow wells were drilled near some of the oil springs in the province of Salta, on La Plata drainage near the northern boundary. All these early wells were short-lived, however.

Argentina's largest field, in the south at Comodoro Rivadavia on the east coast of Patagonia, was discovered in 1907 during the drilling of a water well, and it gave the country its first recorded commercial production, 101 barrels for the year. The field, really a series of fields, has been developed by government operation, and its success is indicated by its 1960 production, over 34 million barrels—more than half the country's total for the year.

The development of Comodoro Rivadavia led to the re-exploration of Mendoza and Salta; deeper drilling by the government and by private companies has been finding good fields under and near the old ones.

Oil fields are now almost equally spaced along the full length of Argentina. Starting from the Tierra del Fuego fields, south of the Strait of Magellan and near Cape Horn, the next fields to the north are the Santa Cruz discoveries near Rio Galligos, then the south and north flanks of the Comodoro Rivadavia area. Next come the Neuquen and Mendoza fields in the west, and finally the Salta fields in the extreme north. These fields cover 1,800 miles—about as far as New York City is from Denver.

Argentina's production doubled during the four-year period from 1956 to 1960, adding a phenomenal sixteen million barrels in 1960, to bring the year's total to 64 million barrels. If exploration drilling continues at the 852-well pace of 1960, we can expect Argentina to be a world leader a few years hence.

Brazil

Brazil has great possibilities; one of them, however, is in the upper Amazon basin where remoteness, jungle, disease-carrying insects, and hostile Indians make exploration difficult and expensive. Shipment of supplies into the country and of oil out to foreign markets is not so difficult as one might think because of the Amazon and its navigable tributaries. Two fields were producing in Amazonas in 1960.

Convenience scored with the finding of the first and of some of the largest fields right on the shore of the Bay of All Saints, one of Brazil's best harbors. The Lobato-Joanes field was found in 1939, practically at the city limits of Salvador. Two years later the prolific Candeias field was brought-in on the north shore of the Bay; then a field was tapped on the tip of Itaparica Island. Itaparica Island lies across the mouth of the Bay like a door ajar at the entrance to a large room. North of the Itaparica, across the Bay, the Dom Joao field was found in 1947, suggesting that the two fields might be connected beneath the harbor. The Dom Joao field now extends eleven miles out under the Bay; it has produced 5½ million barrels of oil. Twenty years of exploration have brought to light twenty fields in the Bacia do Reconcavo basin, as it is now called.

Exploration was at a high pitch in four other areas in 1960, but only one was credited with production. The Sergipe-Alagoas Basin had two fields, one on either side of the mouth of the Rio San Francisco between Recife and Salvador.

In all, Brazil produced 27.8 million barrels in 1960 and was accelerating the rate rapidly late in the year.

Chile

Oil seeps have been known on the west coast of Chile near the Strait of Magellan for many years, but Chile had no production until the Manantiales field was found on Tierra del Fuego in 1945. Later stepouts found oil both to the north and to the south. The northern trend crosses the Strait where directionally drilled holes have extended fields under Hernando Magellan's very course of 1520.

The southern trend extends to the Argentinian border on Tierra del Fuego, thus connecting the producing trend of Argentina with that of Chile.

Chile's southern tip is the only productive area so far in all her 286,397 square miles; her Gaviota field is farther south than Punta Arenas, the southernmost city in the world.

Exploration in northern Chile is picking up enthusiasm among oil men; so Chile bears watching. One of these days her 7.2 million barrels a year may be multiplied into a major factor in the world petroleum picture.

Other Latin-American Countries

The other Latin-American countries undergo continuous exploration, but the results warrant little talk.

Oil indications have been reported from nearly all the Central-

American countries, and some holes have been drilled in Honduras, Guatemala, Costa Rica, and Panama. No commercial oil has been found but several American companies have maintained active interest during the last twenty years.

Paraguay, which lies in the La Plata basin, may have possibilities, although the character of the terrain has made it difficult to find them. Much the same may be said of Uruguay, although here the possibilities, and the difficulties of the search, may be somewhat less.

Homeward Bound

And now, having seen the New World from Point Barrow to the Strait of Magellan and oil fields from Fort Norman to Los Patos, suppose we take passage on an oil tanker and go home.

A Dios

Here ends our journey, home again by the fire. We hope you have enjoyed the trip as much as we have.

Give us credit for one thing: Not once have we handed you a table or a chart or a form. You will find a few, not many, in the appendices that follow. You can use them for home work on winter evenings should the spirit move you.

And so good night and every good fortune. May all your joints hold and all your wildcats come in.

APPENDIX A

Organization of an Oil Company

DIFFERENT oil companies organize and administer their work in different ways. The following chart, though not representing any particular company, is fairly typical of the general organization of an integrated company:

	Departments		*Divisions*
	Exploration[1]		Geologic Geophysics Scouting Land and Leasing[2]
	Production		Drilling Producing
	Transportation		
		Pipeline	Right of Way[3] Construction Maintenance and Operation
		Marine Vehicular	
President and Directors	Manufacturing		Refining Natural Gasoline
	Marketing		Traffic Sales Advertising[4]
	Legal		
	Accounting		
	Tax[5]		
	Public Relations		

---

[1] In many companies Exploration comes under the Production Department.

[2] The title work of the Land and Leasing Division may be done by the Legal Department.

[3] The work of acquiring rights of way may be done by the Land and Leasing Division or by men attached to the Legal Department. Title work in connection with rights of way may also be done by the Legal Department.

[4] Advertising may be handled by the Public Relations Department or, on the other hand, there may be no Public Relations Department except the Advertising division of the Marketing Department.

[5] Tax matters may be handled by the Legal Department.

413

APPENDIX B

Geologic Timetables

THE geologic history of the earth, as recorded in the rocks that are exposed at its surface, is divided into *eras*, and the rocks that were deposited in the various eras are divided into *systems*. The oldest system is the Archean or Archeozoic, and it probably represents a longer time than all the systems of all the eras since. Its rocks, and those of the next younger system, the Algonkian or Proterozoic, are almost entirely igneous or metamorphic, and are therefore not oil bearing. The youngest system, the Quaternary, is by far the shortest, but it is still going on. Because its rocks are so young, it, like the two oldest systems, contains no commercial oil deposits. Almost every system between the two oldest and the youngest has been found somewhere in the world to contain enough oil to be of commercial value.

When you ask a geologist, "How old is oil?" or "How many years ago was that rock made?" he usually hedges a little, talks about geologic time, and answers something like this. "Well, this is a Cretaceous sandstone so it is about one hundred million years old—plus or minus a few million years."

He is not being flippant or evasive. He speaks of time in terms of million-year-units, as the astronomer talks of light years when he measures distance. Geologic events cannot be dated closer even though the physicists have recently devised better methods for measuring time.[1]

---

[1] To give an example, let's climb New York City's Empire State Building. If we start from the street level and call it the beginning of earthly time, then whisk to the 102nd floor in one of its 74 elevators, then climb the 222-foot television tower on top, we have finally come to the time of man's first appearance on earth. From our wallet we pick one United States postage stamp, lick it thoroughly and stick it on the top of the red beacon light. The thickness of the stamp represents the period of recorded history of man on earth. The first evidence of simple vegetable life (algae) would have been seen as we passed the 96th floor on the way up.

If we grasp nothing more than the idea of how infinitesimal our own life span is to that of geologic time, we have gained some appreciation of true antiquity.

If you would like to see all this in terms of numbers of years, examine Table I.

414

The following table gives the names of the eras and systems, and of some of the major subdivisions of systems, which are called *series*.[2]

TABLE I

Main Divisions of Geologic Time

Era	System	Series (Considered Systems by some geologists)	Living Things	Years Ago
Cenozoic (recent life)	Quaternary	Recent	Man	
		Pleistocene	Mammoths	1,000,000
	Tertiary	Pliocene	Ancient horses	
		Miocene	Apes	
		Oligocene	Monkeys	
		Eocene	Camels	
		Paleocene	Rodents	70,000,000
Mesozoic (middle life)	Cretaceous		Turtles	130,000,000
	Jurassic		Dinosaurs	
	Triassic		Ammonites	200,000,000
Paleozoic (ancient life)	Permian	often	Insects	230,000,000
	Pennsylvanian	together	Forests	250,000,000
	Mississippian	called	Sharks	280,000,000
	Devonian	Carboniferous	Shellfish	330,000,000
	Silurian		Corals	360,000,000
	Ordovician		Snails	430,000,000
	Cambrian		Sponges	520,000,000
Proterozoic	often lumped into		Algae	
Archeozoic	Pre-Cambrian		None	2,000,000,000

A succession of sedimentary beds that were deposited continuously and under about the same general conditions is called a *formation*. It may be a few feet or several thousand feet thick. It may be all one type of rock, such as shale or sandstone or limestone, or it may be made up of alternations of types, such as interbedded sandstones and shale, or sandstones and shales and limestones. If an individual bed or group of beds within a

---

[2] Not all geologists agree regarding the subdivisions of geologic time or the names of the eras, systems, and series. The classification and nomenclature in the table are widely though not universally accepted by American geologists.

formation is rather distinct in character from the rest of the formation and its character is sufficiently persistent to make it recognizable throughout a considerable area it is called a *member* of the formation.

Further study of what has been called a formation may show that all of its members are distinct and persistent throughout wide areas. The members are then named as formations and what was originally called a formation is called a "group." The same succession of beds may be a formation in one locality and a group in another. In Wyoming, for example, the Niobrara formation is a somewhat limy shale, the lime content being greater in the lower part than in the upper. The amount of lime in the

Table II

Generalized Geologic Section in Eastern Texas, Western Louisiana, and Southwestern Arkansas

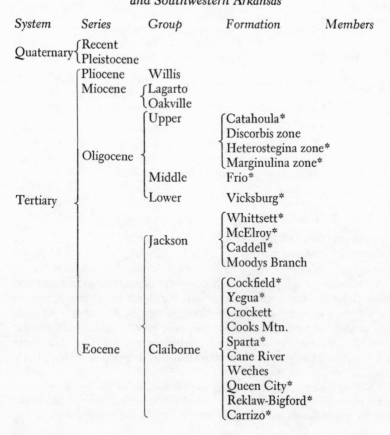

System	Series	Group	Formation	Members
Quaternary	Recent			
	Pleistocene			
Tertiary	Pliocene	Willis		
	Miocene	Lagarto		
		Oakville		
	Oligocene	Upper	Catahoula*	
			Discorbis zone	
			Heterostegina zone*	
			Marginulina zone*	
		Middle	Frio*	
		Lower	Vicksburg*	
	Eocene	Jackson	Whittsett*	
			McElroy*	
			Caddell*	
			Moodys Branch	
		Claiborne	Cockfield*	
			Yegua*	
			Crockett	
			Cooks Mtn.	
			Sparta*	
			Cane River	
			Weches	
			Queen City*	
			Reklaw-Bigford*	
			Carrizo*	

TABLE II—*Continued*

Generalized Geologic Section in Eastern Texas, Western Louisiana, and Southwestern Arkansas

	Wilcox* Midway	Wilcox Midway	⎧ Upper ⎨ Middle ⎩ Lower

Cretaceous (Upper Cretaceous) Gulf		⎧ Navarro* ⎪ Taylor* ⎨ Austin chalk* ⎪ Eagleford* ⎪ Woodbine* ⎩ Tuscaloosa*	Nacatoch sand* Blossom sand* Tokio*

Lower Cretaceous Comanchean Trinity		⎧ Washita-Fredericksburg ⎨ Paluxy sand* ⎧ Mooringsport* ⎩ Glenrose ⎨ Ferry Lake* ⎩ Rodessa*
	Coahuila	⎧ Sligo* ⎩ Travis Peak*

	Cotton Valley	⎧ Schuler ⎩ Bossier

Jurassic or Permian	Haynesville* Buckner Smackover lime*

Altered (metamorphosed) Paleozoic rocks	(Eagle Mills)	⎧ Norphlet ⎨ Louann salt ⎩ Wermer anhydrite

*Formations and members from which oil has been obtained.

lower part increases toward the south, until in southern Colorado the lower part is predominantly limestone. There the Niobrara is a group, the shaly upper part being the Apishapa formation and the predominantly limestone lower part the Timpas formation.

Groups, formations, and members are generally named from the localities, called *type localities*, at which they were first recognized and described.

The following table gives the formations that primarily interest the oil geologist in eastern Texas, western Louisiana, southwestern Arkansas, and the Gulf Coast. Production in the Gulf Coast fields is confined to beds above the Midway group. Farther north the beds above the Mt. Selman group are not present.

The following table shows the formations that primarily interest the oil geologist in the San Joaquin Valley, California:

TABLE III

Geologic Section in San Joaquin Valley, California

System	Series	Group	Formation	Members
Quaternary { Recent Pleistocene			Tulare	
Tertiary	Pliocene		San Joaquin* Etchegoin Jacalitos*	
	Miocene	Santa Margarita	{ Reef Ridge { McLure	Hallmark sand* Stevens sand*
			Temblor Vaqueros	Kettleman sand* Veddar sand*
	Oligocene and Upper Eocene		{ Kreyenhagen { Tejon	
	Middle and Lower Eocene		{ Domengine { Capay { Meganos { Cantua	
	Paleocene		{ Martinez { Moreno { Panoche	
Cretaceous (Upper Cretaceous)			{ Moreno and Panoche* { Chico⁴	
Comanchean (Lower Cretaceous)			{ Horsetown⁵ { Knoxville⁵	

*Formations and members from which oil has been obtained.
⁴ Yields some gas at Tracy.   ⁵ Showings only.

APPENDIX C

Oil and Gas Lease

Form 88 (Producers)

B

Colo., Neb. 1960 Rev.

OIL AND GAS LEASE

Courtesy of
Tatlock's Inc.
Majestic Bldg.: AM 6-1681
Denver 2, Colorado

THIS AGREEMENT, Entered into this the_____day

of_____, 19___

between_____

_____

_____

_____hereinafter called lessor,

and_____hereinafter called lessee, does witness:

1. That lessor, for and in consideration of the sum of_____
Dollars in hand paid and of the covenants and agreements hereinafter
contained to be performed by the lessee, has this day granted, leased, and
let and by these presents does hereby grant, lease, and let exclusively unto
the lessee the hereinafter described land, and with the right to unitize this
lease or any part thereof with other oil and gas leases as to all or any part
of the lands covered thereby as hereinafter provided, for the purpose of
carrying on geological, geophysical and other exploratory work, including
core drilling, and the drilling, mining, and operating for, producing, and
saving all of the oil, gas, casinghead gas, casinghead gasoline and all other
gases and their respective constituent vapors, and for constructing roads,
laying pipe lines, building tanks, storing oil, building powers, stations,
telephone lines and other structures thereon necessary or convenient for
the economical operation of said land alone or conjointly with neighboring lands, to produce, save, take care of, and manufacture all of such
substances, and for housing and boarding employees, said tract of land
with any reversionary rights therein being situated in the County of
_____, State of_____, and
described as follows:

419

in Section_____, Township_____
Range_____, and containing_____acres, more or less.

2. This lease shall remain in force for a term of_____years and as long thereafter as oil, gas, casinghead gas, casinghead gasoline or any of the products covered by this lease is or can be produced.

3. The lessee shall deliver to lessor as royalty, free of cost, on the lease, or into the pipe line to which lessee may connect its wells the equal one-eighth part of all oil produced and saved from the leased premises, or at the lessee's option may pay to the lessor for such one-eighth royalty the market price for oil of like grade and gravity prevailing on the day such oil is run into the pipe line or into storage tanks.

4. The lessee shall monthly pay lessor as royalty on gas marketed from each well where gas only is found, one-eighth ($\frac{1}{8}$) of the proceeds if sold at the well, or if marketed by lessee off the leased premises, then one-eighth ($\frac{1}{8}$) of its market value at the well. The lessee shall pay the lessor: (a) one-eighth ($\frac{1}{8}$), of the proceeds received by the lessee from the sale of casinghead gas, produced from any oil well; (b) one-eighth ($\frac{1}{8}$) of the value, at the mouth of the well, computed at the prevailing market price, of the cashinghead gas, produced from any oil well and used by lessee off the leased premises for any purpose or used on the leased premises by the lessee for purposes other than the development and operation thereof. Lessor shall have the privilege at his own risk and expense of using gas from any gas well on said land for stoves and inside lights in the principal dwelling located on the leased premises by making his own connections thereto.

Where gas from a well or wells, capable of producing gas only, is not sold or used for a period of one year, lessee shall pay or tender as royalty, an amount equal to the delay rental as provided in paragraph (5) hereof, payable annually at the end of each year during which such gas is not sold or used, and while said royalty is so paid or tendered this lease shall be held as a producing property under paragraph numbered two thereof.

5. If operations for the drilling of a well for oil or gas are not commenced on said land on or before the_____ day of_____, 19__, this lease shall terminate as to both parties, unless the lessee shall on or before said date pay or tender to the lessor or for the lessor's credit in the _____Bank at_____, or its successors, which Bank and its successors are the lessor's agent and shall continue as the depository of any and all sums payable under this lease regardless of changes of ownership in said land or in the oil and gas

or in the rentals to accrue hereunder, the sum of_____
Dollars, which shall operate as a rental and cover the privilege of deferring
the commencement of operations for drilling for a period of one year. In
like manner and upon like payments or tenders the commencement of
operations for drilling may further be deferred for like periods successively.
All payments or tenders may be made by check or draft of lessee or any
assignee thereof, mailed or delivered on or before the rental paying date,
either direct to lessor or assigns or to said depository bank, and it is under-
stood and agreed that the consideration first recited herein, the down payment, covers not only the privilege granted to the date when said first
rental is payable as aforesaid, but also the lessee's option of extending that
period as aforesaid and any and all other rights conferred. Notwithstanding
the death of the lessor or his successors in interest, the payment or tender
of rentals in the manner above shall be binding on the heirs, devisees,
executors, and administrators of such persons.

6. If at any time prior to the discovery of oil or gas on this land and
during the term of this lease, the lessee shall drill a dry hole, or holes on
this land, this lease shall not terminate, provided operations for the drilling
of a well shall be commenced by the next ensuing rental paying date, or
provided the lessee begins or resumes the payment of rentals in the manner
and amount hereinabove provided, and in this event the preceding para-
graphs hereof governing the payment of rentals and the manner and effect
thereof shall continue in force.

7. In case said lessor owns a less interest in the above described land than
the entire and undivided fee simple estate therein then the royalties and
rentals herein provided for shall be paid the said lessor only in the proportion which his interest bears to the whole and undivided fee. However,
such rental shall be increased at the next succeeding rental anniversary after
any reversion occurs to cover the interest so acquired.

8. The lessee shall have the right to use, free of cost, gas, oil and water
found on said land for its operations thereon, except water from the wells
of the lessor. When required by lessor, the lessee shall bury its pipe lines
below plow depth and shall pay for damage caused by its operations to
growing crops on said land. No well shall be drilled nearer than 200 feet
to the house or barn now on said premises without written consent of the
lessor. Lessee shall have the right at any time during, or after the expiration of, this lease to remove all machinery, fixtures, houses, buildings and
other structures placed on said premises, including the right to draw and
remove all casing, but lessee shall be under no obligation to do so, nor shall
lessee be under any obligation to restore the surface to its original condition, where any alterations or changes were due to operations reasonably
necessary under this lease.

9. If the estate of either party hereto is assigned (and the privilege of assigning in whole or in part is expressly allowed), the covenants hereof shall extend to the heirs, devisees, executors, administrators, successors, and assigns, but no change of ownership in the land or in the rentals or royalties or any sum due under this lease shall be binding on the lessee until it has been furnished with either the original recorded instrument of conveyance or a duly certified copy thereof or a certified copy of the will of any deceased owner and of the probate thereof, or certified copy of the proceedings showing appointment of an administrator for the estate of any deceased owner, whichever is appropriate, together with all original recorded instruments of conveyance or duly certified copies thereof necessary in showing a complete chain of title back to lessor to the full interest claimed, and all advance payments of rentals made hereunder before receipt of said documents shall be binding on any direct or indirect assignee, grantee, devisee, administrator, executor, or heir of lessor.

10. If the leased premises are now or shall hereafter be owned in severalty or in separate tracts, the premises nevertheless shall be developed and operated as one lease, and all royalties accruing hereunder shall be treated as an entirety and shall be divided among and paid to such separate owners in the proportion that the acreage owned by each separate owner bears to the entire leased acreage. There shall be no obligation on the part of the lessee to offset wells on separate tracts into which the land covered by this lease may be hereafter divided by sale, devise, descent or otherwise or to furnish separate measuring or receiving tanks. It is hereby agreed that in the event this lease shall be assigned as to a part or as to parts of the above described land and the holder or owner of any such part or parts shall make default in the payment of the proportionate part of the rent due from him or them, such default shall not operate to defeat or affect this lease insofar as it covers a part of said land upon which the lessee or any assignee hereof shall make due payment of said rentals. If at any time there be as many as four parties entitled to rentals or royalties, lessee may withhold payments thereof unless and until all parties designate, in writing, in a recordable instrument to be filed with the lessee, a common agent to receive all payments due hereunder, and to execute division and transfer orders on behalf of said parties, and their respective successors in title.

11. Lessor hereby warrants and agrees to defend the title to the land herein described and agrees that the lessee, at its option, may pay and discharge in whole or in part any taxes, mortgages, or other liens existing, levied, or assessed on or against the above described lands and, in event it exercises such option, it shall be subrogated to the rights of any holder or holders thereof and may reimburse itself by applying to the discharge of

any such mortgage, tax or other lien, any royalty or rentals accruing here-under.

12. Notwithstanding anything in this lease contained to the contrary, it is expressly agreed that if lessee shall commence operations for drilling at any time while this lease is in force, this lease shall remain in force and its terms shall continue so long as such operations are prosecuted and, if production results therefrom, then as long as production continues.

13. If within the primary term of this lease, production on the leased premises shall cease from any cause, this lease shall not terminate provided operations for the drilling of a well shall be commenced before or on the next ensuing rental paying date; or, provided lessee begins or resumes the payment of rentals in the manner and amount hereinbefore provided. If, after the expiration of the primary term of this lease, production on the leased premises shall cease from any cause, this lease shall not terminate provided lessee resumes operations for drilling a well within sixty (60) days from such cessation, and this lease shall remain in force during the prosecution of such operations and, if production results therefrom, then as long as production continues.

14. Lessee may at any time surrender or cancel this lease in whole or in part by delivering or mailing such release to the lessor, or by placing same of record in the proper county. In case said lease is surrendered and can-celed as to only a portion of the acreage covered thereby, then all payments and liabilities thereafter accruing under the terms of said lease as to the portion canceled shall cease and determine and any rentals thereafter paid may be apportioned on an acreage basis, but as to the portion of the acreage not released the terms and provisions of this lease shall continue and remain in full force and effect for all purposes.

15. All provisions hereof, express or implied, shall be subject to all fed-eral and state laws and the orders, rules, or regulations (and interpretations thereof) of all governmental agencies administering the same, and this lease shall not be in any way terminated wholly or partially nor shall the lessee be liable for damages for failure to comply with any of the express or implied provisions hereof if such failure accords with any such laws, orders, rules or regulations (or interpretations thereof). If lessee should be pre-vented during the last six months of the primary term hereof from drilling a well hereunder by the order of any constituted authority having juris-diction thereover, or if lessee should be unable during said period to drill a well hereunder due to equipment necessary in the drilling thereof not being available on account of any cause, the primary term of this lease shall continue until six months after said order is suspended and/or said equipment is available, but the lessee shall pay delay rentals herein pro-vided during such extended time.

16. Lessee is hereby expressly granted the right and privilege (which Lessee may exercise at any time either before or after production has been obtained upon this premises or any premises consolidated herewith) to consolidate the gas leasehold estate created by the execution and delivery of this lease, or any part or parts thereof, with any other gas leasehold estate or estates to form one or more gas operating units of approximately 640 acres each. If such operating unit or units is so created by Lessee, Lessor agrees to accept and shall receive out of the gas production from each such unit such portion of the gas royalty as the number of acres out of this lease placed in such unit bears to the total number of acres included in such unit. The commencement or completion of a well, or the continued operation or production of gas from an existing well, on any portion of an operating unit shall be construed and considered as the commencement or completion of a well, or the continued operation of, or production of gas from, a well on each and all of the lands within and comprising such operating unit; provided, that the provisions of this paragraph shall not affect the payment or non-payment of delay rentals with respect to portions of this premises not included in a unit, but this lease as to such portions of this premises not included in a unit, shall be deemed to be a separate lease. In the event portions of the above described lands are included in several units, each portion so included shall constitute a separate lease, and the particular owner or owners of the lands under each separate lease shall be solely entitled to the benefits of and be subject to the obligations of lessor under each separate lease. Lessee shall execute in writing and record in the conveyance records of the county in which the land herein leased is situated an instrument identifying and describing the consolidated acreage.

17. This lease and all its terms, conditions, and stipulations shall extend to and be binding on all successors of said lessor and lessee.

IN WITNESS WHEREOF, we sign the day and year first above written.

_____(SEAL)	_____(SEAL)
_____(SEAL)	_____(SEAL)
_____(SEAL)	_____(SEAL)
_____(SEAL)	_____(SEAL)

World Oil Production

WHEN we review oil production statistics, we should bear in mind the inevitable fact that reporting agencies differ in their sources, choices of emphasis, manners of presentation and scopes of detail. For this reason one seldom finds exact agreement between data presented in several publications. The differences, however, are usually less than a tenth of a per cent of the total.

The following bar graph and statistical tables show many interesting and a few disturbing situations of the present and trends for the future.

Table I—"Production of Crude Oil and 'Lease Condensate' in the United States by States"—shows dramatically the dominant role of Texas and its Gulf Coast neighbor, Louisiana, in oil production in the United States. In 1960, these two states accounted for 51.4 per cent of the nation's total production and 17.2 per cent of all the world's production. The first five states accounted for 80 per cent of the national total and the first ten states produced 92 per cent of the nation's domestic supply of oil.

The Ten Leading Oil States

Starting from the pioneer producing areas in the Appalachians, the trend of United States production has been steadily moving to the west and more recently to the southwestern states. Pennsylvania produced the most until 1895, when it was topped by Ohio. California moved up to first place from 1903 to 1906 and then was displaced by Oklahoma in 1907. For the next two decades California and Oklahoma traded positions as the leading producing state. In 1928 Texas moved into first place and has remained there ever since. One by one several older producing states have disappeared from the lists, to be replaced by western and southwestern states with prolific new production.

TABLE I

Production of Crude Oil and "Lease Condensate" in the United States by States[1]

Rank in 1960	State	1960 Production	1955 Production	Rank in 1955	Bbls Daily 1960 (Avg)	Producing Wells 1960	Daily Average per Well 1960	% of U.S. Total 1960	% of World Total 1960	Year of 1st Recorded Production
1	Texas	931,996,000	1,053,297,000	1	2,546,437	197,256	12.9	36.2	12.1	1889
2	Louisiana	392,202,000	271,010,000	3	1,071,590	25,025	42.8	15.2	5.1	1902
3	California	304,847,000	354,812,000	2	832,915	37,927	22.0	11.8	4.0	1861
4	Oklahoma	191,948,000	202,817,000	4	524,448	80,409	6.5	7.4	2.5	1891
5	Wyoming	135,778,000	99,483,000	6	370,978	7,423	50.0	5.3	1.8	1894
6	Kansas	113,500,000	121,669,000	5	310,109	40,297	7.7	4.4	1.5	1889
7	New Mexico	108,424,000	82,958,000	7	296,240	12,991	22.8	4.2	1.4	1913
8	Illinois	79,432,000	81,423,000	8	217,027	32,182	6.7	3.1	1.0	1889
9	Mississippi	52,681,000	37,741,000	10	143,937	2,621	54.9	2.5	less than 1.0	1939
10	Colorado	47,207,000	52,655,000	9	128,981	2,390	54.0	1.8	"	1887
11	Utah	37,606,000	2,227,000	22	102,749	792	129.7	1.4	"	1907
12	Montana	30,200,000	15,654,000	12	82,514	3,973	20.8	1.2	"	1916
13	Arkansas	28,739,000	28,369,000	11	78,522	6,370	12.3	1.1	"	1920
14	Nebraska	24,488,000	11,203,000	15	66,907	1,672	40.0	less than 1.0	"	1939
15	No. Dakota	21,865,000	11,143,000	16	59,740	1,436	41.6	"	"	1951
16	Kentucky	21,498,000	15,518,000	13	58,738	19,732	3.0	"	"	1860
17	Michigan	15,675,000	11,266,000	14	42,828	4,410	9.7	"	"	1900
18	Indiana	11,600,000	10,988,000	17	31,694	5,388	5.9	"	"	1889
19	Alabama	7,322,000	1,411,000	23	20,005	375	53.3	"	"	1944
20	Pennsylvania	6,276,000	8,531,000	18	17,148	56,176	0.3	"	"	1859

21	Ohio	5,005,000	4,353,000	19	13,675	15,765	0.9	"	"	"	"	1860
22	West Virginia	2,306,000	2,320,000	21	6,301	14,116	0.4	"	"	"	"	1860
23	New York	1,811,000	2,904,000	20	4,948	17,232	0.3	"	"	"	"	1865
24	Florida	377,000	495,000	24	1,030	11	93.6	"	"	"	"	1943
	Other States²	908,000	183,000		2,481	2,366	1.0					
TOTAL		2,573,691,000	2,484,428,000		7,031,942	588,335	12.0					

¹ The inclusion of "Lease Condensate" in the totals of produced oil raises each figure in this Table to somewhat above those given for the states in other parts of the book. These variations are not large enough to change any status rank or position.

² Includes Alaska, Arizona, Missouri, Nevada, South Dakota, Tennessee, Virginia and Washington.

Source: O & G Journal
January 31, 1961
(1955 figures compare with API
Petroleum Facts and Figures)

In 1959, Louisiana displaced California as the second-place producing state. Table II shows how the same few states held the lead from 1910 to 1960; only their places within the first ten shifted.

TABLE II

The Ten Leading Oil States in the United States
1910 to 1960, at Five-Year Intervals

Rank	1910	1915	1920	1925	1930
1	California	Oklahoma	Oklahoma	California	Texas
2	Oklahoma	California	California	Oklahoma	California
3	Illinois	Texas	Texas	Texas	Oklahoma
4	W. Virginia	Illinois	Kansas	Arkansas	Kansas
5	Ohio	Louisiana	Louisiana	Kansas	Louisiana
6	Texas	W. Virginia	Wyoming	Wyoming	Arkansas
7	Pennsylvania	Pennsylvania	Illinois	Louisiana	Wyoming
8	Louisiana	Ohio	Kentucky	Pennsylvania	Pennsylvania
9	Indiana	Wyoming	W. Virginia	Illinois	New Mexico
10	Kansas	Kansas	Pennsylvania	Ohio	Kentucky

Rank	1935	1940	1945	1950	1955	1960
1	Texas	Texas	Texas	Texas	Texas	Texas
2	California	California	California	California	California	Louisiana
3	Oklahoma	Oklahoma	Oklahoma	Louisiana	Louisiana	California
4	Kansas	Illinois	Louisiana	Oklahoma	Oklahoma	Oklahoma
5	Louisiana	Louisiana	Kansas	Kansas	Kansas	Wyoming
6	New Mexico	Kansas	Illinois	Illinois	Wyoming	Kansas
7	Pennsylvania	New Mexico	New Mexico	Wyoming	New Mexico	New Mexico
8	Michigan	Arkansas	Wyoming	New Mexico	Illinois	Illinois
9	Wyoming	Wyoming	Arkansas	Mississippi	Colorado	Mississippi
10	Arkansas	Michigan	Mississippi	Arkansas	Mississippi	Colorado

World Production by Regions and Countries

Fifty-eight nations (or political entities) produced the world's oil in 1960. In many more the search for oil continued with results that ranged from spectacular success to quiet failure. In South America, Brazil struggled to expand its modest reserves; and Argentina, after only three years of highly concentrated exploration and development drilling, became self-sufficient in crude oil and natural gas. In Europe, large gas reserves in The Netherlands were discovered that already loom large in the energy requirements of densely populated Western Europe.

In Northern Africa, Libya moved toward becoming a major producing nation with continuing spectacular discoveries. Algeria's growing gas and oil resources became one of the major factors in the uncertain political future of the country. Exploration increased in Tunisia and Spanish Sahara, and Morocco sought to increase its depleting domestic reserves. Elsewhere in Africa, exploration activity increased in the new nations and colonial region of western and eastern Africa, and Nigeria and Gabon recorded significant production.

The Middle East is now the world's most prolific oil region. Here, average daily production per well is measured in thousands or tens of thousands of barrels of oil per day, wealth that causes the fabled riches of the East, of older days, to pale in comparison.

Farther East in Asia and in Oceania, oil production rates are more modest. Even Communist China's vaunted claims of production seem puny in comparison with other areas of similar potential. Only Indonesia, still recuperating from the devastation of World War II and post-war dislocation, showed an impressive total production for 1960.

TABLE III

Production in the World by Regions and Countries*
(In bbls.)

	Annual Production 1960	Av. Daily Production 1960	Producing Wells 1960	Av. Daily Prod./ Well
NORTH AMERICA	2,865,824,092	7,830,120	616,749	
Canada	191,841,815	524,158	14,618	35.9
United States	2,574,933,000	7,035,336	599,977	11.7
Mexico	99,049,277	270,626	2,154	125.6
WEST INDIES	42,518,369	116,170	3,258	
Cuba	161,040	440	110	4.0
Trinidad	42,357,329	115,730	3,148	36.8
SOUTH AMERICA	1,222,815,117	3,341,040	21,243	
Argentina	62,898,000	171,852	3,750	45.8
Bolivia	3,574,132	9,765	127	76.9
Brazil	29,612,676	80,909	686*	117.9*
Chile	7,230,779	19,756	125	158.0
Colombia	55,770,445	152,378	2,143	71.1
Ecuador	2,798,920	7,664	1,703	4.5

	Annual Production 1960	Av. Daily Production 1960	Producing Wells 1960	Av. Daily Prod./ Well
Peru	19,255,034	52,609	2,209	23.8
Venezuela	1,041,675,131	2,846,107	10,500*	271.0*
WESTERN EUROPE	97,501,920	266,399	6,170	
Austria	16,619,694	45,409	913	49.7
France	14,116,246	38,569	525	73.5
West Germany	39,262,169	107,274	3,881	27.6
Italy	846,010	2,312	107	21.6
Sicily	12,870,042	35,164	83	423.7
Netherlands	13,165,583	35,971	416	86.4
United Kingdom	622,176	1,700	245	6.9
EASTERN EUROPE	1,188,582,852	3,254,202	41,575	
Albania	1,567,920	10,994	300*	36.6*
Bulgaria	1,460,000*	3,990*	150*	26.6*
Czechoslovakia	929,134	2,538	175*	14.5*
East Germany	1,460,000*	3,990*	10*	399.0*
Hungary	9,270,450	25,329	530*	47.8*
Poland	1,442,448	3,941	3,000	1.3*
Rumania	85,169,000*	232,702	4,660*	49.9*
USSR	1,080,400,000*	2,951,912	32,350*	91.2*
Yugoslavia	6,883,900	18,808	400*	47.0*
NORTHERN AFRICA	90,481,045	247,211	592	
Algeria	67,226,388	183,674	285	644.5
Egypt	22,559,182	61,637	227	271.5
Morocco	695,475	1,900	80	23.8
WESTERN AFRICA	13,027,464	35,640	161	
Angola	481,045	1,314	11	119.4
Congo Republic	388,023	1,060	9	117.8
Gabon	5,856,175	16,000	100	160.0
Nigeria	6,289,684	17,232	40	430.8
Senegal	12,537	34	1	34.0
ASIA—MIDDLE EAST	1,929,787,897	5,272,649	1,194	
Bahrain	16,500,424	45,083	150	300.6
Iran	390,754,900	1,067,636	102	10467.0
Iraq	354,591,851	968,830	99	9786.1
Israel	900,000	2,459	24	102.4
Kuwait	594,278,196	1,623,711	367	4424.3
Neutral Zone	49,830,195	136,148	202	674.0
Qatar	63,907,684	174,611	51	3423.7

	Annual Production 1960	Av. Daily Production 1960	Producing Wells 1960	Av. Daily Prod./ Well
Saudi Arabia	456,453,173	1,247,140	196	6363.0
Turkey	2,573,474	7,031	3	2343.7
Asia—Far East	26,579,761	72,623	3,715	
Burma	4,065,000	11,107	410	27.1
India	3,284,850	8,975	500*	18.0*
Pakistan	2,639,204	7,211	79	91.3
China	12,810,000	35,000	—	—
Japan	3,730,019	10,191	2,682	3.8
Taiwan	14,188	39	34	1.1
Thailand	36,500	100	10	10.0
Oceania	186,063,876	508,369	2,963	
Indonesia	150,510,300	411,228	2,390	172.1
New Zealand	4,500	12	4	3.0
North Borneo	34,004,787	92,909	524	177.3
Philippines	1,000	3	12	.2
West New Guinea	1,543,289	4,217	43	98.1
WORLD TOTAL	7,663,184,393	20,944,423	697,620	30.0

* See footnote and preliminary discussion to Table I.

Table IV

Comparison of Crude Oil Production in United States and Rest of World 1951–1960

	World		United States		Rest of World	
Year	Bbls Daily	% Diff. Prior Yr.	Bbls Daily	% Diff. Prior Yr.	Bbls Daily	% Diff. Prior Yr.
1951	11,733,507	+12.6	6,158,112	+13.9	5,575,395	+11.2
1952	12,380,093	+ 5.5	6,256,382	+ 1.6	6,123,711	+ 9.8
1953	13,145,356	+ 6.2	6,457,759	+ 3.2	6,687,597	+ 9.2
1954	13,744,775	+ 4.6	6,342,433	− 1.8	7,402,342	+10.7
1955	15,413,378	+12.1	6,806,652	+ 7.3	8,606,726	+16.3
1956	16,732,708	+ 8.6	7,151,046	+ 5.1	9,581,662	+11.3
1957	17,887,179	+ 6.9	7,169,592	+ 0.3	10,717,587	+11.8
1958	18,090,947	+ 1.1	6,709,553	− 6.4	11,381,394	+ 6.2
1959	19,480,215	+ 7.7	7,053,671	+ 5.1	12,426,544	+ 9.2
1960	20,944,423	+ 7.5	7,035,336	− 0.3	13,909,087	+11.9
est. 1961	22,166,000	+ 5.8	7,144,000	+ 1.5	15,022,000	+ 8.0

From 1930 to 1945 the United States produced three-fifths to two-thirds of the world's oil total each year. During World War II the percentage reached the peak—66.7 in 1943—then began a steady decline, ascending again briefly in 1951 during the Korean War. Since 1951 the percentage the United States produces has declined as world production has nearly doubled, production outside the United States has nearly tripled, and United States production has about held its own.

TABLE V

World Production by Countries and States
(First Ten in Each Category)[1]

Rank in 1960	Country or State	Production in 1960 (thousand bbls)
1	United States	2,573,691
2	Russia	1,080,400
3	Venezuela	1,041,675
4	Texas	931,996
5	Kuwait	594,278
6	Saudi Arabia	456,453
7	Louisiana	392,202
8	Iran	390,754
9	Iraq	354,591
10	California	304,847
11	Oklahoma	191,948
12	Canada	191,841
13	Indonesia	150,510
14	Wyoming	135,778
15	Kansas	113,500
16	New Mexico	108,424
17	Mexico	99,049
18	Illinois	79,432
19	Mississippi	52,861
20	Colorado	47,207

[1] Oil & Gas Journal, January 30, 1961

YEARLY
UNITED STATES AND WORLD
OIL PRODUCTION COMPARED

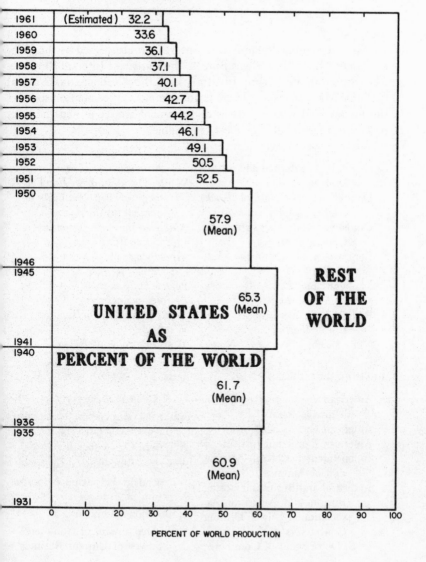

PERCENT OF WORLD PRODUCTION

References

Figures are compiled mainly from information furnished by the U. S. Bureau of Mines, the Census Bureau, the Bureau of Internal Revenue, The American Petroleum Institute, and the "Refiner and Natural Gasoline Manufacturer." There is not complete agreement on some of the figures, and where there is disagreement we have used the ones that seem most reliable or most reasonable.

We borrowed heavily from:

Crane, R. J., *Bitumen and Petroleum in Antiquity*

Danby, Arthur, *Natural Rock Asphalts and Bitumens*

Goldstein, R. R., *The Petroleum Chemicals Industry*

Longrigg, S. H., *Oil in the Middle East*

Magoffin and Davis, *Romance of Archaeology*

Redwood, Sir Boverton, A *Treatise on Petroleum*

Ver Wiebe, W. A., *North*

American Petroleum

White, David, *Outstanding Features of Petroleum Development in America*

American Journal of Science

International Oilman

Mines Magazine

Oil and Gas Journal

Pétrole Informations

Petroleum Engineer

Rocky Mountain Oil Reporter

World Oil

World Petroleum

and from the publications of:

American Association of Petroleum Geologists

American Gas Association

American Petroleum Institute

Independent Petroleum Association of America

Industrial and Business Training Bureau and American Association of Oilwell Drilling Contractors

U. S. Department of Commerce

U. S. Geological Survey

American Association of Petroleum Geologists

American Gas Association

American Petroleum Institute

Independent Petroleum Association of America

International Oilman

U. S. Department of Commerce Survey of Current Business

Index

Abadan, Persia, refinery, 312
Abasand, Canada, oil sand mining,173
Abqaiq oil field, Saudi Arabia, 309
Absorption plant, 155
Abu Dhabi, oil production in, 309
Abu Durba field, Egypt, 343
Accountants, 21
Acetates, 289
Acetone, 289
Acetylene series, 207
Achi-Su field, Russia, 329
Acidizer, 14
Acidizing wells, 106
Acreage factor in proration, 146-148, 149-150
Adena field, 381
Adhesion as force opposing oil production, 139
Adriatic Sea, coastal exploration, 339
Adsorption as force opposing oil production, 141
Advertising, gasoline, 266
Advertising men, 20
Afghanistan, oil possibilities in, 313-314
Agent, sales, 18, 267-268
Agha Jari oil field, Iran, 312
Agitators, refinery, 250
Agriculture, oil industry and, 282
Agrigento, Sicily, oil spring near, 339
Agrigentum, oil of, 339
Aguas Calientes field, Peru, 409
Ahmad oil field, Kuwait, 312
Ahwoz oil field, Iran, 312
Ain Dar oil field, Saudi Arabia, 309
Ain Zeft, Algeria, oil springs, 343
Air drive, 169
Air lift, 161
Airplanes, in U.S., 257
  gasoline consumption by, 258
Alabama, oil fields in, 387
Alaska, oil in, 393-395
Albania, oil occurrences in, 338
  oil production in, 432
Alberta, oil in, 399-400
  oil sands in, 32-33, 173, 399
Alberz oil field, Iran, 312
Alcohol, 288, 297-300

Aldan River, Siberia, 330
Aleppo, Syria, 308
Alexander the Great, 304, 314
Alexandretta, Turkey, 308
Algae as a source of oil, 30
Algeria, oil in, 343-344
Algonkian system, 415-416
Algonquins, use of oil, 351
Alleghenies, oil springs in, 351
Allowable production, 148
Alluvial fans, 28
Alps, core of, 30
Amarillo field, Texas, 359, 375
Amarillo Mts., Texas, 53, 375
Amazing decade, the, 359
Amazon Basin, 407, 408, 409
Amazonas, Brazil, oil production in, 410-411
Amerada Petroleum Co., oil discovery in North Dakota, 390
American Assoc. of Petroleum Geologists, 47, 280, 284
American Gilsonite Co., 174
American Indians, use of oil, 351, 401, 406, 408
American Petroleum Institute, 280, 281, 283, 285
American Society for Testing Materials, 223, 280
Ammonium hydroxide, use in treating oils, 249
Ammonium nitrate, fertilizer, 295
Ammonium sulphate, fertilizers, 295
Amory oil field, Mississippi, 372
Amu Darya, Russia, 330
Amur River, Siberia, 330
Anchorage, Alaska, oil center, 394
Anderson oil field, New Mexico, 377
Anesthetics, 293
Aneth oil field, Utah, 393
Anglo-Iranian Oil Co.; see British Petroleum Co.
Anglo-Persian Oil Co.; see British Petroleum Co.
Angola (Africa), oil fields in, 346
  "asphalt rock" mines in, 346
Aniline, 281

435

Rocks—*Cont.*
 source, 22, 30, 32, 283
Rocky Mountain field, 350
Rocky Mountains, care of, 30
 erosion of, 25
Rodessa field, Ark., Tex., 370
Rods, sucker, 163
Rogers, H. D., 47
Rohri, Pakistan, 316
Romans, use of oil products, 339
Romi salt basin, Russia, 329
Room and pillar mining, 176
Rope socket, 88
Rotary drilling, 12, 83, 109-137
 coring, 126
 crew, 129
 mud, 115-120
 passing up sands, 83
Rotary helpers, 13
Rotary practice, improvement in, 109-110
Rotary rig, 109-111
 weight indicator on, 127
Rotary runner, 13
Rotary tools and equipment, cost of, 83
 description, 110-115
Rotary well, cleaning, 118
 drilling, 109-137
 drilling-time logging, 127
 electric logging, 127-129
 first, 373-374
 logging, 125-126
 testing, 131-133
Rotterdam, Holland, oil production near, 332
Roughneck, 13
Roustabouts, 13, 15, 17
Royal Dutch Co. for the Exploitation of
 Petroleum Wells in the Nether-
 lands, 318-319
Royal Dutch-Shell group, 319
Royalties, as investments, 74
 landowners, amount of, 74-75
 landowners, sale of, 78
 overriding, 74
 under oil lease, 74
Royalty acre, 78
Royalty companies, 74
Rubbers, buna, 91
 synthetic, 291-292, 298
Rudeis field, Egypt, 343
Rudrasagar oil field, India, 316
Ruffner, David, 352
Rumania, oil in, 337-338
 oil production in, 432
Run ticket, 182
Running casing, 95-96
Running pipe, 125
Running to coke, 248

Running tools, 86
Russia, oil in, 327-331
 oil production in, 432, 434
 pipelines in, 195
 unit operations in, 150

Saban, oilfield area, Venezuela, 406
Sabine uplift, 376
Sacramento Valley, Calif., 387
Sacred fire, 301, 318, 327
Saddle, structural geologic, 50
SAE specifications for lubricants, 235
Sag, structural geologic, 50
Saginaw field, Mich., 359, 364
St. Lawrence Valley, sinking of, 25
Sakhalin, oil in, 326-327
 production compared with Oklahoma, 368
Sales, open-market, gasoline, 270-271
 staff, 18
 struggle for, 262-272
Sales efficiency as weapon, 267-268
Salt, composition of, 204
 relation to search for oil, 352-353
Salt, beds (Congo), 346
 series, 419
Salt Creek field, Wyoming, 358, 381-383
Salt domes, 43-45
 discoveries of, 358, 359
 See also Colorado, Louisiana, Texas,
 Utah, Gulf Coast, Germany, Rus-
 sia, Republic of Congo
Salt springs, as source of oil, 352-353
 location of wells because of, 46
Salta Province, Argentina, oil in, 410
Saltville, Virginia, limit of oil produc-
 tion, 396
Salvador, city of, Brazil, oil fields near, 411
Sample log, 93
Samson post, 85
Samuel, Sir Marcus, 319
San Antonio district, Texas, 373
San Bernardino Mts., Calif., 385
San Joaquin formation, 419
San Joaquin Valley, Calif., geologic sec-
 tion, 419
 heavy oils in, 33
 oil in, 358, 385-387
San Juan Basin, N. M., 377
 helium gas in, 378
 natural gas in, 378
 oil fields in, 377
San Juan County, Utah, oil in, 393
San Juan River, Mexico, 401
San River area oil fields, Poland, 332
Sand, compaction into sandstone, 26
 deposition of, 26